Modeling, Control and Diagnosis of Electrical Machines and Devices

Modeling, Control and Diagnosis of Electrical Machines and Devices

Editors

Moussa Boukhnifer
Larbi Djilali

Basel • Beijing • Wuhan • Barcelona • Belgrade • Novi Sad • Cluj • Manchester

Editors
Moussa Boukhnifer
Université de Lorraine
Metz
France

Larbi Djilali
Tecnológico Nacional
de Mexico
Chihuahua
Mexico

Editorial Office
MDPI
St. Alban-Anlage 66
4052 Basel, Switzerland

This is a reprint of articles from the Special Issue published online in the open access journal *Energies* (ISSN 1996-1073) (available at: https://www.mdpi.com/journal/energies/special_issues/2531K82S5R).

For citation purposes, cite each article independently as indicated on the article page online and as indicated below:

Lastname, A.A.; Lastname, B.B. Article Title. *Journal Name* **Year**, *Volume Number*, Page Range.

ISBN 978-3-7258-1339-1 (Hbk)
ISBN 978-3-7258-1340-7 (PDF)
doi.org/10.3390/books978-3-7258-1340-7

© 2024 by the authors. Articles in this book are Open Access and distributed under the Creative Commons Attribution (CC BY) license. The book as a whole is distributed by MDPI under the terms and conditions of the Creative Commons Attribution-NonCommercial-NoDerivs (CC BY-NC-ND) license.

Contents

About the Editors . . . . **vii**

Moussa Boukhnifer and Larbi Djilali
Modeling, Control and Diagnosis of Electrical Machines and Devices
Reprinted from: *Energies* **2024**, *17*, 2250, doi:10.3390/en17102250 . . . . **1**

**Olaoluwa Demola Aladetola, Mondher Ouari, Yakoub Saadi, Tedjani Mesbahi,
Moussa Boukhnifer and Kondo Hloindo Adjallah**
Advanced Torque Ripple Minimization of Synchronous Reluctance Machine for Electric Vehicle
Application
Reprinted from: *Energies* **2023**, *16*, 2701, doi:10.3390/en16062701 . . . . **5**

Hamad Alharkan
Torque Ripple Minimization of Variable Reluctance Motor Using Reinforcement Dual NNs
Learning Architecture
Reprinted from: *Energies* **2023**, *16*, 4839, doi:10.3390/en16134839 . . . . **35**

Piotr Gnaciński, Marcin Pepliński, Adam Muc, Damian Hallmann and Piotr Jankowski
Effect of Ripple Control on Induction Motors
Reprinted from: *Energies* **2023**, *16*, 7831, doi:10.3390/en16237831 . . . . **48**

Yuhua Sun, Nicola Bianchi, Jinghua Ji and Wenxiang Zhao
Improving Torque Analysis and Design Using the Air-Gap Field Modulation Principle for
Permanent-Magnet Hub Machines
Reprinted from: *Energies* **2023**, *16*, 6214, doi:10.3390/en16176214 . . . . **60**

Mariam Saeed, Daniel Fernández, Juan Manuel Guerrero, Ignacio Díaz and Fernando Briz
Insulation Condition Assessment in Inverter-Fed Motors Using the High-Frequency Common
Mode Current: A Case Study
Reprinted from: *Energies* **2024**, *17*, 470, doi:10.3390/en17020470 . . . . **75**

**Jose A. Ruz-Hernandez, Larbi Djilali, Mario Antonio Ruz Canul, Moussa Boukhnifer and
Edgar N. Sanchez**
Neural Inverse Optimal Control of a Regenerative Braking System for Electric Vehicles
Reprinted from: *Energies* **2022**, *15*, 8975, doi:10.3390/en15238975 . . . . **91**

**Yasser Damine, Noureddine Bessous, Remus Pusca, Ahmed Chaouki Megherbi,
Raphaël Romary and Salim Sbaa**
A New Bearing Fault Detection Strategy Based on Combined Modes Ensemble Empirical Mode
Decomposition, KMAD, and an Enhanced Deconvolution Process
Reprinted from: *Energies* **2023**, *16*, 2604, doi:10.3390/en16062604 . . . . **110**

Ahmed Belkhadir, Remus Pusca, Driss Belkhayat, Raphaël Romary and Youssef Zidani
Analytical Modeling, Analysis and Diagnosis of External Rotor PMSM with Stator Winding
Unbalance Fault
Reprinted from: *Energies* **2023**, *16*, 3198, doi:10.3390/en16073198 . . . . **137**

Junqing Li, Chengzhi Zhang, Yuling He, Xiaodong Hu, Jiya Geng and Yapeng Ma
Impact of Inter-Turn Short Circuit in Excitation Windings on Magnetic Field and Stator Current
of Synchronous Condenser under Unbalanced Voltage
Reprinted from: *Energies* **2023**, *16*, 5695, doi:10.3390/en16155695 . . . . **160**

Przemyslaw Pietrzak, Piotr Pietrzak and Marcin Wolkiewicz
Microcontroller-Based Embedded System for the Diagnosis of Stator Winding Faults and Unbalanced Supply Voltage of the Induction Motors
Reprinted from: *Energies* **2024**, *17*, 387, doi:/10.3390/en17020387 . . . . **178**

About the Editors

Moussa Boukhnifer

Moussa Boukhnifer (Senior Member, IEEE) received his M.Sc. degree in Electrical Engineering from the Institut National des Sciences Appliquées de Lyon, Lyon, France, in 2002, and a Ph.D. degree in Control and Engineering from the Université d'Orléans, Orléans, France, in December 2005. He received a habilitation for heading research (HDR Habilitation à Diriger des Recherches) from the Université de Paris Sud, France, in December 2015. He is currently an Associate Professor HDR at ENIM (Ecole Nationale d'Ingénieurs de Metz), Université de Lorraine, France. His main research interests are in the fields of diagnosis, FTC control, and energy management, alongside their applications in electrical and autonomous systems. He is the author or coauthor of more than 150 journal and conference papers. He has served as a programme committee member or session chair for many international conferences and as an Editorial Board Member of many international journals. He is an Associate Editor of *IEEE Transactions on Vehicular Technology*, *IEEE Access*, and the *Transactions of the Institute of Measurement and Control* (*TIMC*) journal.

Larbi Djilali

Larbi Djilali was born in Algeria in 1987. He received his B.Eng. degree in Maintenance in Instrumentation from the University of Oran, Algeria in 2010. In 2014, he received his M.Sc. in Electrical Engineering—Control Automatic from the National Polytechnic School of Oran (ENPO). In 2019, he obtained his PhD. Degree in Electrical Engineering from the University of Laghouat, Algeria, and in 2020, his Ph.D. in Electrical Engineering from the Center for Research and Advanced Studies, National Polytechnic Institute (CINVESTAVIPN), Guadalajara campus, Mexico. He worked as a Professor of Electrical Engineering graduate programs at the Autonomous University of Carmen, Campeche, Mexico from 2020 to 2022. He has worked as an Invited Professor at TecNM Chihuahua since 2023. Larbi has published different articles in several research areas, including linear and nonlinear control, robust control, intelligent-based neural control, and their applications in renewable power systems, micro-grids, and power electronics.

Editorial

Modeling, Control and Diagnosis of Electrical Machines and Devices

Moussa Boukhnifer [1],* and Larbi Djilali [2]

1. Université de Lorraine, LCOMS, F-57000 Metz, France
2. Faculty of Technology, University of Ciudad del Carmen, Campeche 24130, Mexico; larbidjar@chihuahua.tecnm.mx
* Correspondence: moussa.boukhnifer@univ-lorraine.fr

Citation: Boukhnifer, M.; Djilali, L. Modeling, Control and Diagnosis of Electrical Machines and Devices. *Energies* **2024**, *17*, 2250. https://doi.org/10.3390/en17102250

Received: 8 April 2024
Revised: 17 April 2024
Accepted: 28 April 2024
Published: 7 May 2024

Copyright: © 2024 by the authors. Licensee MDPI, Basel, Switzerland. This article is an open access article distributed under the terms and conditions of the Creative Commons Attribution (CC BY) license (https://creativecommons.org/licenses/by/4.0/).

1. Introduction

Nowadays, the increasing use of electrical machines and devices in more critical applications has driven the research in condition monitoring and fault tolerance. Condition monitoring of electrical machines has a very important impact in the field of electrical system maintenance, mainly because of its potential functions of failure prediction, fault identification, and dynamic reliability estimation. Fault diagnosis of electrical machines and devices has received a great deal of attention due to its benefits in reducing maintenance costs, preventing unplanned downtime, and, in many cases, preventing damage and failure. Fault-tolerant design offers a solution that combines fault occurrence conditions, fault detection and location tools, and the reconfiguration of control functions. On the other hand, recent advances in intelligent technology using artificial intelligence and advanced machine learning capabilities provide new perspectives for meaningful fault diagnosis and fault-tolerant control. These outstanding advances can improve the performance of condition monitoring and have significant potential for fault detection in electrical machines and equipment.

Based on the above premises, this Special Issue, titled "Modeling, Control and Diagnosis of Electrical Machines and Devices", aims to highlight the recent trends, research, development, applications, solutions, and challenges related to condition monitoring and fault diagnosis of electrical machines and devices. Topics of interest include the following:

- Modeling of electrical machines and devices.
- Robust control strategies of electrical machines and devices.
- Failure detection and diagnosis of electrical machines and devices.
- Fault-tolerant control of electrical machines and devices.
- Condition monitoring techniques and applications in electrical machines and devices.
- AI techniques for electrical machine fault diagnosis and fault-tolerant control.
- Machine learning techniques for electrical machine fault diagnosis and fault-tolerant control.

There are 10 scientific research articles published in this Special Issue. A summary of the articles published in this Special Issue is outlined in the following section.

2. Highlights of Published Papers

This section provides a summary of this Special Issue of *Energies*, which includes published articles [1–10] covering various topics related to the modeling, control, and diagnosis of electrical devices.

Saeed et al., in ref. [1], extensively studied the use of the common mode current for a stator winding insulation condition assessment. Two main approaches were followed. The first modeled the electric behavior of ground–wall insulation as an equivalent RC circuit; these methods have been successfully applied to high-voltage, high-power machines. The second used the high frequency of the common mode current, which results from the voltage pulses applied by the inverter. This approach has mainly been studied for the case of low-voltage, inverter-fed machines and has not yet reached the level of maturity of the first one. One fact noticed after a literature review is that, in most cases, the faults detected were induced by connecting external elements between the winding and stator magnetic cores. The paper presented a case study on the use of the high-frequency common mode current to monitor the stator insulation condition. Insulation degradation occurred progressively with the machine operating normally; no exogenous elements were added.

Pietrzak et al. [2] proposed a low-cost embedded system based on a microcontroller with the ARM Cortex-M4 core for the extraction of stator winding faults (inter-turn short circuits) and an unbalanced supply voltage of the induction motor drive. The voltage induced in the measurement coil by the axial flux was used as a source of diagnostic information. The process of signal measurement, acquisition, and processing using a cost-optimized embedded system (NUCLEO-L476RG), with the potential for industrial deployment, was described in detail. In addition, the analysis of the possibility of distinguishing between inter-turn short circuits and unbalanced supply voltage was carried out. The effect of motor operating conditions and fault severity on the symptom extraction process was also studied. The results of the experimental research conducted on a 1.5 kW IM confirmed the effectiveness of the developed embedded system in the extraction of these types of faults.

Gnaciński et al. [3] described the effect of RC on low-voltage induction motors through the use of experimental and finite element methods. One method for the remote management of electrical equipment is ripple control (RC), based on the injection of voltage inter-harmonics into the power network to transmit information. The disadvantage of this method is its negative impact on energy consumers, such as light sources, speakers, and devices counting zero crossings. The results showed that the provisions concerning RC included in the European Standard EN 50160 Voltage Characteristics of Electricity Supplied by Public Distribution Network are imprecise, failing to protect induction motors against excessive vibrations.

Sun et al. [4] investigated the torque generation mechanism and its improved design in Double Permanent Magnet Vernier (DPMV) machines for hub propulsion based on the field modulation principle. Firstly, the topology of the proposed DPMV machine was introduced, and a commercial PM machine was used as a benchmark. Secondly, the rotor PM, stator PM, and armature magnetic fields were derived and analyzed considering the modulation effect. Meanwhile, the contribution of each harmonic to average torque was pointed out. It can be concluded that the 7th-, 12th-, 19th- and 24th-order flux density harmonics are the main source of average torque. Thanks to the multi-working harmonic characteristics, the average torque of DPMV machines has significantly increased by 31.8% compared to the commercial PM machine while also reducing the PM weight by 75%. Thirdly, the auxiliary barrier structure and dual three-phase winding configuration were proposed from the perspective of optimizing the phase and amplitude of working harmonics, respectively.

Li et al. [5] analyzed the fault characteristics of inter-turn short circuits in the excitation windings of synchronous condensers under unbalanced grid voltage. Mathematical models were developed to represent the air gap flux density and stator parallel currents for four operating conditions: normal operation and inter-turn short-circuit fault under balanced voltage, as well as a process without a fault and with an inter-turn short-circuit fault under unbalanced voltage. By comparing the harmonic contents and amplitudes, various aspects of the fault mechanism of synchronous condensers were revealed, and the operating characteristics under different conditions were analyzed. Considering the four aforementioned operating conditions, finite element simulation models were created for the TTS-300-2 synchronous condenser in a specific substation as a case study. The results demonstrate that the inter-turn short-circuit fault in the excitation windings under unbalanced voltage leads to an increase in even harmonic currents in the stator parallel currents, particularly in the second and fourth harmonics.

Alharkan, in ref. [6], developed a novel reinforcement neural network learning approach based on machine learning to find the best solution for the tracking problem of the switched reluctance motor (SRM) device in real time. The reference signal model, which minimizes torque pulsations, was combined with a tracking error to construct the augmented structure of the SRM device. A discounted cost function for the augmented SRM model was described to assess the tracking performance of the signal. To track the optimal trajectory, a neural network (NN)-based RL approach was developed. This method achieved the optimal tracking response to the Hamilton–Jacobi–Bellman (HJB) equation for a nonlinear tracking system. Simulation findings were undertaken for SRM to confirm the viability of the suggested control strategy.

Belkhadir et al. [7] presented an analytical model of the stator winding unbalance fault represented by lack of turns. Here, mathematical approaches were used by introducing a stator winding parameter for the analytical modeling of the faulty machine. This model can be employed to determine the various quantities of the machine under different fault levels, including the magnetomotive force, the flux density in the air-gap, the flux generated by the stator winding, the stator inductances, and the electromagnetic torque. On this basis, a corresponding link between the fault level and its signature was established. The feasibility and efficiency of the analytical approach were validated by finite element analysis and experimental implementation.

Aladetola et al. [8] developed a control approach to minimize the issue of torque ripple effects in synchronous reluctance machines (SynRMs). This work was performed in two steps: Initially, the reference current calculation bloc was modified to reduce the torque ripple of the machine. A method for calculating the optimal reference currents based on the stator joule loss was proposed. The proposed method was compared to two methods used in the literature, the FOC and MTPA methods. A comparative study between the three methods based on the torque ripple rate showed that the proposed method allowed for a significant reduction in the torque ripple. The second contribution to the minimization of the torque ripple was to propose a sliding mode control. This control suffers from the phenomenon of "Chattering", which affects the torque ripple. To solve this problem, a second-order sliding mode control was proposed.

Damine et al. [9] introduced a robust process for extracting rolling bearing defect information based on combined mode ensemble empirical mode decomposition (CMEEMD) and an enhanced deconvolution technique. Firstly, the proposed CMEEMD extracts all combined modes (CMs) from adjoining intrinsic mode functions (IMFs) decomposed from the raw fault signal via ensemble empirical mode decomposition (EEMD). Then, a selection indicator known as kurtosis median absolute deviation (KMAD) was created in this research to identify the combination of the appropriate IMFs. Finally, the enhanced deconvolution process minimized noise and improved defect identification in the identified CM. Analyzing real and simulated bearing signals demonstrated that the developed method showed excellent performance in extracting defect information. Comparing the results between selecting the sensitive IMF using kurtosis and selecting the sensitive CM using the proposed KMAD showed that the identified CM contained rich fault information in many cases.

Ruz-Hernandez et al. [10] presented the development of a neural inverse optimal control (NIOC) for a regenerative braking system installed in electric vehicles (EVs), which is composed of a main energy system (MES), including a storage system and an auxiliary energy system (AES). The latter one is composed of a supercapacitor and a buck–boost converter. To build up the NIOC, a neural identifier was trained with an extended Kalman filter (EKF) to estimate the real dynamics of the buck–boost converter. The NIOC was implemented to regulate the voltage and current dynamics in the AES. For testing the drive system of the EV, a DC motor was considered, with speed controlled using a PID controller to regulate the tracking source in regenerative braking. Simulation results illustrated the efficiency of the proposed control scheme to (1) track time-varying references of the AES voltage and current dynamics measured at the buck–boost converter and (2) guarantee that charging and discharging operation modes of the supercapacitor would be initiated. In addition, it was demonstrated that the proposed control scheme enhances the EV storage system's efficacy and performance when the regenerative braking system is working.

Author Contributions: Investigation, M.B. and L.D.; Writing—original draft, M.B. and L.D.; Writing—review and editing, M.B. and L.D. All authors have read and agreed to the published version of the manuscript.

Funding: This research received no external funding.

Conflicts of Interest: The authors declare no conflicts of interest.

References

1. Saeed, M.; Fernandez, D.; Guerrero, J.M.; Diaz, I.; Briz, F. Insulation Condition Assessment in Inverter-Fed Motors Using the High-Frequency Common Mode Current: A Case Study. *Energies* **2024**, *17*, 470. [CrossRef]
2. Pietrzak, P.; Pietrzak, P.; Wolkiewicz, M. Microcontroller-Based Embedded System for the Diagnosis of Stator Winding Faults and Unbalanced Supply Voltage of the Induction Motors. *Energies* **2024**, *17*, 387. [CrossRef]
3. Gnacinski, P.; Peplinski, M.; Muc, A.; Hallmann, D.; Jankowski, P. Effect of Ripple Control on Induction Motors. *Energies* **2023**, *16*, 7831. [CrossRef]
4. Sun, Y.; Bianchi, N.; Ji, J.; Zhao, W. Improving Torque Analysis and Design Using the Air-Gap Field Modulation Principle for Permanent-Magnet Hub Machines. *Energies* **2023**, *16*, 6214. [CrossRef]
5. Li, J.; Zhang, C.; He, Y.; Hu, X.; Geng, J.; Ma, Y. Impact of Inter-Turn Short Circuit in Excitation Windings on Magnetic Field and Stator Current of Synchronous Condenser under Unbalanced Voltage. *Energies* **2023**, *16*, 5695. [CrossRef]
6. Alharkan, H. Torque Ripple Minimization of Variable Reluctance Motor Using Reinforcement Dual NNs Learning Architecture. *Energies* **2023**, *16*, 4839. [CrossRef]
7. Belkhadir, A.; Pusca, R.; Belkhayat, D.; Romary, R.; Zidani, Y. Analytical Modeling, Analysis and Diagnosis of External Rotor PMSM with Stator Winding Unbalance Fault. *Energies* **2023**, *16*, 3198. [CrossRef]
8. Aladetola, O.D.; Ouari, M.; Saadi, Y.; Mesbahi, T.; Boukhnifer, M.; Adjallah, K.H. Advanced Torque Ripple Minimization of Synchronous Reluctance Machine for Electric Vehicle Application. *Energies* **2023**, *16*, 2701. [CrossRef]
9. Damine, Y.; Bessous, N.; Pusca, R.; Megherbi, A.C.; Romary, R.; Sbaa, S. A New Bearing Fault Detection Strategy Based on Combined Modes Ensemble Empirical Mode Decomposition, KMAD, and an Enhanced Deconvolution Process. *Energies* **2023**, *16*, 2604. [CrossRef]
10. Ruz-Hernandez, J.A.; Djilali, L.; Canul, M.A.R.; Boukhnifer, M.; Sanchez, E.N. Neural Inverse Optimal Control of a Regenerative Braking System for Electric Vehicles. *Energies* **2022**, *15*, 8975. [CrossRef]

Disclaimer/Publisher's Note: The statements, opinions and data contained in all publications are solely those of the individual author(s) and contributor(s) and not of MDPI and/or the editor(s). MDPI and/or the editor(s) disclaim responsibility for any injury to people or property resulting from any ideas, methods, instructions or products referred to in the content.

Article

Advanced Torque Ripple Minimization of Synchronous Reluctance Machine for Electric Vehicle Application

Olaoluwa Demola Aladetola [1], Mondher Ouari [1], Yakoub Saadi [2], Tedjani Mesbahi [2], Moussa Boukhnifer [3,*] and Kondo Hloindo Adjallah [1]

1 Laboratoire de Conception, Optimisation et Modélisation des Systèmes, Université de Lorraine, 57000 Metz, France
2 ICube, CNRS (UMR 7357) INSA Strasbourg, University of Strasbourg, 67000 Strasbourg, France
3 Université de Lorraine, LCOMS, 57000 Metz, France
* Correspondence: moussa.boukhnifer@univ-lorraine.fr

Abstract: The electric machine and the control system determine the performance of the electric vehicle drivetrain. Unlike rare-earth magnet machines such as permanent magnet synchronous machines (PMSMs), synchronous reluctance machines(SynRMs) are manufactured without permanent magnets. This allows them to be used as an alternative to rare-earth magnet machines. However, one of the main drawbacks of this machine is its high torque ripple, which generates significant acoustic noise. The most typical method for reducing this torque ripple is to employ an optimized structural design or a customized control technique. The objective of this paper is the use of a control approach to minimize the torque ripple effects issue in the SynRM. This work is performed in two steps: Initially, the reference current calculation bloc is modified to reduce the torque ripple of the machine. A method for calculating the optimal reference currents based on the stator joule loss is proposed. The proposed method is compared to two methods used in the literature, the FOC and MTPA methods. A comparative study between the three methods based on the torque ripple rate shows that the proposed method allows a significant reduction in the torque ripple. The second contribution to the minimization of the torque ripple is to propose a sliding mode control. This control suffers from the phenomenon of "Chattering" which affects the torque ripple. To solve this problem, a second-order sliding mode control is proposed. A comparative study between the different approaches shows that the second-order sliding mode provides the lowest torque ripple rate of the machine.

Keywords: electric vehicle; synchronous reluctance machine; field-oriented control; maximum torque per ampere; optimal current calculation; sliding mode control; torque ripple minimization

Citation: Aladetola, O.D.; Ouari, M.; Saadi, Y.; Mesbahi, T.; Boukhnifer, M.; Adjallah, K.H. Advanced Torque Ripple Minimization of Synchronous Reluctance Machine for Electric Vehicle Application. *Energies* **2023**, *16*, 2701. https://doi.org/10.3390/en16062701

Academic Editor: King Jet Tseng

Received: 10 February 2023
Revised: 5 March 2023
Accepted: 7 March 2023
Published: 14 March 2023

Copyright: © 2023 by the authors. Licensee MDPI, Basel, Switzerland. This article is an open access article distributed under the terms and conditions of the Creative Commons Attribution (CC BY) license (https://creativecommons.org/licenses/by/4.0/).

1. Introduction

The rapid increase in the number of conventional vehicles has led to a significant increase in greenhouse gas emissions, the depletion of fossil fuels, and various negative consequences for the people living in these environments [1]. Unlike conventional vehicles, which face the problem of fuel poverty, electric vehicles (EVs) can have significant emissions and environmental benefits over conventional vehicles. As well, they can significantly reduce fuel costs due to the high efficiency of electric drive components [2].

The electrification of the automotive sector is accelerating, and carmakers and equipment manufacturers are reinventing electric machines to adapt them to the constraints of electric drivetrains. A high power density, high torque density, wide speed range, and efficiency are critical factors in the selection of electric motor technology for this application [3]. Permanent magnet synchronous machines (PMSMs) are by far the most widely utilized electric machine technology in the electric vehicle (EV) market [4].

In 2021, PMSMs accounted for 84% of the electric car market [5]. However, the magnets used in these machines are typically rich in rare-earth materials (REMs), primarily

Neodymium, but also often contain a range of heavy rare earth, such as Dysprosium [6,7]. Nevertheless, the cost of REM-based machines has increased over several years. Furthermore, due to restricted resources, the use of REM-based machines in EV applications is now being challenged [8].

The above factors have prompted several equipment manufacturers to design rare-earth-free machines, such as Renault's wound rotor synchronous machine (WRSM) in the ZOE and the Audi induction machine (IM) in its e-tron models [9,10]. Due to the robustness, simplicity of fabrication, small size, and compatibility with the requirements of the EV electric machine, the synchronous reluctance machine (SynRM) is an alternative for REM-based machines. The SynRM has a wound stator that has neither conductors nor magnets like the IM and it operates like a WRSM without a DC field winding in its rotor [11]. Moreover, the power converter used to supply this machine is a three-phase inverter, which facilitates the replacement of the IMs and PMSMs without a specific power converter. Figure 1 illustrates a simplified architecture of the essential components of an electric vehicle propelled using a synchronous reluctance machine.

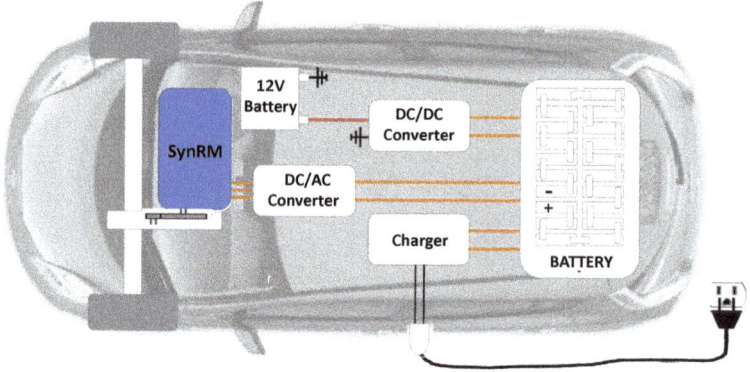

Figure 1. Simplified electric vehicle drivetrain architecture propelled by a SynRM.

However, the nonlinear magnetic path of the SynRM and the operational saturation of the rotor core segments cause significant torque ripples and acoustic noise [12,13]. However, these drawbacks can be significantly reduced with an optimal structure design [14,15] and/or a good control strategy [16–18]. The optimization of flux barriers [19,20], rotor ribs [21,22], rotor skewing [23,24], and adding permanent magnets [25,26] is the most common strategy for an optimal structural design. Although optimizing the SynRM structure may offer satisfactory results, the design procedure is typically time-consuming.

The focus of this research is on control-based strategies for minimizing the torque ripple of the SynRM. Several strategies to reduce the torque ripple effect of the machine have been investigated in the literature.

The authors of the publication [11] present a general review of various control scheme strategies for SynRM's current regulation. This research examines the designs, techniques, benefits, and drawbacks of synchronous reluctance machine control systems, such as direct torque control (DTC), field-oriented control (FOC), predictive control, and many others. This study demonstrates the limitations of each method for reducing the torque ripple effect in a synchronous reluctance machine. The DTC method provides high dynamic control, which makes it superior to other methods [11]. Because it does not use a current controller, this approach achieves a substantially superior transient torque control performance. Furthermore, it controls the machine solely by stator resistance, resulting in reasonably robust machine control with quick dynamics. This method is appropriate for specifications that require a better transient response rather than a steady-state response for control [27]. Nevertheless, this method generates significant torque ripples as compared to other approaches, and its implementation necessitates the use of a torque sensor or an extra sensorless torque block solution [28]. This adds significant computing time. Furthermore, because this method uses a variable switching frequency to control the flux, it produces a relatively high harmonic current and high torque ripples, causing significant noise levels in the machine. In reference [29], the torque ripples were handled satisfactorily in the DTC technique using multilevel inverters. Moreover, a mechanism is created and employed to limit the torque in [27–29]. In this technique, the torque-limiting mechanism adjusts the flux reference with respect to the torque error sign to ensure a steady machine operation.

A field-oriented control (FOC) strategy is proposed in [30] to achieve convenient control of the SynRM. This method controls the SynRM in the d, q reference frame, representing the machine as a direct current (DC) machine. In this review, FOC is categorized into two techniques for controlling decoupling currents i_d, i_q in the synchronous reference frame: direct field-oriented control (DFOC) and indirect field-oriented control (IFOC). This method features a precise control method, reduced torque ripples in comparison to the DTC method, improved steady-state responsiveness, and a consistent switching frequency, which makes it attractive to researchers because of its high steady-state performance [30].

Another control-based method for reducing the SynRM torque ripple is to add specified current harmonics to the original sinusoidal stator currents. The authors of [31] investigated the average torque of a two-phase SynRM and defined the optimal current using different stator inductance harmonics. Each torque harmonic requires multiple current harmonics to be reduced. When numerous dominant torque harmonics are taken into account, the process of determining the link between each torque harmonic and the corresponding current harmonics can become lengthy and difficult. Therefore, the suggested method makes determining the appropriate currents for a multiphase SynRM extremely challenging. The stator inductances and low-order harmonics are measured in [32] to determine the optimal currents using the electromagnetic torque equation. But nonetheless, measuring the high-order stator inductance harmonics accurately is extremely difficult. This means the optimal currents determined by measured inductances may not result in the most effective torque ripple reduction. Some strategies for reducing torque ripples rely on a reference currents calculation bloc. This bloc's purpose is to generate reference currents via the reference torque [16]. To minimize the torque ripple, the active torque ripple cancellation control technique is examined in [33]. To provide a smooth output torque, the active torque ripple cancellation method actively regulates the excitation of current waveforms using torque to the current function. The term "active" refers to a method for canceling the torque ripple of the machine while it is functioning at a varied torque-speed range.

This paper will address the problem of the torque ripple minimization of a synchronous reluctance machine used in electric vehicle propulsion. Based on a velocity/current cascade control strategy, we first suggest changing the reference currents calculation bloc, which transforms the reference torque into reference currents via a stator current optimization method. In other words, the torque ripple can be reduced by optimizing the reference currents because stator currents represent the machine's torque. To assess the efficacy of the suggested method, we will replicate the reference currents calculation investigated in the literature, namely the control by flux-oriented control (FOC) and maximum torque per ampere (MTPA) with PI control. The torque ripple ratio of each method is then examined in a comparative study.

Secondly, based on the velocity/current cascade control, the optimal currents calculations method from the first study will be chosen. We propose nonlinear controls to replace the PI control, notably the classical sliding mode control, and the second-order sliding mode control, to improve the stator current control and hence the torque ripple minimization. The performance of each control approach is then compared, along with the torque ripple ratio.

The structure of this article is as follows: Section 2 explains the modeling and behavior of the synchronous reluctance machine, as well as the velocity/current cascade control strategy. The reference currents calculation bloc description utilizing the FOC, the MTPA control, and the suggested optimal current computation approach are covered in Section 3. In Section 4, the proposed classical and second-order sliding mode controllers are combined with the optimal reference currents calculation method. Section 5 contains the conclusions that bring this article to a close.

2. SynRM Modeling and Description of the Velocity/Currents Cascade Control Strategy

In this section, the description of the modeling, as well as the velocity/currents control strategy used in driving the synchronous reluctance machine in this work are presented.

2.1. SynRM Modeling

The synchronous reluctance machine is a pure AC machine that requires a polyphase sinusoidal AC current. The torque of this machine is produced by a difference in magnetic conductivities along the direct axes of the rotor, as well as by the quadrature, which lacks permanent magnets and field windings [34,35]. The SynRM used in this work is a three-phase flux barrier type with four-pole machine as shown in Figure 2a.

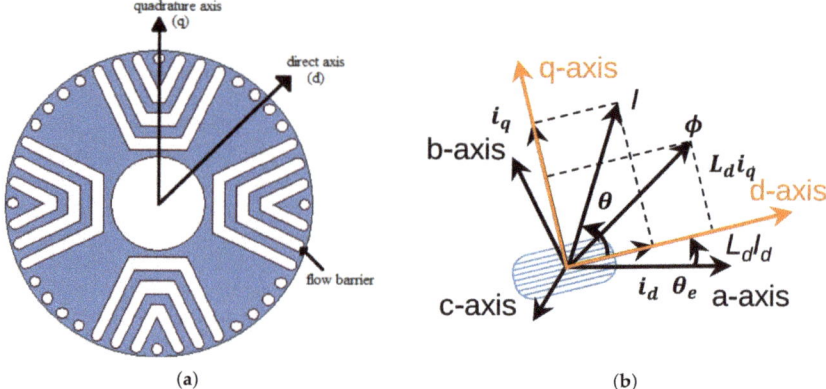

Figure 2. (a): Cross section of an exemplary SynRM with four poles. (b): Transformation of the system in synchronous (dq) reference frame [36].

2.1.1. Electric Model

SynRM's electrical model is based on the following assumptions [37]:

- Magnetic materials are isotropic and non-saturable.
- The hysteresis effect and iron losses are neglected.
- The inductance variations are sinusoidal (first harmonic hypothesis).
- The capacitive coupling between the machine's windings is ignored.

Given the assumptions, the voltage v applied to a phase is equal to the resistive voltage drop across the phase winding plus the flux change beneath a rotor pole and is denoted by

$$\begin{cases} v = R_s i + \dfrac{d\Phi}{dt} \\ \Phi = L(p\theta) \cdot i \end{cases} \quad (1)$$

where

- $v = \begin{bmatrix} v_a & v_b & v_c \end{bmatrix}^T$: the stator voltage vector;
- $i = \begin{bmatrix} i_a & i_b & i_c \end{bmatrix}^T$: the stator current vector;
- $\Phi = \begin{bmatrix} \Phi_a & \Phi_b & \Phi_c \end{bmatrix}^T$: the vector of the total fluxes through the windings $a - b - c$;
- R_s: the resistance of a stator phase;
- θ and p: the mechanical position and the number of pole pairs, respectively;

- $L(p\theta)$: the stator inductance matrix given by [38]

$$L(p\theta) = \begin{bmatrix} L_a(p\theta) & M_{ab}(p\theta) & M_{ac}(p\theta) \\ M_{ba}(p\theta) & L_b(p\theta) & M_{bc}(p\theta) \\ M_{ca}(p\theta) & M_{cb}(p\theta) & L_c(p\theta) \end{bmatrix} \quad (2)$$

With L_i is the stator inductance of phase i and M_{ij} is the mutual inductance between phases i and j ($i, j = (a, b, c)$) [38,39].

The electrical equations in the $d - q$ frame (see Figure 2b), in the absence of a zero sequence current component, are given by [40–42]

$$\begin{cases} v_{ds} = R_s i_{ds} + \dfrac{d\Phi_{ds}}{dt} - p\Omega\Phi_{qs} \\ v_{qs} = R_s i_{qs} + \dfrac{d\Phi_{qs}}{dt} + p\Omega\Phi_{ds} \end{cases} \quad (3)$$

with the following:
- v_{ds} and v_{qs} are the stator voltage in the d and q axes.
- Ω is the machine velocity.
- Φ_s, Φ_{qs}, and Φ_{ds} are the total stator and flux linkage in the d and q axes given by

$$\begin{cases} \Phi_{ds} = L_d i_{ds} \\ \Phi_{qs} = L_q i_{qs} \\ \Phi_s = \sqrt{\Phi_{ds}^2 + \Phi_{qs}^2} \end{cases} \quad (4)$$

- L_d, L_q are the d and q-axes stator inductances.

Finally, the voltage equations can be written as follows:

$$\begin{bmatrix} v_{ds} \\ v_{qs} \end{bmatrix} = R_s \begin{bmatrix} i_{ds} \\ i_{qs} \end{bmatrix} + \begin{bmatrix} L_d \\ L_q \end{bmatrix} \frac{d}{dt} \begin{bmatrix} i_{ds} \\ i_{qs} \end{bmatrix} + p\Omega \begin{bmatrix} 0 & -L_q \\ L_d & 0 \end{bmatrix} \begin{bmatrix} i_{ds} \\ i_{qs} \end{bmatrix} \quad (5)$$

2.1.2. Electromechanical Model

The electromagnetic torque of the SynRM can be expressed by [42]

$$T_e = p(L_d - L_q) i_{ds} i_{qs} \quad (6)$$

From the electromagnetic torque equation, the fundamental relation of the dynamics of the rotating part of the machine is given by [40–42]

$$\frac{d\Omega}{dt} = \frac{1}{J}(T_e - T_r - f_r \Omega) \quad (7)$$

- Ω: rotational velocity of the machine, in rad/s.
- T_e: electromagnetic torque produced by the machine, in Nm.
- T_L: load torque, in Nm.
- f_r: viscous friction coefficient, in Ns^2/m^2.

The SynRM state model in $d-q$ is finally written as follows:

$$\frac{d}{dt}\begin{bmatrix}i_{ds}\\i_{qs}\\\Omega\\\theta\end{bmatrix}=\begin{bmatrix}-\frac{R_s}{L_d}i_{ds}+p\Omega\frac{L_q}{L_d}i_{qs}\\-\frac{R_s}{L_q}i_{qs}+p\Omega\frac{L_d}{L_q}i_{ds}\\\frac{\frac{3}{2}p(L_d-L_q)i_{ds}i_{qs}}{J}-\frac{f_r}{J}\Omega-\frac{T_r}{J}\\\Omega\end{bmatrix}+\begin{bmatrix}\frac{1}{L_d}&0\\0&\frac{1}{L_q}\\0&0\\0&0\end{bmatrix}\begin{bmatrix}V_{ds}\\v_{qs}\end{bmatrix} \qquad (8)$$

2.1.3. Vehicle Load Torque Modeling

Figure 3 shows the driving force and the mean forces resistant to the advance of a vehicle in a slope α [43].

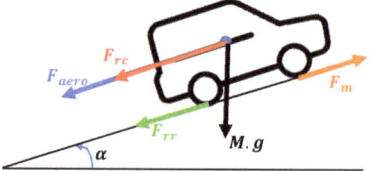

Figure 3. The typical driving force and resisting forces components of a vehicle [43].

where
- F_m: the slope force or tractive force that is required to drive the vehicle up.
- F_{aero}: the aerodynamic force created by the friction of the vehicle's body moving through the air.
- F_{rr}: the rolling resistance force.
- F_{rc}: the resistance force exerted by the vehicle weight as it goes up and down a hill.
- M: the vehicle mass.
- g: the acceleration due to gravity on Earth.

The expression of each resisting force is given by [44]

$$\begin{cases}F_{aero}=\frac{1}{2}\rho c_x s_f V^2\\F_{rr}=f_{rr}mg\cos(\alpha)\\F_{rc}=mg\sin(\alpha)\end{cases} \qquad (9)$$

where
- ρ: the density of the air, in kg/m^3.
- c_x: the drag coefficient.
- s_f: frontal cross-sectional area, in m^2.
- f_{rr}: rolling resistance value, in N.

From [44–46], the linear speed of a vehicle V can be expressed using different forces as follows:

$$M\frac{dV}{dt}=F_m-F_{aero}-F_{rr}-F_{rc} \qquad (10)$$

Because $V=R_{sc}\Omega$ where R_{sc} is the resistance of the EV in a slope.

The total load torque of the vehicle T_r in the steady state can be written from the Equation (9) by

$$T_r=\frac{1}{2}\rho c_x s f R_{sc}^3 \Omega_{sc}^2+mgR_{sc}[\sin(\alpha)+f_{rr}\cos(\alpha)] \qquad (11)$$

2.2. SynRM Cascade Control Strategy

Figure 4 shows the cascade velocity/currents control strategy used in this study [47]. The EV driver is presented by a velocity controller that provides the reference torque T_e^*. An indirect torque control approach is used to regulate the machine's torque by regulating the stator currents given by i_d^* and i_q^*. The reference currents calculation bloc is used to transform the reference torque into reference currents i_d^* and i_q^*. These currents are then controlled in the internal control loop.

This strategy allows torque to be controlled indirectly by controlling the currents and provides a separation between the electrical and mechanical variables.

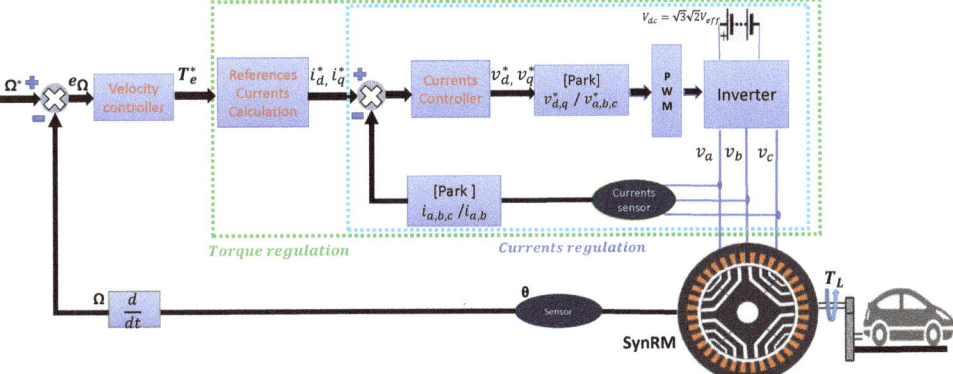

Figure 4. Block diagram of the cascade velocity/currents control strategy [47].

3. Torque Ripple Minimization by Using the Currents References Calculation

In this section, different techniques for calculating current reference in the reference currents calculation bloc have been adopted, to study the effect of currents on torque ripple and control output.

3.1. Conventional Field-Oriented Control (FOC)

The conventional field-oriented control of the synchronous machine controls the current with respect to the reference current which automatically controls the torque by using only one component of the current and by setting the other to a constant (zero in the case of a permanent magnet synchronous machine) [48]. Analogously, this command consists of imposing a constant value on one component of the current and allowing the other component to regulate the torque given that the expression of the electromagnetic torque of the machine in the reference $(d - q)$ is

$$T_e = p(L_d - L_q)i_{d_s}i_{q_s} \qquad (12)$$

By imposing the reference of the component i_d to a constant,

$$i_d^* = c^{te} \qquad (13)$$

From (12) and (13), the reference of the component i_q can be calculated as follows:

$$i_q^* = \frac{T_e^*}{p(L_d - L_q)i_d^*} \qquad (14)$$

3.2. Maximum Torque per Ampere (MTPA)

MTPA or maximum torque per ampere operation is the most preferred operating mode for any motor operating with the vector control [49].

This method provides the maximum torque for a given operating current. This method controls both currents i_d and i_q. The operating condition at the maximum point can be deduced from the electromagnetic torque equation:

$$T_e = p(L_d - L_q)i_{d_s}i_{q_s} \tag{15}$$

Assuming sinusoidal stator currents, Park's transformation allows us to write

$$\begin{cases} i_{ds} = \sqrt{\dfrac{3}{2}}\, I_s \sin\gamma \\ i_{qs} = \sqrt{\dfrac{3}{2}}\, I_s \cos\gamma \end{cases} \tag{16}$$

with I_s the amplitude of the stator current and $\gamma = \omega t + \varphi$, where ω and φ are the electrical network pulsation in rad/s and phase at the reference origin in rad, respectively.

Thus, from Equations (15) and (16), the expression of the electromagnetic couple becomes

$$T_e = \frac{3}{2}p(L_d - L_q)I_s^2 \sin\gamma \cos\gamma \tag{17}$$

Knowing that $\sin\gamma\cos\gamma = \dfrac{\sin 2\gamma}{2}$, the expression of the electromagnetic torque becomes

$$T_e = \frac{3}{2}p(L_d - L_q)I_s^2 = \frac{\sin 2\gamma}{2} \tag{18}$$

Then, the condition for the maximization of torque per ampere can be written as

$$\left.\frac{dT_e}{d\gamma}\right|_{I_s = cte} = \frac{3}{2}p(L_d - L_q)I_s^2 \cos 2\gamma = 0 \tag{19}$$

Solving the Equation (15) allows finding the expression of the components of the current as follows [49]:

$$i_d = i_q = \sqrt{\frac{T_e}{\frac{3}{2}p(L_d - L_q)}} \tag{20}$$

By replacing the measured values by the reference values in Equation (20), we can write

$$i_d^* = i_q^* = \sqrt{\frac{T_e^*}{\frac{3}{2}p(L_d - L_q)}} \tag{21}$$

3.3. Optimal Currents Calculations

The electromagnetic torque of the machine can be written in the form [50]

$$T_e = \frac{1}{2}i^T \frac{\partial L}{\partial \theta} i \tag{22}$$

The currents in the $a - b - c$ frame can be written as the following:

$$i = \begin{bmatrix} i_a \\ i_b \\ i_c \end{bmatrix} = P(p\theta)\begin{bmatrix} i_d \\ i_q \\ i_h \end{bmatrix} = T_{32}.\, R(\theta).\begin{bmatrix} i_d \\ i_q \end{bmatrix} \tag{23}$$

with

- i_h: zero sequence current assumed to be null;
- $P(p\theta)$: Park's matrix;
- $R(\theta)$: rotation matrix;

- T_{32}: Concordia matrix.

The machine torque can be written as:

$$T_e = \begin{bmatrix} i_d \\ i_q \end{bmatrix}^T \cdot R^T \cdot T_{32}^T \cdot \frac{\partial L(p\theta)}{\partial \theta} \cdot T_{32} \cdot R \begin{bmatrix} i_d \\ i_q \end{bmatrix} \quad (24)$$

We suppose

$$\begin{bmatrix} a(p\theta) & c(p\theta) \\ c(p\theta) & b(p\theta) \end{bmatrix} = R^T \cdot T_{32}^T \cdot \frac{\partial L(p\theta)}{\partial \theta} \cdot T_{32} \cdot R \quad (25)$$

By replacing (25) in (24), the torque is given as follows:

$$\Gamma_{em} = a(p\theta)i_d^2 + b(p\theta)i_q^2 + 2c(p\theta)i_d i_q \quad (26)$$

The problem is to determine the currents i_d and i_q which will provide a constant torque. This problem has an infinite number of solutions. To remedy this, the solution which generates the least stator loss by joule effect is sought. The stator joule losses are defined by

$$P_j = R_s(i_d^2 + i_q^2) \quad (27)$$

The search for the solution becomes an optimization problem with the stator loss equation as an objective function of two variables and the torque Equation (27) as a constraint [51]:

$$\begin{cases} \Gamma_{em} = a(p\theta)i_d^2 + b(p\theta)i_q^2 + 2c(p\theta)i_d i_q \\ (i_d^2 + i_q^2) \quad \text{to minimize} \end{cases} \quad (28)$$

In order to solve the problem, the Lagrangian function (Δ) is used. It can be written as

$$\Delta = (i_d^2 + i_q^2) + \mu\left(\Gamma_{em} - \left(a(p\theta)i_d^2 + b(p\theta)i_q^2 + 2c(p\theta)i_d i_q\right)\right) \quad (29)$$

with μ being the Lagrange multiplier.

The derivation of Δ with respect to i_d, i_d, and μ gives

$$\begin{cases} 2i_q + \mu(-2ai_d - 2ci_q) = 0 \\ 2i_d + \mu(-2bi_q - 2ci_d) = 0 \\ T_e = a(p\theta)i_d^2 + b(p\theta)i_q^2 + 2c(p\theta)i_d i_q \end{cases} \quad (30)$$

By solving the system of Equations (28), we can write

$$\begin{cases} i_q = \frac{(1-\mu a)i_d}{\mu c} \\ i_d = \sqrt{\frac{|T_e|}{\frac{\mu^2(a^2b-ac^2)+\mu(2c^2-2ab)+b}{\mu^2 c^2}}} \end{cases} \quad (31)$$

$$\mu = \begin{cases} \frac{a+b+\sqrt{(a-b)^2+4c^2}}{2(ab-c^2)} & \text{if } T_e < 0 \\ \frac{a+b+\sqrt{(a-b)^2-4c^2}}{2(ab-c^2)} & \text{if } T_e > 0 \end{cases} \quad (32)$$

By replacing the measured values by the reference values in Equation (31), we can write

$$\begin{cases} i_q^* = \frac{(1-\mu a)i_d}{\mu c} \\ i_d^* = \sqrt{\frac{|T_e^*|}{\frac{\mu^2(a^2b-ac^2)+\mu(2c^2-2ab)+b}{\mu^2 c^2}}} \end{cases} \quad (33)$$

In this section, we are interested in the reduction in the torque ripple through the optimization of the reference current calculations used in the reference currents calculation bloc. In order to examine the developed method, the optimization of the reference currents are compared with the two methods of the literature, namely FOC and MTPA. For that, we will integrate the three methods in the reference currents calculation bloc in the cascade control strategy adopted in this study and presented in Section 2.2.

3.4. Simulation Results of Different Techniques of Current Calculation with PI Regulators

The reference currents calculation bloc will be used in this section to implement the three reference currents calculations that were previously described. A PI controller is used to regulate the velocity and current using the cascade velocity/currents control strategy. The simulation results were achieved using the Matlab/Simulink software tools, with the SynRM parameters utilized listed in Appendix A. The chosen velocity profile presented in Figure 5a covers multiple operating points: low velocity (300 rpm), nominal velocity (1500 rpm), and negative velocity (-300 rpm). As depicted in Figure 5b, various torque loads were applied at various points during the steady and transient states. The PI velocity controller parameters used are $K_p = 2.31$, $K_i = 387$, and PI currents controllers parameters used in the simulation are $K_p^{'} = 1400$ and $K_i^{'} = 10^6$.

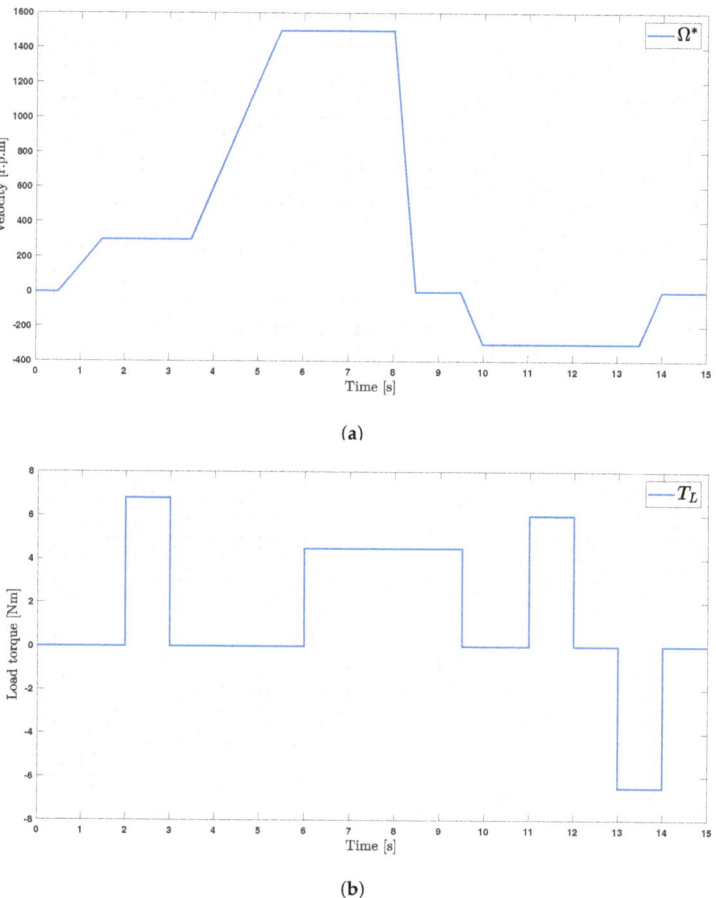

(a)

(b)

Figure 5. Reference velocity and load torque profiles applied in the simulation: (**a**) reference velocity profile and (**b**) load torque profile.

3.4.1. Simulation with the Conventional Field-Oriented Control

The FOC reference currents calculation method is implemented in the reference currents calculation bloc as follows:

$$\begin{cases} i_d^* = 3 & \text{rated RMS current} \\ i_q^* = \frac{T_e^*}{p(L_d - L_q)i_d^*} \end{cases} \quad (34)$$

Figure 6 displays the velocity response using the conventional field-oriented control with respect to the selected profile. With no static error, the velocity closely matches the reference. Nevertheless, a tracking error results from the PI controller property. It is also important to note that the velocity has no overshoot thanks to the controllers' parameters chosen in the simulation.

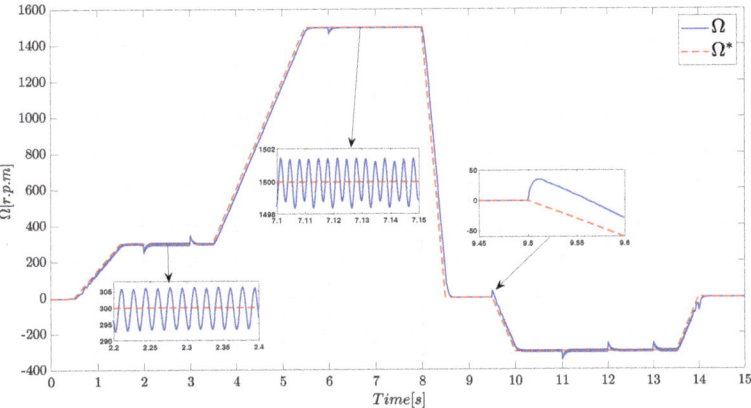

Figure 6. Velocity response obtained by PI controllers and conventional field-oriented control.

During the different moments of the application of the load torque, small decreases in machine velocity are observed compared to the reference. In the same way, small overshoots appear when the load is canceled. These drops and overshoots disappear gradually to return to the reference, as shown by the zooms in the figure.

For the current control, Figure 7 shows the evolution of the current components i_d and i_q. The evolution of the currents shows a good control of both current components over the different velocity and torque ranges, as shown in Figure 7a. Figure 7b demonstrates that even at low velocity the current is constant at the $id = 3A$ because of the value of $id^* = 3A$.

Although this method ensures the correct operation of the machine over the entire velocity/torque operating range, it can be seen that the reference current i_d^* is somewhat undulating, which has an impact on the current i_d and therefore on the torque ripple, as shown in the different zooms of Figure 7a.

The torque evolution is shown in Figure 8. In this strategy, the torque has been controlled from the calculation of the reference current using the classical field-oriented control. It can be seen that the torque is able to convince the load torque (T_L) and the intrinsic torque of the machine ($f_r\Omega$). Nevertheless, as illustrated in the zooms in Figure 8, this technique results in significant machine torque ripples at low and high machine velocities, which are quantified at 41.07 and 48.08% with the torque load, respectively.

The torque ripple rate is calculated at a steady state as follows:

$$\Delta T_e(\%) = \frac{T_{e_{max}} - T_{e_{min}}}{T_{e_{avg}}} \times 100 \quad (35)$$

- $T_{e_{max}}$: the maximum torque value.
- $T_{e_{min}}$: the minimum torque value.
- $T_{e_{avg}}$: the average torque value.

These ripples are caused by the control of stator currents, which are directly related to torque, as given in Equation (12). In other words, the conventional field-oriented control does not provide optimal reference currents for optimizing stator currents and, as a result, the reduction in torque ripples.

In the following part, the maximum torque per ampere (MTPA) method is tested under the same simulation conditions to determine its capacity to reduce the machine's torque ripple.

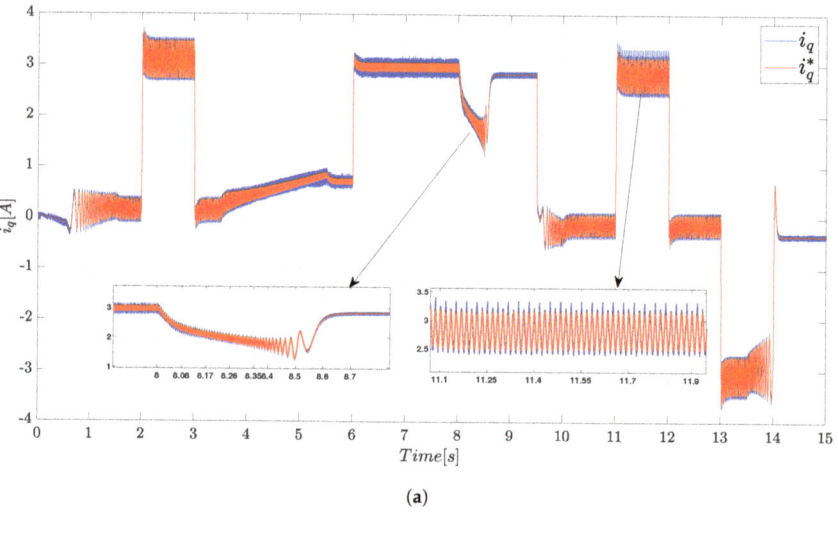

(a)

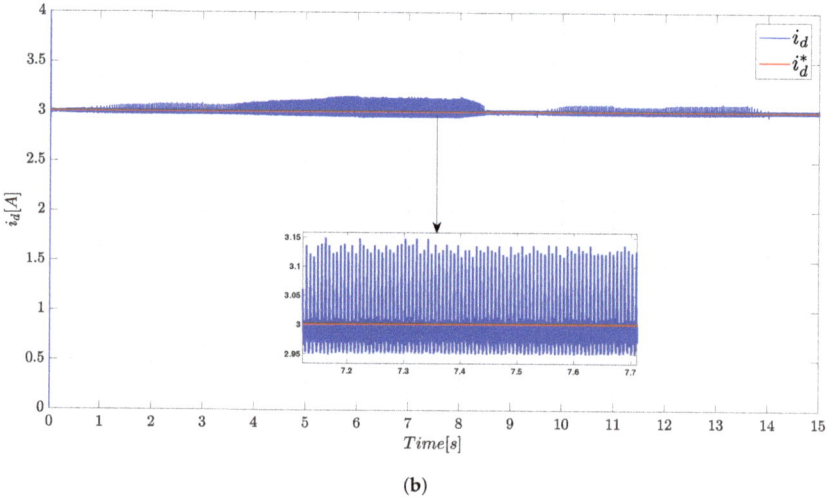

(b)

Figure 7. Response of the current components obtained by the PI controllers and the conventional field-oriented control: (**a**) i_q response and (**b**) i_d response.

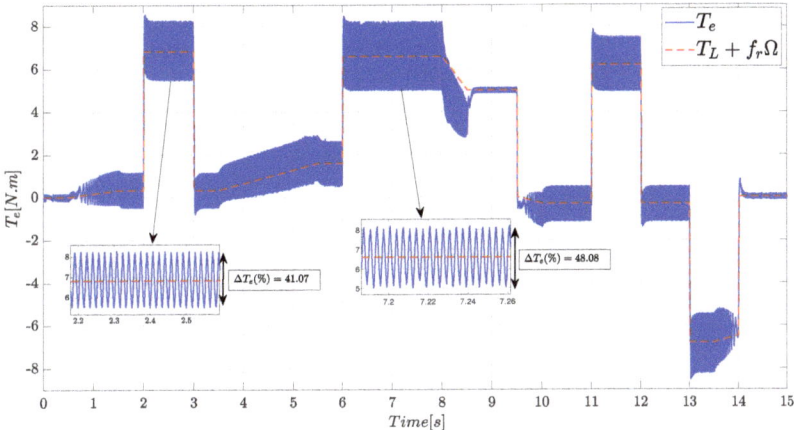

Figure 8. Torque response obtained by PI controllers and conventional field-oriented control.

3.4.2. Simulation with the Maximum Torque per Ampere Method

Applying the same simulated conditions as in the previous section. Similarly, the MTPA approach will be implemented in the reference currents calculation bloc as follows:

$$i_d^* = i_q^* = \sqrt{\frac{T_e^*}{\frac{3}{2}p(L_d - L_q)}} \qquad (36)$$

Figure 9 depicts the velocity response to the profile selected using the maximum torque per ampere approach. The velocity closely follows the reference with no static error. However, there is a tracking error caused by the PI regulator's property. It should also be observed that the velocity has no overshoot. This is due to the regulator parameters chosen.

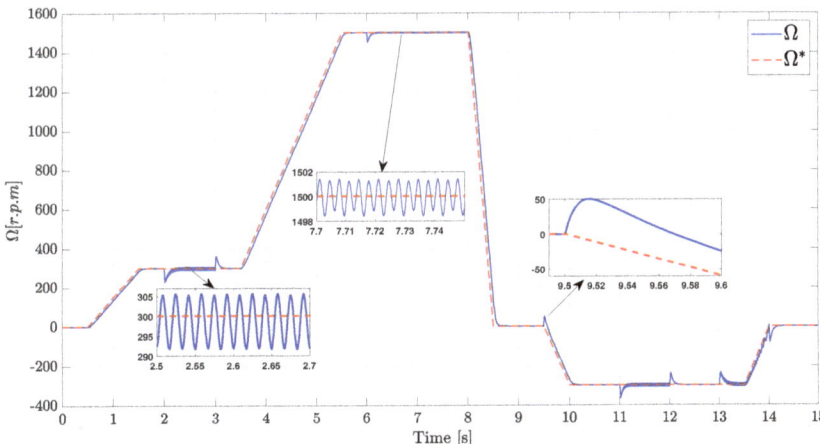

Figure 9. Velocity response obtained by PI controllers and maximum torque per ampere method.

Figure 10 illustrates the evolution of the current components i_d and i_q for current regulation. The evolution of the currents also shows a satisfactory regulation of the two current components over a wide range of velocity and torque. Similarly, the zooms in Figure 10a,b show that the reference currents exhibit undulations that affect the stator currents and therefore the torque ripple.

Figure 11 shows the machine's torque. From this illustration, we can see that the maximum torque per ampere method provides a machine torque that can overcome the load torque (T_L) and the intrinsic torque ($f_r \Omega$) of the machine. Although the torque ripple rate is a bit lower compared to the FOC method, the machine's torque still contains a significant ripple. The torque ripple rates at low and high machine velocities are quantified at 47.07 and 47.02% with the torque load, respectively, as illustrated in the zooms in Figure 11.

In the following section, we will put the optimal currents calculations method to the test under identical simulation conditions in order to evaluate whether it can reduce the machine's torque ripple.

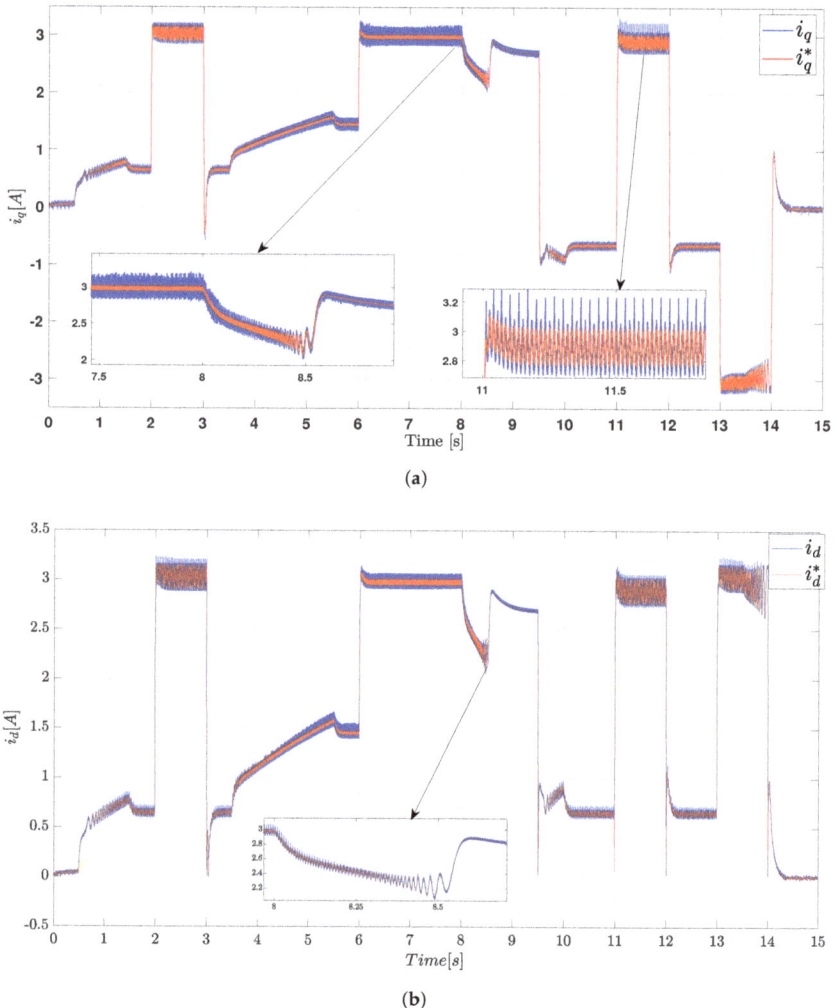

Figure 10. Response of the current components obtained by the PI controllers and the maximum torque per ampere method: (**a**) i_q response and (**b**) i_d response.

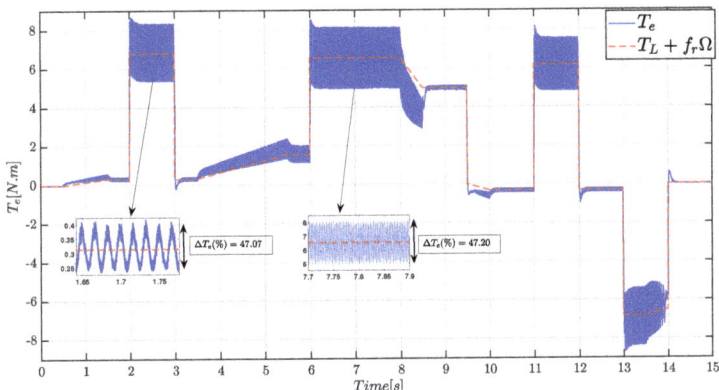

Figure 11. Torque response obtained by PI controllers and maximum torque per ampere method.

3.4.3. Simulation with the Optimal Currents Calculations Method (OCCM)

The same simulated conditions used for FOC and MTPA is also implemented for the optimal currents calculation method (OCCM). Similarly, in the reference currents calculation bloc, the optimal currents calculations method (OCCM) will be implemented as follows:

$$\begin{cases} i_q^* = \frac{(1-\mu a)i_d}{\mu c} \\ i_d^* = \sqrt{\frac{|T_e^*|}{\frac{\mu^2(a^2b-ac^2)+\mu(2c^2-2ab)+b}{\mu^2 c^2}}} \end{cases} \quad (37)$$

Figure 12 shows the velocity response of the system for the selected profile. The velocity onset has the same dynamic characteristics as in the previous case. However, the velocity fluctuations are very small compared to the previous case.

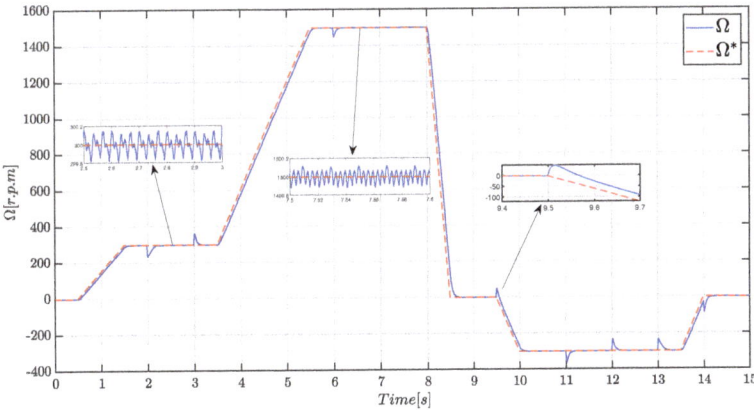

Figure 12. Velocity response obtained by PI controllers and optimal currents calculations method.

Figure 13 shows a good control of the current components i_d and i_q. The current components show a reduction in fluctuations, as shown by the zooms in Figure 13a,b. This can be justified by optimizing the reference currents, which necessarily has an impact on minimizing the torque ripple of the machine.

Figure 14 shows the machine's torque. The optimal currents calculations method as shown in Figure 14 assures a machine torque that can convince the machine's load torque (TL) and intrinsic torque ($f_r\Omega$). Moreover, there is a minimization of the machine torque ripple 9.08 and 10.8% with the torque load. This can be justified by the optimization in the reference currents; therefore, the optimization in the currents regulation impacts on the minimization of the torque ripple.

The different reference current calculation strategies presented similar dynamic performances of the currents and velocity control. However, the optimization of the reference current can reduce the torque ripple of the machine. The optimum current method solves the problem of excessive consumption and minimizes torque ripples over the entire operating range. The minimization is performed with a harmonized stator current.

Table 1 summarizes the torque ripple rate for the three examined methods.

Table 1. Comparison of torque ripple rates of the three strategies

	FOC		MTPA		OCCM	
ΔT_e	Without Load	With Load	Without Load	With Load	Without Load	With Load
At 300 r.p.m	383%	41.7%	50%	40.7%	38.08%	9.08%
At 1500 r.p.m	130%	48.8%	53.7%	47.2%	17.3%	10.8%

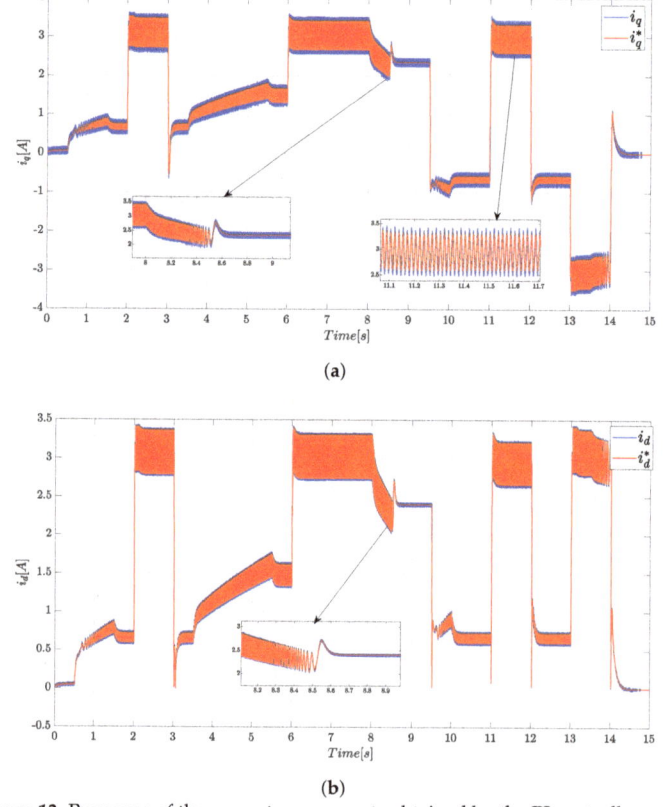

Figure 13. Response of the current components obtained by the PI controllers and optimal currents calculations method: (**a**) i_q response and (**b**) i_d response.

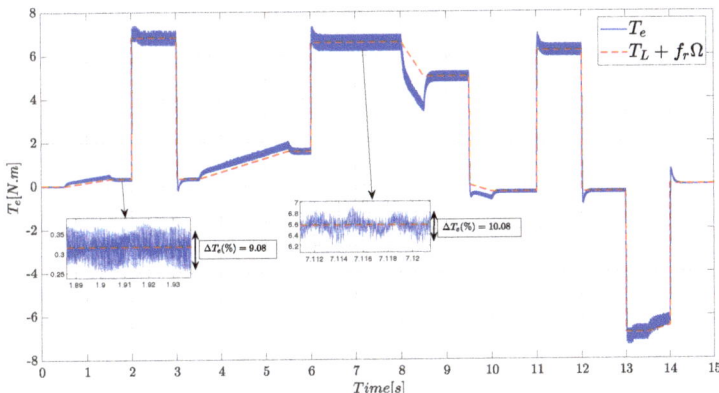

Figure 14. Torque response obtained by PI controllers and optimal currents calculations method.

4. Torque Ripple Minimization by Using Advanced Control Techniques

The reference currents calculation bloc can be used to minimize the machine torque ripple, as we have observed in this study. The optimal current calculation method (OCCM), when compared to the FOC and MTPA methods, was proven to be effective at reducing the torque ripple of the synchronous reluctance machine.

In this part, the torque ripple is reduced by using advanced controls. Indeed, the torque is directly linked to the stator currents of the machine. Therefore, better control of these currents can impact the torque ripple. Our objective is to investigate this hypothesis by improving the control of the currents with advanced controls based on the theory of sliding mode control in order to improve the dynamic performance (suppression of the tracking error) and robustness. The optimal current calculation method will be combined with the sliding mode control that replaces the PI velocity and currents controllers.

A conventional sliding mode control (SMC) is proposed to replace the PI controllers. However, due to the disadvantages of the control, particularly the chattering phenomenon [52], a second-order sliding mode control based on a super-twisting algorithm (STA) to minimize the chattering and improve the current response, and thereby reducing the torque ripple, is proposed.

4.1. Sliding Mode Control

Sliding mode control (SMC) is a class of a variable structure system (VSS) that targets decreasing the complexity of high-order systems to first-order state variables, defined as a sliding function and its derivative [53]. It is characterized by its simplicity of implementation, very good dynamic responsiveness, and, most importantly, its robustness with respect to internal uncertainties, as manifested by an insensitivity to variations in the parameters of the system to be controlled, as well as external disruptions [54–56].

This section will synthesize a conventional sliding mode control for velocity and currents control to replace PI controllers in the velocity/currents cascade control strategy.

4.1.1. Synthesis of a Conventional Sliding Mode for Velocity Controller

The selected sliding surface depends on the velocity tracking error (e_Ω) as follows:

$$e_\Omega(t) = \Omega^*(t) - \Omega(t) \tag{38}$$

The expression of the sliding surface (s_1) is

$$s_1(t) = e_\Omega(t) + \lambda_1 \int_0^t e_\Omega(\tau)d\tau \tag{39}$$

where λ_1 is a positive constant ($\lambda_1 > 0$).

This choice of sliding surface results in an error that tends to zero (if $s_1 = 0$, then $e_\Omega = 0$).

Thus, the following state variables are used:

$$\begin{cases} x_1(t) = \int_0^t e_\Omega(\tau) \, d\tau \\ x_2(t) = e_\Omega(t) \end{cases} \quad (40)$$

With x_1 and x_2 representing, respectively, the error and its integral. From (40), we deduce that

$$\dot{x}_1(t) = x_2(t) \quad (41)$$

Thus, the sliding surface can be defined as

$$s_1(t) = x_2(t) + \lambda_1 x_1(t) \quad (42)$$

Therefore, the electromechanical Equation (7) of the SynRM model is rewritten as follows:

$$\begin{aligned} \dot{x}_2 &= \dot{\Omega}^* - \frac{1}{J}T_e + \frac{f_r}{J}\Omega + \frac{1}{J}T_L \\ &= \dot{\Omega}^* - \frac{1}{J}T_e + \frac{f_r}{J}(\Omega - \Omega^* + \Omega^*) + \frac{1}{J}T_L \\ &= \dot{\Omega}^* - \frac{1}{J}T_e + \frac{f_r}{J}\Omega^* - \frac{f_r}{J}x_2 + \frac{1}{J}T_L \end{aligned} \quad (43)$$

Then, the system can be put in the form of a state space representation:

$$\begin{cases} \dot{x}_1 = x_2 \\ \dot{x}_2 = f(x) + gu + d \end{cases} \quad (44)$$

With $f(x) = \dot{\Omega}^* + \frac{f_r}{J}\Omega^* - \frac{f_r}{J}x_2$, $g = \frac{1}{J}$, $u = T_e$, , $d = \frac{1}{J}T_L$.

In order to determine the continuous component (u_c) of the SMC [57,58],

$$\dot{s}_1(x) = s(x) = 0 \quad (45)$$

Thus, considering (42) and (44), assuming that $u = u_c$,

$$u_c = g^{-1}(-f(x) - d - \lambda_1 x_2) \quad (46)$$

The existence and convergence condition [58] is used to determine the discontinuous component (u_d) as follows:

$$s_1(x) \cdot \dot{s}_1(x) < 0 \quad (47)$$

Given that,

$$\dot{s}_1(x) = \dot{x}_2(t) + \lambda_1 \dot{x}_1(t) \quad (48)$$

After simplification and by posing $u = u_d$,

$$u_d = -c_1 sign(s_1) \quad (49)$$

Knowing that the sliding mode control law is the sum of the continuous control and the discontinuous components [58],

$$u = u_{eq} + u_d \quad (50)$$

By replacing (46) and (49) in (50),

$$u = J\dot{\Omega}^* + f_r\Omega^* + (\lambda_1 J - f_r)e_\Omega + T_L - c_1 sign(s_1) \tag{51}$$

where the control u represents the total reference torque T_e^* provided by the velocity controller. The term T_L is considered as a disturbance to be compensated by the controller. The final sliding mode control law for velocity is

$$T_e^* = J\dot{\Omega}^* + f_r\Omega^* + (\lambda_1 J - f_r)e_\Omega - c_1 sign(s_1) \tag{52}$$

4.1.2. Synthesis of a Conventional Sliding Mode for Currents Controllers

The same method of synthesizing the velocity control law is used for the two current components i_d and i_q. Similarly, the sliding surfaces (s_2 and s_3) are defined in terms of the current tracking error (e_d and e_q) as follows:

$$\begin{cases} s_2(t) = e_d(t) + \lambda_2 \int_0^t e_d(\tau)d\tau \\ s_3(t) = e_q(t) + \lambda_3 \int_0^t e_q(\tau)d\tau \end{cases} \tag{53}$$

where λ_2 and λ_3 are positive constants.

The electrical Equation (3) can be rewritten in the following form:

$$\begin{cases} \frac{di_d}{dt} = v_d - \frac{L_d}{R_s}i_d + pL_q\Omega\, i_q \\ \frac{di_q}{dt} = v_q - \frac{L_q}{R_s}i_q - pL_d\Omega\, i_d \end{cases} \tag{54}$$

The terms $pL_q\Omega\, i_q$ and $pL_d\Omega\, i_d$ are considered as disturbances to be compensated by the current regulators. Thus, (54) becomes

$$\begin{cases} \frac{di_d}{dt} = v_d - \frac{L_d}{R_s}i_d \\ \frac{di_q}{dt} = v_q - \frac{L_q}{R_s}i_q \end{cases} \tag{55}$$

In the same way as the SMC velocity controller, to find the continuous components, the following condition is used:

$$\begin{cases} \dot{s}_2(x) = s_2(x) = 0 \\ \dot{s}_3(x) = s_3(x) = 0 \end{cases} \tag{56}$$

Thus, the continuous components (v_{cd} and v_{cq}) have the following form:

$$\begin{cases} v_{cd} = L_d\frac{di_d^*}{dt} + R_s i_d^* + (\lambda_1 L_d - R_s)e_{i_d} \\ v_{cw} = L_q\frac{di_q^*}{dt} + R_s i_q^* + (\lambda_1 L_q - R_s)e_{i_q} \end{cases} \tag{57}$$

In order to determine the discontinuous components, the convergence condition is used as follows:

$$\begin{cases} \dot{s}_2(x) < s_2(x) = 0 \\ \dot{s}_3(x) < s_3(x) = 0 \end{cases} \tag{58}$$

After simplification, the discontinuous components (v_{d_d} and v_{d_q}) are written as

$$\begin{cases} v_{d_d} = -c_2 sign(s_2) \\ v_{d_q} = -c_3 sign(s_3) \end{cases} \quad (59)$$

The sliding mode control law for the i_d and i_q currents controllers is the sum of the continuous and discontinuous components:

$$\begin{cases} u_d = v_{dc} + v_{d_d} \\ u_q = v_{q_c} + v_{q_d} \end{cases} \quad (60)$$

where u_d and u_q represent the voltages generated by SMC i_d and i_q currents controllers. Finally, the conventional sliding mode control law of the currents i_d and i_q can be written as

$$\begin{cases} v_d = v_{dc} + v_{d_d} \\ v_q = v_{q_c} + v_{q_d} \end{cases} \quad (61)$$

4.1.3. Simulation Results

In this section, the conventional sliding mode controllers developed in the cascade control strategy (see Section 2.2) are implemented using the optimal current calculation method (OCCM). To evaluate the conventional SMC against the PI controllers, the same torque and velocity profiles presented in Section 3.4 were used. The controller parameters used are $\lambda_1 = 3, c_1 = 1, \lambda_2 = \lambda_3 = 2$, and $c_2 = c_3 = 5$.

Figure 15 shows a very good response over the entire velocity and torque range. In contrast to the PI control (see Figure 12), the static error and tracking error are almost zero. It should also be noted that the velocity fluctuations are very small or even negligible. The velocity drops and overshoots when applying or removing the load have been significantly reduced, as shown by the zooms in Figure 15.

From this result, the static and dynamic performance of the velocity response is significantly improved by using an SMC velocity controller.

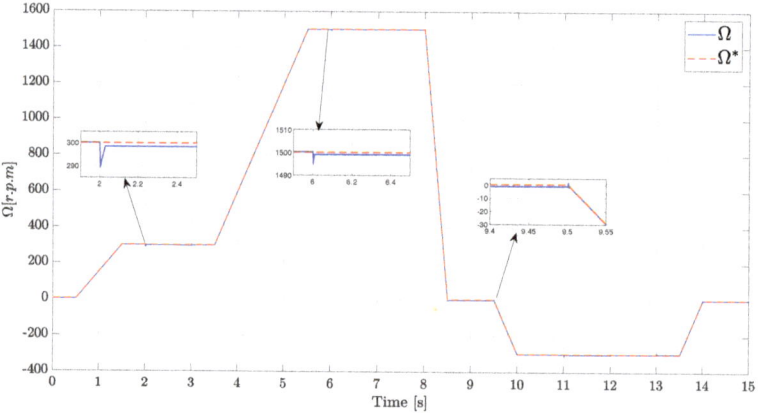

Figure 15. Velocity response obtained by SMC controller and optimal currents calculations method.

For the currents control, Figure 16 shows good control of the i_d and i_q current components. The current curve shows large current peaks compared to the PI controller (see Figure 13). These peaks are due to the high dynamics of the velocity controller to eliminate the tracking error. Moreover, it is caused by the discontinuous theme of the SMC or, as it is called in the literature, the chattering phenomenon.

The torque response of the SynRM employing the SMC is shown in Figure 17. It can be seen from this figure that the SMC with the optimal current calculation method ensures a machine torque that can convince the resistive torque (T_L) and the intrinsic torque ($f_r\Omega$) of the machine.

When compared to the PI controllers, at a very low speed and under load, for example, the torque response in the sliding mode shows a slight increase in torque ripples (10.07% compared to PI 9.8%). These ripples are a consequence of the chattering effect which is a drawback of the sliding mode control.

The next section suggests a control strategy based on the theory of higher-order sliding mode control to address this flaw.

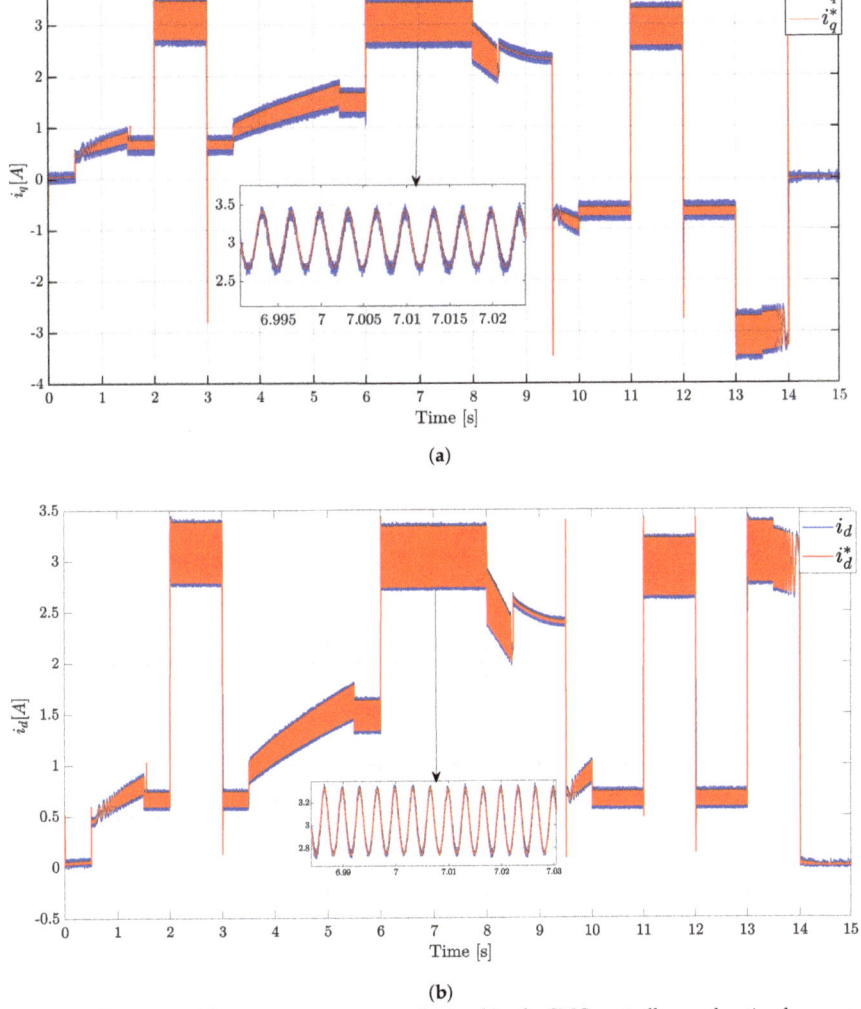

Figure 16. Response of the current components obtained by the SMC controllers and optimal currents calculations method: (**a**) i_q response and (**b**) i_d response.

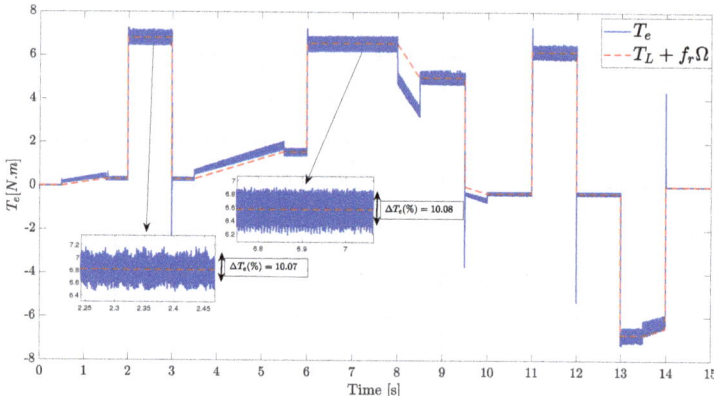

Figure 17. Torque response obtained by SMC controllers and optimal currents calculations method.

4.2. Higher-Order Sliding Mode Control

The control of systems by the classical sliding mode has shown that it presents an undesirable chattering phenomenon. In order to reduce or eliminate these phenomena, many solutions have been proposed in the literature [59].

The higher-order sliding mode has been chosen, and this method is based on the theory of the classical sliding mode control presented previously. In addition, it ensures that the desired performance is maintained and that there is a better convergence accuracy. The discontinuous control term is applied to the higher-order derivatives of the sliding variable while maintaining the advantages of the classical sliding mode control, because the discontinuity does not appear directly in the control but rather in one of its higher derivatives [59,60].

To lessen the chattering phenomenon, we suggest using a second-order sliding mode control built on the super-twisting algorithm (STA) in this work. By minimizing this phenomenon, the excellent static and dynamic performances offered by the conventional sliding mode control are maintained while reducing the torque ripple. In order to determine its effect on the torque ripple of the synchronous reluctant machine, we will thus propose the STA for the velocity and currents controllers and evaluate it against the SMC and the PI.

4.2.1. Synthesis of the Velocity Controller by Super-Twisting Algorithm

For the synthesis of the speed controller, the sliding surface s_4 is defined by

$$s_4(t) = y_1(t) = e_\Omega(t) + \lambda_1 \int_0^t e_\Omega(\tau) d\tau \tag{62}$$

This sliding surface is chosen similarly to that of the conventional sliding mode control. From the mechanical Equation (7) of the SynRM,

$$\dot{\Omega} = \frac{1}{J}T_e - \frac{f}{J}\Omega - \frac{1}{J}T_L \tag{63}$$

Using (62) and (63), the derivative of the surface is expressed as

$$\dot{\Omega} = \frac{1}{J}T_e - \frac{f}{J}\Omega - \frac{1}{J}T_L \tag{64}$$

It should be noted that the super-twisting algorithm system is specifically developed for systems with a relative degree $n = 1$ whose goal is to reduce chattering problems. This algorithm does not require the knowledge of the second derivative of the sliding variable as in the case of other algorithms. Thus, the algorithm guarantees that the trajectories of the system twist around the origin in the phase portrait [61] which brings about having the model of the system in relative order one:

$$\dot{y}_1 = \phi(y_1, t) + Y(x, t)u_{ST}(t) \tag{65}$$

with $\dot{y}_1$ being the sliding surface, and ϕ and Y being bounded functions [62,63]:

$$\begin{cases} |\phi| \leq \Phi \\ 0 < Y_m \leq Y(x,t) \leq Y_M \end{cases} \tag{66}$$

From Equations (65) and (64), we can deduce

$$\dot{y}_1 = \dot{\Omega}^* + \frac{f}{J}\Omega^* + (\frac{-f_r}{J} + \lambda_1)e_\Omega + \frac{1}{J}T_L - \frac{1}{J}T_e \tag{67}$$

The definition of the upper and lower bounds of the previously defined functions is chosen as follows [47]:

$$\begin{cases} \phi_{ST} = \dot{\Omega}^* + \frac{f}{J}\Omega^* + (\frac{-f}{J} + \lambda_1)e_\Omega + \frac{1}{J}T_L \\ Y_{ST} = 1 \\ u_{ST} = -\frac{1}{J}T_e \end{cases} \tag{68}$$

So, the sufficient conditions of convergence can be chosen as

$$\begin{cases} W = 3\Phi_{ST} \\ \lambda = 5\sqrt{2\Phi_{ST}} \\ \rho = 0.5 \end{cases} \tag{69}$$

By choosing $S_0 = s_4^2$, the order can be written as

$$u_{ST} = u_1 + u_2 \tag{70}$$

With

$$\begin{cases} \dot{u}_1 = -W sign(s_1) \\ \dot{u}_2 = -\lambda |s_1|^{0.5} sign(s_1) \end{cases} \tag{71}$$

4.2.2. Synthesis of Current Controllers by Super-Twisting Algorithm

In a similar way to the surface used in the classical sliding mode, the sliding surfaces of the currents (s_5 and s_6) are defined by

$$\begin{cases} s_5(t) = y_2(t) = e_d(t) + \lambda_2 \int_0^t e_d(\tau)d\tau \\ s_6(t) = y_3(t) = e_q(t) + \lambda_3 \int_0^t e_q(\tau)d\tau \end{cases} \tag{72}$$

From the electrical equations of the machine, the surface derivatives can be expressed as

$$\begin{cases} \dot{y}_2 = \dot{i}_d^* + \frac{R}{L_d}i_d^* + (\frac{-R}{L_d} + \lambda_1)e_d + \frac{1}{L_d}E_q - \frac{1}{L_d}v_d \\ \dot{y}_3 = \dot{i}_q^* + \frac{R}{L_q}i_q^* + (\frac{-R}{L_q} + \lambda_1)e_q + \frac{1}{L_q}E_d - \frac{1}{L_q}v_q \end{cases} \tag{73}$$

From Equations (65) and (73), we can deduce

$$\begin{cases} \phi_{d_{ST}} = i_d^* + \frac{R}{L_d}\Omega^* + (\frac{-R}{L_d} + \lambda_1)e_d + \frac{1}{L_d}E_q \\ Y_{d_{ST}} = 1 \\ u_{d_{ST}} = -\frac{1}{L_d}v_d \end{cases} \begin{cases} \phi_{q_{ST}} = i_q^* + \frac{R}{L_q}i_q^* + (\frac{-R}{L_q} + \lambda_1)e_q + \frac{1}{L_q}E_d \\ Y_{q_{ST}} = 1 \\ u_{q_{ST}} = -\frac{1}{L_q}v_q \end{cases} \quad (74)$$

The upper and lower bounds of the previously defined functions are chosen as follows:

$$\begin{cases} \Phi_{d_{ST}} = \left|i_d^* + \frac{R}{L_d}i_d^* + (\frac{-R}{L_d} + \lambda_1)e_d\right| + \left|\frac{1}{L_d}E_q\right| \\ Y_{d_{mST}} = 0.5 \;,\; Y_{d_{MST}} = 1 \\ U_d = -\frac{1}{L_d}v_{d_{max}} \end{cases} \begin{cases} \Phi_{q_{ST}} = \left|i_q^* + \frac{R}{L_q}i_q^* + (\frac{-R}{L_q} + \lambda_1)e_q\right| + \left|\frac{1}{L_q}E_d\right| \\ Y_{q_{mST}} = 0.5 \;,\; Y_{q_{MST}} = 1 \\ U_q = -\frac{1}{L_q}v_{q_{max}} \end{cases} \quad (75)$$

4.2.3. Simulation Results

Using the OCCM reference currents calculation bloc with the identical velocity and torque profile in the PI and SMC case, the STA sliding mode controllers of the velocity and currents were implemented in the cascade control strategy.

Figure 18 demonstrates excellent tracking over the whole velocity range. A zero static error and zero tracking error are displayed in the velocity response. Overshoot and undershoot are extremely rare, and there are barely any velocity fluctuations, as shown in the zooms of the figure.

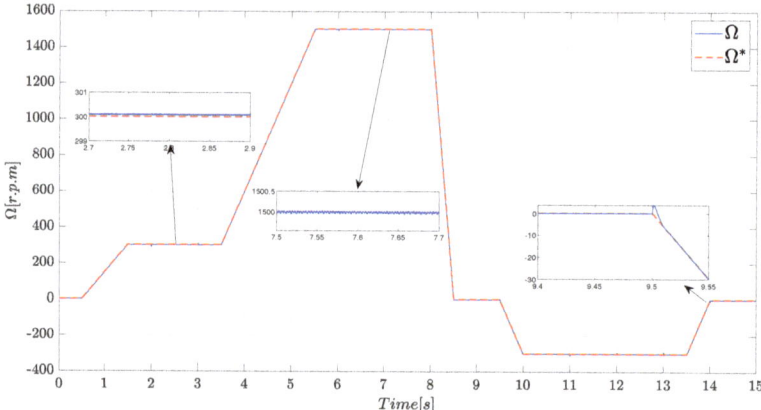

Figure 18. Velocity response obtained by STA controller and optimal currents calculations method.

Figure 19 shows good regulation of the current components i_d and i_q. The waveform of the currents is similar to the SMC case with less fluctuation, as shown in the zooms.

Figure 20 depicts the torque response of the SynRM using the SMC. It shows that the SMC with the optimal current calculation method assures a machine torque that can convince the machine's load torque (T_L) and intrinsic torque ($f_r\Omega$). When compared to the PI and SMC control, the torque response of the STA control demonstrates a reduction in the torque ripple. As an illustration, at a low speed with load, the torque ripple rate in the PI control is 9.8%, 10.07% in the SMC control, and 5% in the STA control.

Table 2 summarizes the machine torque ripple rate for low and high speeds with and without load torque.

Table 2. Comparison of torque ripple rates for the three control modes.

	PI		SMC		STA	
ΔT_e	Without Load	With Load	Without Load	With Load	Without Load	With Load
At 300 r.p.m	38.08%	9.8%	43%	10.07%	29%	5%
At 1500 r.p.m	17.3%	10.8%	21.7%	10.8%	12.7%	8%

This table makes it abundantly evident that applying STA control considerably reduces the torque ripple of the synchronous reluctance machine.

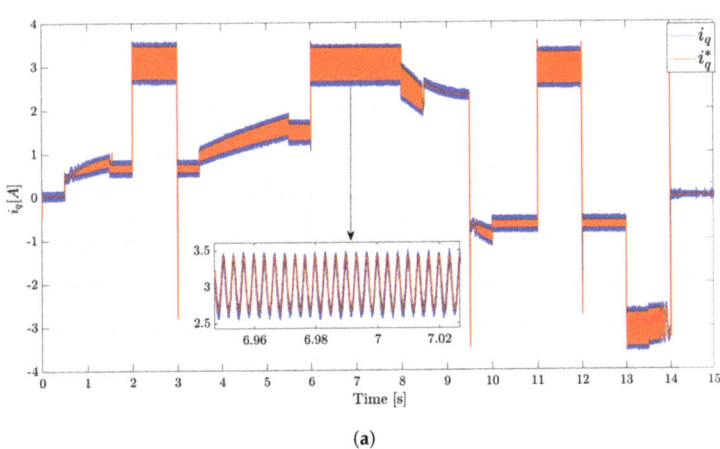

(a)

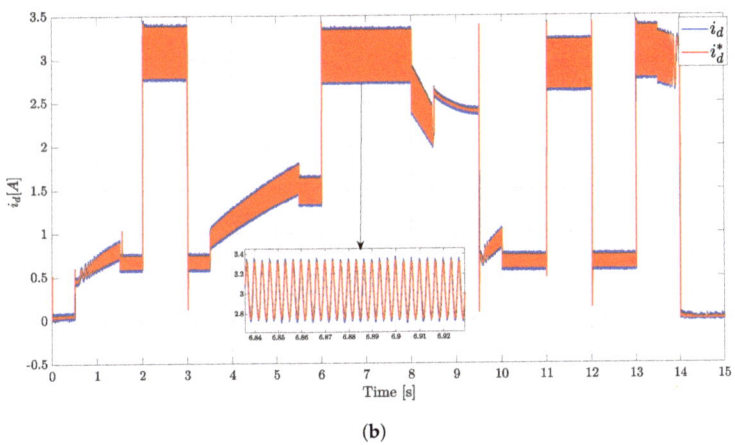

(b)

Figure 19. Response of the current components obtained by the STA controllers and optimal currents calculations method: (a) i_q response and (b) i_d response.

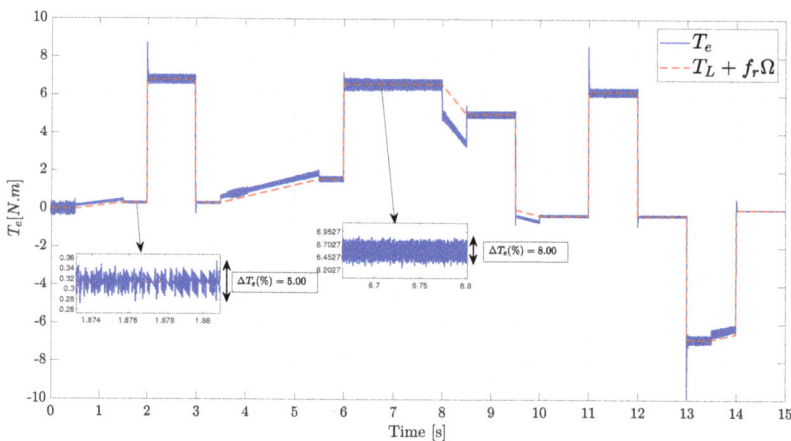

Figure 20. Torque response obtained by STA controllers and optimal currents calculations method.

5. Conclusions

In this work, we addressed the problem of torque ripple reduction in a synchronous reluctance machine for an electric vehicle drivetrain application.

We have based our approach on control-based solutions. In a cascade velocity/currents control strategy, we first proposed a new reference currents calculation bloc based on the optimization of the stator joule loss. In order to examine the contribution of this method on torque ripple reduction, we compared it to two methods used in the literature, namely the conventional field-oriented control and maximum torque per ampere. The simulation results clearly show the effectiveness and superiority of this proposed method in reducing the torque ripple of the machine.

To improve the system's static and dynamic performance, in the second part, we synthesized advanced velocity and current controllers based on the variable structure theory. Classical sliding mode controllers have been proposed using the method provided in the first part, namely the calculation of optimal currents. When compared to the PI controllers, the simulation results demonstrate a gain in performance but a minor increase in torque ripple. This is related to the chattering phenomenon, which constitutes the drawback of conventional sliding mode control. We then presented second-order sliding mode controllers based on the super-twisting algorithm to avoid this problem. The simulation results show that in addition to the improvement in the drivetrain performance, the torque ripples are significantly reduced.

In conclusion, a combination of the optimal current calculation method and the second-order sliding mode control produces the most effective combination for improving the synchronous reluctance machine's performance and torque ripple reduction.

Author Contributions: Methodology, O.D.A., M.O., Y.S., T.M., M.B. and K.H.A.; Software, O.D.A., M.O. and Y.S.; Validation, O.D.A., M.B. and Y.S.; Formal analysis, Y.S., T.M. and M.B.; Data curation, Y.S.; Writing—original draft, O.D.A. and Y.S.; Writing—review & editing, M.B. and Y.S.; Visualization, Y.S. and M.B.; Supervision, Y.S. and M.B. All authors have read and agreed to the published version of the manuscript.

Funding: This research received no external funding

Conflicts of Interest: The authors declare no conflict of interest.

Abbreviations

The following abbreviations are used in this manuscript:

EV	Electric vehicle
SynRM	Synchronous reluctance machines
PMSM	Permanent magnet synchronous machine
WRSM	Wound rotor synchronous machine
REMs	Rare-earth materials
FOC	Field-oriented control
DFOC	Direct field-oriented control
IFOC	Indirect field-oriented control
MTPA	Maximum torque per ampere
OCCM	Optimum current calculation method
PI	Proportional integral
SMC	Sliding mode control
STA	Super-twisting algorithm
$\phi_s, \phi_{d_s}, \phi_{q_s}$	Stator flux linkage in the d and q axes
i_d, i_q	Stator current in the d and q axes
V_d, V_q	Voltages in the d and q axes
L_{d_s}, L_{q_s}	Inductance in the d and q axes
L_i	Stator inductance of phase i
M_{ij}	Mutual inductance between phases i and j
Ω	Rotational velocity of the machine, in rad/s.
Ω^*	Rotational velocity reference of the machine, in rad/s.
T_e	Electromagnetic torque produced by the machine, in Nm
T_L	Load torque, in Nm
f_r	Viscous friction coefficient, in Ns^2/m^2
F_m	The slope force or tractive force that is required to drive the vehicle up
f_{aero}	Aerodynamic force created by the friction of the vehicle's body moving through the air
F_{rr}	Rolling resistance force
F_{rc}	Resistance force exerted by the vehicle weight as it goes up and down a hill
M	Vehicle mass
g	The acceleration due to gravity on Earth
ρ	Density of the air, in kg/m^3
C_x	Drag coefficient
S_f	Frontal cross-sectional area, in m^2
R_{sc}	Rolling resistance opposing the slope
i_d^*, i_q^*	Reference current in the d and q axis
Δ	Lagrangian function used to optimize the currents

Appendix A

Table A1. The synchronous reluctance machine's parameters.

Parameter	Value
Rated power	$P_n = 1.1$ kW
Number of pole pairs	$p = 2$
Rated RMS current	$I = 3$ A
Power supply voltage	220/380 V
Phase resistance	$R_s = 6.2$ Ohm
Direct inductance	$L_d = 0.34$ H
Quadrature inductance	$L_q = 0.105$ H
Rated speed	1500 r.p.m
Maximum velocity	1800 r.p.m
Torque at rated velocity	7 Nm
Torque at maximum velocity	5.8 N m
Machine inertia	$J = 0.005$ kg · m^2
Viscous friction coefficient	$f = 0.01$ Nm/s

References

1. Ashok, B.; Chidambaram, K.; Muhammad Usman, K.; Rajasekar, V.; Deepak, C.; Ramesh, R.; Narendhra, T.; Chellappan, K. Transition to Electric Mobility in India: Barriers Exploration and Pathways to Powertrain Shift through MCDM Approach. *J. Inst. Eng. (India) Ser.* **2022**, *103*, 1251–1277. [CrossRef]
2. Upadhyay, A.; Dalal, M.; Sanghvi, N.; Nair, S.; Scurtu, I.C.; Dragan, C. Electric Vehicles over Contemporary Combustion Engines. *Iop Conf. Ser. Earth Environ. Sci.* **2021**, *635*, 012004. [CrossRef]
3. El Hadraoui, H.; Zegrari, M.; Chebak, A.; Laayati, O.; Guennouni, N. A Multi-Criteria Analysis and Trends of Electric Motors for Electric Vehicles. *World Electr. Veh. J.* **2022**, *13*, 65. [CrossRef]
4. Rodríguez, E.; Rivera, N.; Fernández-González, A.; Pérez, T.; González, R.; Battez, A.H. Electrical compatibility of transmission fluids in electric vehicles. *Tribol. Int.* **2022**, *171*, 107544. [CrossRef]
5. Mohammad, K.S.; Jaber, A.S. Comparison of electric motors used in electric vehicle propulsion system. *Indones. J. Electr. Eng. Comput. Sci.* **2022**, *27*, 11–19. [CrossRef]
6. Gielen, D.; Lyons, M. *Critical Materials for the Energy Transition: Rare Earth Elements*; International Renewable Energy Agency: Abu Dhabi, United Arab Emirates, 2022; Volume 48.
7. Luo, X.; Qiu, Q.; Jing, L.; Zhang, D.; Huang, P.; Jia, S. Heavy Rare Earth Doped Nd-Fe-B Permanent Magnetic Material Performance Enhancement Methods and Their Motor Application Research. In Proceedings of the 2022 IEEE 5th International Electrical and Energy Conference (CIEEC), Nanjing, China, 27–29 May 2022; IEEE: New York, NY, USA, 2022; pp. 3121–3126.
8. Alves Dias, P.; Bobba, S.; Carrara, S.; Plazzotta, B. *The Role of Rare Earth Elements in Wind Energy and Electric Mobility*; European Commission: Luxembourg, 2020.
9. Singh, B.; Chowdhury, A.; Dixit, A.K.; Mishra, V.; Jain, A.; Kumar, N. Investigation on electric vehicle motor challenges, solutions and control strategies. *J. Inf. Optim. Sci.* **2022**, *43*, 185–191. [CrossRef]
10. Maciejewska, M.; Fuć, P.; Kardach, M. Analysis of electric motor vehicles market. *Combust. Engines* **2019**, *58*, 169–175. [CrossRef]
11. Heidari, H.; Rassõlkin, A.; Kallaste, A.; Vaimann, T.; Andriushchenko, E.; Belahcen, A.; Lukichev, D.V. A review of synchronous reluctance motor-drive advancements. *Sustainability* **2021**, *13*, 729. [CrossRef]
12. Mohanarajah, T.; Rizk, J.; Hellany, A.; Nagrial, M.; Klyavlin, A. Torque Ripple Improvement in Synchronous Reluctance Machines. In Proceedings of the 2018 2nd International Conference On Electrical Engineering (EECon), Colombo, Sri Lanka, 28 September 2018; pp. 44–50. [CrossRef]
13. Muteba, M.; Twala, B.; Nicolae, D.V. Torque ripple minimization in synchronous reluctance motor using a sinusoidal rotor lamination shape. In Proceedings of the 2016 XXII International Conference on Electrical Machines (ICEM), Lausanne, Switzerland, 1–4 September 2016; pp. 606–611. [CrossRef]
14. Liang, J.; Dong, Y.; Sun, H.; Liu, R.; Zhu, G. Flux-Barrier Design and Torque Performance Analysis of Synchronous Reluctance Motor with Low Torque Ripple. *Appl. Sci.* **2022**, *12*, 3958. [CrossRef]
15. Li, X.; Wang, Y.; Qu, R. Design of synchronous reluctance motors with asymmetrical flux barriers for torque ripple reduction. In Proceedings of the 2021 IEEE 4th Student Conference on Electric Machines and Systems (SCEMS), Virtually, 2–3 December 2021; IEEE: New York, NY, USA, 2021; pp. 1–6.
16. Wu, H.; Depernet, D.; Lanfranchi, V.; Benkara, K.E.K.; Rasid, M.A.H. A novel and simple torque ripple minimization method of synchronous reluctance machine based on torque function method. *IEEE Trans. Ind. Electron.* **2020**, *68*, 92–102. [CrossRef]
17. Singh, A.K.; Raja, R.; Sebastian, T.; Rajashekara, K. Torque ripple minimization control strategy in synchronous reluctance machines. In Proceedings of the IECON 2021–47th Annual Conference of the IEEE Industrial Electronics Society, Toronto, ON, Canada, 13–16 October 2021; pp. 1–6.
18. Abou-ElSoud, A.M.; Nada, A.S.A.; Aziz, A.A.M.A.; Sabry, W. Synchronous Reluctance Motors Torque Ripples Reduction using Feedback Cascaded PII Controller. In Proceedings of the 2022 23rd International Middle East Power Systems Conference (MEPCON), Cairo, Egypt, 13 December 2022; IEEE: New York, NY, USA, 2022; pp. 1–5.
19. Moghaddam, H.A.; Vahedi, A.; Ebrahimi, S.H. Design Optimization of Transversely Laminated Synchronous Reluctance Machine for Flywheel Energy Storage System Using Response Surface Methodology. *IEEE Trans. Ind. Electron.* **2017**, *64*, 9748–9757. [CrossRef]
20. Yan, D.; Xia, C.; Guo, L.; Wang, H.; Shi, T. Design and Analysis for Torque Ripple Reduction in Synchronous Reluctance Machine. *IEEE Trans. Magn.* **2018**, *54*, 1–5. [CrossRef]
21. Gallicchio, G.; Palmieri, M.; Cupertino, F.; Di Nardo, M.; Degano, M.; Gerada, C. Design Methodologies of High Speed Synchronous Reluctance Machines. In Proceedings of the 2022 International Conference on Electrical Machines (ICEM), Valencia, Spain, 5–8 September 2022; pp. 448–454. [CrossRef]
22. Donaghy-Spargo, C.M. Electromagnetic–Mechanical Design of Synchronous Reluctance Rotors With Fine Features. *IEEE Trans. Magn.* **2017**, *53*, 1–8. [CrossRef]
23. Bianchi, N.; Degano, M.; Fornasiero, E. Sensitivity Analysis of Torque Ripple Reduction of Synchronous Reluctance and Interior PM Motors. *IEEE Trans. Ind. Appl.* **2015**, *51*, 187–195. [CrossRef]
24. Bonthu, S.S.R.; Tarek, M.T.B.; Choi, S. Optimal Torque Ripple Reduction Technique for Outer Rotor Permanent Magnet Synchronous Reluctance Motors. *IEEE Trans. Energy Convers.* **2018**, *33*, 1184–1192. [CrossRef]
25. Liu, H.C.; Kim, I.G.; Oh, Y.J.; Lee, J.; Go, S.C. Design of Permanent Magnet-Assisted Synchronous Reluctance Motor for Maximized Back-EMF and Torque Ripple Reduction. *IEEE Trans. Magn.* **2017**, *53*, 1–4. [CrossRef]

26. Huynh, T.A.; Hsieh, M.F.; Shih, K.J.; Kuo, H.F. An Investigation Into the Effect of PM Arrangements on PMa-SynRM Performance. *IEEE Trans. Ind. Appl.* **2018**, *54*, 5856–5868. [CrossRef]
27. Zhang, X.; Foo, G.H.B.; Vilathgamuwa, D.M.; Maskell, D.L. An Improved Robust Field-Weakeaning Algorithm for Direct-Torque-Controlled Synchronous-Reluctance-Motor Drives. *IEEE Trans. Ind. Electron.* **2015**, *62*, 3255–3264. [CrossRef]
28. Zhang, Z.; Liu, X. A Duty Ratio Control Strategy to Reduce Both Torque and Flux Ripples of DTC for Permanent Magnet Synchronous Machines. *IEEE Access* **2019**, *7*, 11820–11828. [CrossRef]
29. Mohan, D.; Zhang, X.; Beng Foo, G.H. Generalized DTC Strategy for Multilevel Inverter Fed IPMSMs with Constant Inverter Switching Frequency and Reduced Torque Ripples. *IEEE Trans. Energy Convers.* **2017**, *32*, 1031–1041. [CrossRef]
30. Li, C.; Wang, G.; Zhang, G.; Xu, D.; Xiao, D. Saliency-Based Sensorless Control for SynRM Drives with Suppression of Position Estimation Error. *IEEE Trans. Ind. Electron.* **2019**, *66*, 5839–5849. [CrossRef]
31. Ortombina, L.; Tinazzi, F.; Zigliotto, M. Adaptive Maximum Torque per Ampere Control of Synchronous Reluctance Motors by Radial Basis Function Networks. *IEEE J. Emerg. Sel. Top. Power Electron.* **2019**, *7*, 2531–2539. [CrossRef]
32. Truong, P.H.; Flieller, D.; Nguyen, N.K.; Mercklé, J.; Sturtzer, G. Torque ripple minimization in non-sinusoidal synchronous reluctance motors based on artificial neural networks. *Electr. Power Syst. Res.* **2016**, *140*, 37–45. [CrossRef]
33. Singh, A.K.; Raja, R.; Sebastian, T.; Rajashekara, K. Torque Ripple Minimization Control Strategy in Synchronous Reluctance Machines. *IEEE Open J. Ind. Appl.* **2022**, *3*, 141–151. [CrossRef]
34. Moghaddam, R.R.; Magnussen, F.; Sadarangani, C. A FEM1 investigation on the Synchronous Reluctance Machine rotor geometry with just one flux barrier as a guide toward the optimal barrier's shape. In *IEEE Eurocon 2009*; IEEE: New York, NY, USA, 2009; pp. 663–670.
35. Agrebi, Y.; Triki, M.; Koubaa, Y.; Boussak, M. Rotor speed estimation for indirect stator flux oriented induction motor drive based on MRAS scheme JES. 2007. Available online: https://www.researchgate.net/profile/Youssef-Agrebi/publication/274892152_Rotor_resistance_estimation_for_indirect_stator_oriented_induction_motor_drive_based_on_MRAS_scheme/links/5a82c7860f7e9bda869fb0ce/Rotor-resistance-estimation-for-indirect-stator-oriented-induction-motor-drive-based-on-MRAS-scheme.pdf (accessed on 9 February 2023) .
36. Xi, T.; Kehne, S.; Fey, M.; Brecher, C. Application of Optimal Control for Synchronous Reluctance Machines in Feed Drives of Machine Tools. In Proceedings of the 2022 International Conference on Electrical, Computer and Energy Technologies (ICECET), Prague, Czech Republic, 20–22 July 2022; pp. 1–6. [CrossRef]
37. Lubin, T. Modélisation et Commande de la Machine Synchrone à Réluctance Variable: Prise en Compte de la Saturation Magnétique. Ph.D. Thesis, Université Henri Poincaré, Nancy, France, 2003.
38. Im, J.B.; Kim, W.; Kim, K.; Jin, C.S.; Choi, J.H.; Lee, J. Inductance Calculation Method of Synchronous Reluctance Motor Including Iron Loss and Cross Magnetic Saturation. *IEEE Trans. Magn.* **2009**, *45*, 2803–2806. [CrossRef]
39. Dilys, J.; Baskys, A. Self-identification of permanent magnet synchronous motor inductance for efficient sensorless control. In Proceedings of the 2017 Open Conference of Electrical, Electronic and Information Sciences (eStream), Vilnius, Lithuania, 27 April 2017; pp. 1–4. [CrossRef]
40. Dursun, D.C.; Yildiz, A.; Polat, M. Modeling of Synchronous Reluctance Motor and Open and Closed Loop Speed Control. In Proceedings of the 2022 21st International Symposium INFOTEH-JAHORINA (INFOTEH), East Sarajevo, Bosnia and Herzegovina, 16–18 March 2022; pp. 1–6. [CrossRef]
41. Bao, C.; Chen, H.; Yang, C.; Zhong, J.; Gao, H.; Song, S. Synchronous reluctance motor flux linkage saturation modeling based on stationary identification and neural networks. In Proceedings of the IECON 2022—48th Annual Conference of the IEEE Industrial Electronics Society, Brussels, Belgium, 17–20 October 2022; pp. 1–6. [CrossRef]
42. Santos, J.; Andrade, D.; Viajante, G.; Freitas, M.; Bernadeli, V. Analysis and Mathematical Modeling Of The Synchronous Reluctance Motor. *IEEE Lat. Am. Trans.* **2015**, *13*, 3820–3825. [CrossRef]
43. Hassan, M.R.M.; Mossa, M.A.; Dousoky, G.M. Evaluation of Electric Dynamic Performance of an Electric Vehicle System Using Different Control Techniques. *Electronics* **2021**, *10*, 2586. [CrossRef]
44. Deuszkiewicz, P.; Radkowski, S. On-line condition monitoring of a power transmission unit of a rail vehicle. *Mech. Syst. Signal Process.* **2003**, *17*, 1321–1334. [CrossRef]
45. Minakawa, M.; Nakahara, J.; Ninomiya, J.; Orimoto, Y. Method for measuring force transmitted from road surface to tires and its applications. *Jsae Rev.* **1999**, *20*, 479–485. [CrossRef]
46. Liang, H.; To Chong, K.; Soo No, T.; Yi, S.Y. Vehicle longitudinal brake control using variable parameter sliding control. *Control. Eng. Pract.* **2003**, *11*, 403–411. [CrossRef]
47. Saadi, Y.; Sehab, R.; Chaibet, A.; Boukhnifer, M.; Diallo, D. Performance Comparison between Conventional and Robust Control for the Powertrain of an Electric Vehicle Propelled by a Switched Reluctance Machine. In Proceedings of the 2017 IEEE Vehicle Power and Propulsion Conference (VPPC), Belfort, France, 14–17 December 2017; pp. 1–6. [CrossRef]
48. Adhavan, B.; Kuppuswamy, A.; Jayabaskaran, G.; Jagannathan, V. Field oriented control of Permanent Magnet Synchronous Motor (PMSM) using fuzzy logic controller. In Proceedings of the 2011 IEEE Recent Advances in Intelligent Computational Systems, Trivandrum, India, 22–24 September 2011; IEEE: New York, NY, USA, 2011; pp. 587–592.
49. Diao, K.; Sun, X.; Bramerdorfer, G.; Cai, Y.; Lei, G.; Chen, L. Design optimization of switched reluctance machines for performance and reliability enhancements: A review. *Renew. Sustain. Energy Rev.* **2022**, *168*, 112785. [CrossRef]

50. Truong, P.H.; Flieller, D.; Nguyen, N.K.; Mercklé, J.; Sturtzer, G. An investigation of Adaline for torque ripple minimization in Non-Sinusoidal Synchronous Reluctance Motors. In Proceedings of the IECON 2013—39th Annual Conference of the IEEE Industrial Electronics Society, Vienna, Austria, 10–13 November 2013; IEEE: New York, NY, USA, 2013; pp. 2602–2607.
51. Truong, P.H.; Flieller, D.; Nguyen, N.K.; Mercklé, J.; Dat, M.T. Optimal efficiency control of synchronous reluctance motors-based ANN considering cross magnetic saturation and iron losses. In Proceedings of the IECON 2015—41st Annual Conference of the IEEE Industrial Electronics Society, Yokohama, Japan, 9–12 November 2015; IEEE: New York, NY, USA, 2015; pp. 004690–004695.
52. Utkin, V.; Lee, H. Chattering problem in sliding mode control systems. In Proceedings of the International Workshop on Variable Structure Systems, VSS'06, Alghero, Italy, 5–7 June 2006; IEEE: New York, NY, USA, 2006; pp. 346–350.
53. Dong, L.; Nguang, S.K. Chapter 5—Sliding mode control for multiagent systems with continuously switching topologies based on polytopic model. In *Consensus Tracking of Multi-Agent Systems with Switching Topologies*; Dong, L., Nguang, S.K., Eds.; Emerging Methodologies and Applications in Modelling; Academic Press: Cambridge, MA, USA, 2020; pp. 87–105. [CrossRef]
54. Shao, K.; Zheng, J.; Wang, H.; Wang, X.; Lu, R.; Man, Z. Tracking Control of a Linear Motor Positioner Based on Barrier Function Adaptive Sliding Mode. *IEEE Trans. Ind. Inform.* **2021**, *17*, 7479–7488. [CrossRef]
55. Edwards, C.; Colet, E.F.; Fridman, L.; Colet, E.F.; Fridman, L.M. In *Advances in Variable Structure and Sliding Mode Control*; Springer: Berlin/Heidelberg, Germany, 2006; Volume 334.
56. Bandyopadhyay, B.; Deepak, F.; Kim, K.S. *Sliding Mode Control Using Novel Sliding Surfaces*; Springer: Berlin/Heidelberg, Germany, 2009; Volume 392.
57. Saadi, Y.; Sehab, R.; Chaibet, A.; Boukhnifer, M.; Diallo, D. Sensorless control of switched reluctance motor for EV application using a sliding mode observer with unknown inputs. In Proceedings of the 2018 IEEE International Conference on Industrial Technology (ICIT), Lyon, France, 19–22 February 2018; pp. 516–521. [CrossRef]
58. Koshkouei, A.J.; Burnham, K.J.; Zinober, A.S. Dynamic sliding mode control design. *IEE-Proc.-Control. Theory Appl.* **2005**, *152*, 392–396. [CrossRef]
59. Lee, H.; Utkin, V.I. Chattering suppression methods in sliding mode control systems. *Annu. Rev. Control.* **2007**, *31*, 179–188. [CrossRef]
60. Dorel, L.; Levant, A. On chattering-free sliding-mode control. In Proceedings of the 2008 47th IEEE Conference on Decision and Control, Cancun, Mexico, 9–11 December 2008; IEEE: New York, NY, USA, 2008; pp. 2196–2201.
61. Levant, A. Sliding order and sliding accuracy in sliding mode control. *Int. J. Control.* **1993**, *58*, 1247–1263. [CrossRef]
62. Barth, A.; Reichhartinger, M.; Reger, J.; Horn, M.; Wulff, K. Lyapunov-design for a super-twisting sliding-mode controller using the certainty-equivalence principle. *IFAC-PapersOnLine* **2015**, *48*, 860–865. [CrossRef]
63. Utkin, V.; Guldner, J.; Shi, J. *Sliding Mode Control in Electro-Mechanical Systems*; CRC Press: Boca Raton, FL, USA, 2017.

Disclaimer/Publisher's Note: The statements, opinions and data contained in all publications are solely those of the individual author(s) and contributor(s) and not of MDPI and/or the editor(s). MDPI and/or the editor(s) disclaim responsibility for any injury to people or property resulting from any ideas, methods, instructions or products referred to in the content.

Article

Torque Ripple Minimization of Variable Reluctance Motor Using Reinforcement Dual NNs Learning Architecture

Hamad Alharkan

Department of Electrical Engineering, College of Engineering, Qassim University, Unaizah 56452, Saudi Arabia; h.alharkan@qu.edu.sa

Abstract: The torque ripples in a switched reluctance motor (SRM) are minimized via an optimal adaptive dynamic regulator that is presented in this research. A novel reinforcement neural network learning approach based on machine learning is adopted to find the best solution for the tracking problem of the SRM drive in real time. The reference signal model which minimizes the torque pulsations is combined with tracking error to construct the augmented structure of the SRM drive. A discounted cost function for the augmented SRM model is described to assess the tracking performance of the signal. In order to track the optimal trajectory, a neural network (NN)-based RL approach has been developed. This method achieves the optimal tracking response to the Hamilton–Jacobi–Bellman (HJB) equation for a nonlinear tracking system. To do so, two neural networks (NNs) have been trained online individually to acquire the best control policy to allow tracking performance for the motor. Simulation findings have been undertaken for SRM to confirm the viability of the suggested control strategy.

Keywords: variable reluctance motor; optimization problems; reinforcement learning (RL); adaptive dynamic programming (ADP); neural network (NN); machine learning method

Citation: Alharkan, H. Torque Ripple Minimization of Variable Reluctance Motor Using Reinforcement Dual NNs Learning Architecture. *Energies* 2023, 16, 4839. https://doi.org/10.3390/en16134839

Academic Editors: Moussa Boukhnifer and Larbi Djilali

Received: 30 April 2023
Revised: 1 June 2023
Accepted: 15 June 2023
Published: 21 June 2023

Copyright: © 2023 by the author. Licensee MDPI, Basel, Switzerland. This article is an open access article distributed under the terms and conditions of the Creative Commons Attribution (CC BY) license (https://creativecommons.org/licenses/by/4.0/).

1. Introduction

Recently, the deployment of Switched Reluctance Motors (SRMs) in a vast scope of car electrification and variable speed systems has garnered significant recognition. The SRM is a flexible contender that might outperform other types of machines due to its inherent durability, fault-tolerant capability, affordable pricing, and natural simplicity due to its lack of magnets, brushes, and winding of a rotor [1,2]. Advancements in power electronic devices and computer programming have increased their efficiency. SRMs are now being considered for a number of applications requiring high-speed performance and dependability, including those involving electric vehicles and aviation [3–7]. SRMs have numerous benefits, but they also have certain drawbacks, such as huge torque ripples that might result in loud noise and vibration when the motor is operating. The system's nonlinear electromechanical characteristic, which depends on current and rotor angle, as well as severe magnetic saturation, to achieve great torque density, is the cause of the torque ripples. As a result, extending the percentage of SRM in high-performance models requires reducing the torque's oscillations.

To limit torque ripples, there are two common approaches that have been employed. The first entails improving the machine's magnetic configuration [8–11]. In one instance, the rotor and stator structures were changed by the SRM's manufacturer to reduce torque ripples; however, this might have degraded efficiency [12]. The second alternative is designing a torque regulator to minimize ripples and address the model's nonlinearity. The SRM's stator current ought to be precisely supplied and adjusted by the controller at the right rotor angle, as well as achieving the current pulses' quick rise and fall times. This can be accomplished be inserting a considerable level of voltage from the DC supply to handle the back electromotive force which occurs during the operating of the machine and simultaneously minify the inductance per phase. That is, as the rotor speed rises, the induced voltage of the motor reaches a point at which the DC voltage pulses currently produced are inadequate to control the torque. In order to reach the highest possible voltage for high performance, the control mechanism would necessarily assume an optimum phase-pulse mode which requires a high switching frequency. As a result, having a controller with reduced torque fluctuations is a technical challenge for the SRM drive.

To relieve torque ripples, many strategies have been proposed. Bang–bang control, sliding mode techniques, and enhanced Proportional-Integral-Derivative (PID) control are several that are often used and simple to apply [13–19]. Bang–bang and delta-modulation regulators have typically been applied to regulate SRMs. For these mechanisms, a number of limitations, including significant torque pulsations, restricted switching frequency due to semiconductor properties, and variable switching frequency, which results in less effective regulation of Electro-Magnetic Interference (EMI) make them impracticable for many applications. In such a model, the current pulse cannot be adjusted speedily enough by the classical PID regulator. Indeed, even more advanced transitioning PID controllers are unable to provide the best response. Additionally, researchers have studied direct torque optimal control approaches. The direct instantaneous torque control (DITC) system can be used to cope with the difficulty to represent the phase current as a function of torque and rotor angles. Although DITC has a straightforward and easy structure, its implementation necessitates complicated switching rules, unrestrained switching frequency, and a very large sample rate [18–26]. Therefore, implementing a controller that can minimize the torque ripples requires a very high dynamic scheme which allows high switching frequency.

In this article, a machine learning algorithm using RL techniques is employed to track the reference signal and reduce pulsations on torque pulses of the SRM drive. This unique approach is able to handle the model variances and produce excellent results even though the SRM experiences nonlinearities dependent on current and rotor angles. In this approach, the SRM tracking problem needs to be solved by optimizing the tracking function and tracing reference trajectory. Dual-stream neural network strategies should be employed and trained to provide optimized duty cycles based on the predetermined utility function [27,28]. The nonlinear tracking Hamilton–Jacobi–Bellman (HJB) equation of SRM is determined by modulating the NNs until convergence, providing the tracking performance for the system model. The fundamental contributions of this research are as follows:

I. Augmenting the SRM drive model to generate the tracking function;
II. Adopting a policy iteration method based on a reinforcement learning algorithm to minimize the torque ripples of the SRM;
III. Deployment of two NNs to optimize the HJB equation and conduct tracking operations for the system.

2. Materials and Methods

The main framework of the proposed model is shown in Figure 1, where the dual neural network architecture using the policy iteration method has been executed to minimize the torque ripples of the variable switched reluctance motor. The internal architecture of the proposed model is further described in the following subsection.

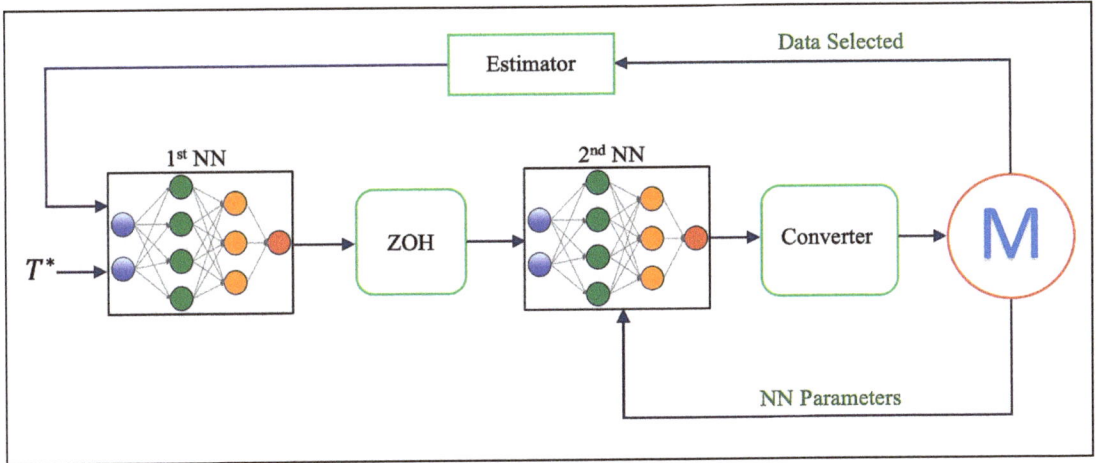

Figure 1. The proposed framework for minimizing torque ripples.

2.1. Modelling the Tracking Function for Srm Drive

In machine learning theory, tracking control is a method used to guide the state of a system to follow a reference path, while the optimal regulation method aims to bring the system's state dynamics to a halt [29]. The tracking control for an SRM drive is designed to align the machine's output torque with the reference torque trajectory. Designing an optimum control system depends on being able to solve the partial differential equation known as the HJB equation, which represents the ideal control strategy for a nonlinear system. Optimal tracking control involves both feedforward and feedback control to accurately guide the system's state towards a reference path while maintaining stability. Using the inverse dynamic technique, one may solve the feedforward portion that achieves tracking performance. By computing the HJB model, it is possible to conduct a feedback function that maintains the system's stability. The authors of [30] discuss the typical responses for both concepts. The disadvantages of utilizing the usual approach are that it needs the inversion of the drive's characteristics in order to derive the control policy and that it uses the full system's parameters. Due to the complex nature of the controller, the typical solutions are consequently not applicable to SRM. To remedy this, the optimum tracking control of the SRM drive is intended to minimize a specified quadratic cost function based on the augmented system model that comprises the machine parameter and reference source model. This enhanced system requires that the reference signal is supplied and generated by a distinct source model. Reinforcement learning consists of a collection of techniques that enable the use of an expanded model in the construction of adaptive tracking control for a nonlinear system. These methods are intended to tackle the tracking issue online and in real time by monitoring data streams [27]. Enabling the controller to calculate the system dynamics and tracking the inaccuracies after each iteration is another method for estimating the inductance surface. All mathematical techniques need an estimator to update the model, which may then be applied to a controller such as model predictive control. Neural networks based on RL methods integrate adaptation and tracking performance simultaneously into a single task. Therefore, to benefit from this advantage while managing the non-linearity of the system, reinforcement dual-NN learning architecture is proposed for minimizing the torque ripples of the machine. By applying the neural network under the concept of value function approximator (VFA), this can approximate the cost function using the least-squares method. In optimal control, there are two techniques to solve the optimal tracking problem online in real time without requiring full knowledge of the system. One approach to RL is based on iterating the Q-function, which is called the Q-learning algorithm. However, this method is only applicable for the linear system. For nonlinear applications, another algorithm should be incorporated with Q-learning to cope with the nonlinearity of the system. The other approach of RL is the dual-neural-network architecture, which can solve the nonlinear system and be implemented for applications such as SRM. Therefore, the dual-neural-network architecture is a fundamental technique of reinforcement learning methods. This method includes two phases. The first NN is responsible for determining the optimum phase voltage of control input in the first stage of the process, which may be executed during the policy improvement phase. The second NN must assess the control input according to the policy evaluation step in the second stage.

Following is a discussion of the tracking issue for the dynamic model of the SRM drive and the derivation of the HJB equation.

2.1.1. Updated Model of SRM Drive

SRM consists of a variable number of salient poles on both the stator and the rotor of the motor. In order to generate the machine's phases, the coils are wound around the stator pole and then installed in pairs that are mirror-opposite to one another. After the phase has been excited, the change in reluctances will cause the torque that is responsible for aligning the rotor pole with the stator poles. Because of its minimal impact on torque generation and dynamics, the mutual inductance between surrounding phases in an SRM is often quite low and has been omitted in the modeling process. In general, the mutual inductance between adjacent winding in an SRM is relatively tiny. As a consequence of this, the voltage and torque equation for one phase of an SRM may be expressed as

$$V = R_s i + \frac{d\lambda(\theta, i)}{dt} \qquad (1)$$

$$T = \frac{1}{2} i^2 \frac{dL(\theta, i)}{d\theta} \qquad (2)$$

where R_s is the phase resistance and λ is the flux linkage per phase computed by $\lambda = L(\theta, i)i$. L is the inductance profile as a function of the rotor position (θ) and the phase current (i). As seen in (2), the electromagnetic torque of a single phase is proportional to the square of the current in this type of machine. For this reason, the fundamental motivation for using the infinite-horizon tracking technique is to find the most suitable scheme for the system of SRM (1) that allows the output torque or the state $x(k)$ to follow the reference trajectory $d(k)$. Subsequently, we can write out the error equation that leads to optimal tracking performance as

$$e_k = x_k - d_k \qquad (3)$$

To develop the enhanced model, it is necessary to make a claim. That is, the reference signal of the machine for the tracking issue is generated by the combination of the reference model and the dynamic model of the motor [31]. The generator model can be formulated as

$$d_{k+1} = \beta d_k \qquad (4)$$

where $\beta \in \mathbb{R}^n$. This reference generator does not account for the fact that it is stable and may offer a broad variety of useful reference signals, including the periodic pulses of the square wave, which is the SRM reference current and torque. The forward method is used to estimate the discrete time domain of the SRM model during discrete execution. Consequently, based on the discrete dynamic model of SRM and the reference generator formulation, the tracking error (3) based on the input voltage signal may be calculated as follows:

$$e_{k+1} = x_{k+1} - d_{k+1} = f(x_k) + g(x_k)u_k - \beta r_k = f(e_k + d_k) - \beta r_k + g(e_k + d_k)u_k \qquad (5)$$

where $f(x_k) = x_k - (tR_s/L_k)x_k$ and $g(x_k) = t/L_k$. $x_k \in \mathbb{R}^n$ is the phase current (i_k), u_k is the DC voltage generated from the DC power supply, t is the discrete sampling time, and

L_k is the phase inductance fluctuation determined by rotor angle and phase current. The reference signal model and the tracking error may be included in the simulation model as an array by incorporating (4) and (5) to create the updated dynamic model:

$$X_{k+1} = \begin{bmatrix} e_{k+1} \\ d_{k+1} \end{bmatrix} = \begin{bmatrix} f(e_k + d_k) - \beta r_k \\ \beta r_k \end{bmatrix} + \begin{bmatrix} g(e_k + d_k) \\ 0 \end{bmatrix} u_k \equiv \Lambda(X_k) + \forall(X_k)u_k \quad (6)$$

where $X_k = \begin{bmatrix} e_k & d_k \end{bmatrix}^T \in \mathbb{R}^{2n}$ is the updated state. Minimizing a quadratic performance index function yields the optimal input signal that minimizes the tracking error. SRM's performance index function is established by weighing the cost of the voltage signal against the tracking inaccuracy and taking the proportion of the two into account as follows:

$$V(X_k) = \sum_{i=k}^{\infty} \gamma^{i-1} \left[(x_i - d_i)^T Q(x_i - d_i) + u_i^T R u_i \right] \quad (7)$$

where Q is a predefined weight matrix for the tracking error and R is a predefined weight matrix for the control policy, and $0 < \gamma \leq 1$ is a discount rate that considerably lowers the long-term cost. The value of γ should be smaller than 1 for SRM situations since $\gamma = 1$ only applies when it is known ahead of time, such as when obtaining the reference signal from an asymptotically stable reference generator model [32]. The value function may be expressed using the updated model (5) as follows:

$$V(X_k) = \sum_{i=k}^{\infty} \gamma^{i-1} \left[X_i^T Q_q X_i + u_i^T R u_i \right] \quad (8)$$

where

$$Q_q = \begin{bmatrix} Q & 0 \\ 0 & 0 \end{bmatrix}, Q > 0 \quad (9)$$

The tracking issue is changed and transformed into a regulating issue by using the updated system and discounted value function (6) [32]. With this improvement, it is feasible to create a reinforcement learning regulator to address the SRM drive's optimum tracking issue without possessing complete information of the machine's specifications.

2.1.2. Formulating the System Using Bellman and Hamilton–Jacobian Equation

One type of RL approach is based on dual neural networks, where the first NN provides the control policy or the action to the machine, and the second NN evaluates the value of that control policy. Different strategies, such as the gradient descent method and least-squares method, may then be utilized to update the control input in the sense that the new input is better than the old input. To allow the use of a RL method for tracking applications such as torque ripple minimization, one can derive the Bellman equation for the SRM drive. One of the adequate RL algorithms used to solve the Bellman equation online in real time and achieve tracking performance is the policy iteration method; that is, updating the policy until convergence leads to the optimal solution of the tracking problem. Following the presentation of the augmented model of the SRM and the performance index in the prior section, the Bellman and HJB equations of the SRM drive will be discussed. This will make it possible for the tracking control to apply the RL online technique in order to solve the issue. (8) may be recast as follows if one makes use of an applicable policy

$$V(X_k) = X_i^T Q_q X_i + u_i^T R u_i + \sum_{i=k+1}^{\infty} \gamma^{i-(k+1)} [X_i^T Q_q X_i + u_i^T R u_i] \quad (10)$$

This can be derived, based on the Bellman equation, as

$$V(X_k) = X_i^T Q_q X_i + u_i^T R u_i + \gamma V(X_{k+1}) \quad (11)$$

The optimum cost function $V^*(X_k)$, based on Bellman's optimality concept for infinite-time conditions, is a time-invariant and satisfies the discretized HJB equation as follows:

$$V^*(X_k) = \min_{u_k}\{X_i^T Q_q X_i + u_i^T R u_i + \gamma V^*(X_{k+1})\} \qquad (12)$$

To obtain the optimal control policy which can minimize the torque ripples, the Hamiltonian function of the Bellman equation can be expressed as

$$H(X_k, u_k) = x_k^T Q_q x_k + u_k^T R u_k + \gamma V^*(X_{k+1}) - V^*(X_k) \qquad (13)$$

At this point, it is crucial to execute the stationary condition $dH(X_k, u_k)/du_k = 0$. This condition is necessary to achieve optimality [33]. Hence, the control policy that can minimize the torque ripples for SRM drive is generated as

$$u_k^* = -\frac{\gamma}{2} R^{-1} \mathcal{G}(X_k)^T \frac{\partial V^*(X_{k+1})}{\partial X_{k+1}} \qquad (14)$$

2.2. Dual-Neural-Network Architecture for Learning the Tracking Problems of SRM Drive

Since the tracking HJB equation is unable to be solved accurately online using a normal approach without incorporating a complete knowledge of the parameter model, the reinforcement dual-neural-network learning methodology was used. Rotor angle and current both have nonlinear effects on phase inductances. This inductance is at its highest value when the stator and rotor poles are lined up, and at its lowest value when the poles are not lined up. Figure 2 displays the inductance surface of the SRM used later in the simulation. This figure is generated by quantizing the inductance profile derived using finite element analysis of the SRM. A table holding the data of the inductance surface may be produced. A 2D grid made up of a selection of several currents and rotor positions is used to quantize this surface. A quantized inductance profile is obtained by recording the inductance in a table at every point of this grid [34,35]. The bearings' age, differences in the airgap, chemical deterioration, and temperature changes may all lead to additional, unidentified alterations in this characteristic. Other changes in the inductance curve might result from inconsistencies between the real and predicted models caused by typical manufacturing defects, such as variances in the permeability, the size of the airgap, or even the quantity of turns in the coil. Adaptive approximation methods to improve the machine's dynamic characteristics may be carried out by utilizing the dual-neural-network procedure. To solve the Bellman problem, the neural network is employed to optimize the cost function values. The second neural network (NN) used in this technique, which accounts for the approximate dynamic programming tracking control, is modified online and in real time using information recorded while the machine is running, such as the torque state, the future augmented state, and the model parameters. The first and second neural networks are developed sequentially in this study, meaning that the first neural network's parameters will stay constant while the parameters of the second network are trained until convergence. These procedures are repeated until the first and second neural networks settle on the ideal trajectory. Using the neural network along with the value function approximation (VFA) concept, an evaluation NN may be created to tune the performance index function using the least-squares technique until convergence [27]. The formulation of the first and second NNs to minimize the torque pulsating is demonstrated in this section.

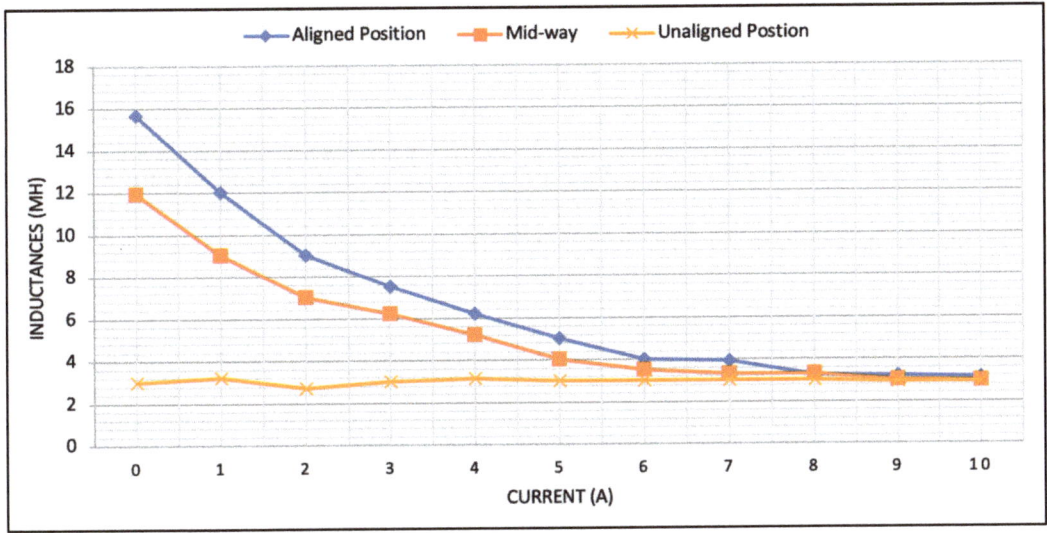

Figure 2. The variation of the base inductance parameters as a function of the current.

2.2.1. Modelling of First Neural Network

This is conducted to develop an observer for the purpose of evaluating the performance index, and as a result, this observer is used in order to generate the feedback control. It is standard practice to use neural networks when attempting to estimate a smooth cost function on a preset data set. The expression that may be used to describe how the weights of the NN, which offer the optimum approximation solution of minimization problem for the SRM drive, work is:

$$V_i(X) = \sum_{j=1}^{N} s_{vi}^j \mathcal{B}_j(X_k) = S_{vi}^T \mathcal{B}(X_k) \qquad (15)$$

where S_{vi} are the approximated quantities of the first NN weights that can be produced in linear system for the machine, shown as

$$S_{vi} = \begin{bmatrix} s_{11} & s_{12} & s_{22} \end{bmatrix} \qquad (16)$$

where $\mathcal{B}(X_k) = X_k \otimes X_k \in \mathbb{R}^N$ is the vector of the convolution operation, and it represents the number of neurons within the hidden layer. The Bellman equation can be reproduced by incorporating the Kronecker concept, which converts the weights matrix (16) into columns of bundling sequences [32].

$$\left[(X_k) \otimes (X_k) - \gamma(X_{k+1}) \otimes (X_{k+1}) \right] \times vec\left(S_{vi+1}^T \right) = X_k^T Q_q X_k + \hat{u}_i^T(X_k) R \hat{u}_i(X_k) \qquad (17)$$

where $\otimes$ is the Kronecker product, and $vec(S_{vi+1}^T)$ is the weights matrix derived by aggregating the entries of matrix W_{vi}. The left side of (17) can be defined as

$$\rho(X_k, \hat{u}_i(X_k)) = X_k^T Q_q X_k + \hat{u}_i^T(X_k) R \hat{u}_i(X_k) \qquad (18)$$

By exploring and gathering sufficient data packets throughout each cycle of the normal running of the motor, including information on the modified state of the motor and the input voltage, the solution of this equation can be can be obtained. The least-squares (LS) approach can be used to improve the weights of the network. This technique is a potent optimizer that does not need any additional model identification unless an observer is required to watch the appropriate data item sets. Thus, the first NN weight's inaccuracy error may be expressed as

$$Err_{vN} = \left(\rho(X_k, \hat{u}_i(X_k)) - S_{vi}^T \mathcal{B}(X_k) \right) \tag{19}$$

Prior to applying LS strategies and to address the policy evaluation method, the total count of individual entities in the data vector should be greater than 3 samples per iteration (17). The sequential least-squares response for the NN weights is then shown as

$$vec\left(S^T\right) = \frac{\mathcal{C}^T \sigma}{\mathcal{C}^T \mathcal{C}} \tag{20}$$

where $\mathcal{C} = [\Delta \overline{X}_k^T \Delta \overline{X}_{k+1}^T \ldots \Delta \overline{X}_{k+N-1}^T]$, $\Delta \overline{X}_k^T = \mathcal{B}^T(X_k) \otimes \mathcal{B}^T(X_k) - \gamma \mathcal{B}^T(X_{k+1}) \otimes \mathcal{B}^T(X_{k+1})$, and $\sigma = [\rho(\overline{X}_1, \hat{u}_i)\rho(\overline{X}_2, \hat{u}_i)\ldots\rho(\overline{X}_N, \hat{u}_i)]^T$. The dynamical parameters of the machine do not need to be inserted in order to tune the weight matrix values, and as $\mathcal{C}$ has a complete rank, $\mathcal{B}(X_k)$ is necessary to satisfy the persistence excitation condition. This can be achieved by adding a modest amount of white noise to the input signal. It will thus be sufficient to attain the persistence excitation condition [31].

2.2.2. Modeling of Second Neural Network

This section aims to create a phase voltage signal that minimizes the approximate amount function of the first NN by approximating the ideal return voltage signal of the machine. The ideal policy to minimize the torque ripples can be expressed as follows:

$$\hat{u}_i(X_k) = \underset{u(0)}{\operatorname{argmin}} \left(X_k^T Q_q X_k + u_i^T(X_k) R \hat{u}_i(X_k) + \gamma S_{vi}^T \mathcal{B}(X_{k+1}) \right) \tag{21}$$

Once the first value matrix has been trained until the parameters settle to their ideal values, the second online NN approximations are applied in order to achieve a result of (14) to fulfill the tracking performance and mitigate the torque ripples. The second NN formulation is described by the equation below.

$$\hat{u}_i(X) = \sum_{j=1}^{P} S_{ui}^j \mathcal{G}_j(X_k) = S_{ui}^T \mathcal{G}(X_k) \tag{22}$$

where $\mathcal{G}(X_k) \in \mathbb{R}^H$ is the parameters of the activation function, where P is the quantity of neurons in the hidden layer. As a result, the actor error may be calculated as the difference between the machine's phase voltage per phase and the control signal that minimizes the predicted performance index in the second NN, which is expressed as

$$err_{u(X_k)} = S_{ui}^T \mathcal{G}(X_k) + \frac{\gamma}{2} R^{-1} \mathcal{G}(X_k)^T \frac{\partial \mathcal{B}(X_{k+1})}{\partial X_{k+1}} S_{vi} \tag{23}$$

The gradient descent strategy may be used to tune the variables of the second NN in real screen time. Because the network only runs a single adjusting sample, this approach is simple to encode in memory. As a result, the second NN value update may be carried out as follows:

$$\begin{aligned} S_{ui}|_{z+1} = S_{ui}|_z & \left. -\Phi \frac{\partial}{\partial S_{ui}} [X_k^T Q_q X_k + \hat{u}_i^T(X_k) R \hat{u}_i(X_k) + \gamma S_{vi}^T \mathcal{B}(X_{k+1})] \right|_{S_{ui}|_z} \\ & = S_{ui}|_z - \Phi \times \Pi(X_k) \left(2R\hat{u}_i + \gamma \mathcal{G}(X_k)^T \frac{\partial \mathcal{B}(X_{k+1})}{\partial X_{k+1}} S_{vi} \right)^T \end{aligned} \tag{24}$$

where $\Phi > 0$ is a training parameter which represents the scaling factor, and z is the repetition number. As demonstrated in (18), only $\mathcal{G}(X_k)$ values of the dynamical model are needed to improve the weight of the NN. The policy iteration (PI) methodology has been utilized extensively for constructing feedback controllers among the RL techniques now in use. Specifically, the linear quadratic tracker (LQT) problem is resolved using PI algorithms. It is common knowledge that resolving an LQT is necessary to solve the Algebraic Riccati equation (ARE). The PI technique starts with an acceptable control policy and iteratively alternates between policy assessment and policy improvement phases until there is no modification to the value or the policy. In contrast to the value iteration (VI) method, In contrast to the value iteration (VI) method, PI is often faster than VI as the control input converges to their optimal solution which achieve torque ripples minimization for the system model. The following Algorithm 1 shows the process which will be executed for the proposed control strategy.

Algorithm 1: Using policy iteration approach, compute the tracking HJB problem of the model online.

Initialization: Launch the computation process with an allowable control policy. Perform and modify the two processes below until convergence is reached.

1st NN:
$$S_{vi}^T \mathcal{B}(X_k) = (X_k^T) Q_q(X_k) + \left(u_k^i\right)^T R \left(u_k\right)^i + \gamma S_{vi}^T \mathcal{B}(X_k + 1)$$

2nd NN:
$$S_{ui}|_{z+1} = S_{ui}|_z - \Phi \times \Pi(X_k) \left(2R\hat{u}_i + \gamma \mathcal{G}\left(X_k\right)^T \frac{\partial \mathcal{B}(X_{k+1})}{\partial X_{k+1}} S_{vi}\right)^T$$

---

3. Simulation Results

To assess the tracking effectiveness of the suggested system, a dual-stream neural network algorithm based on reinforcement learning techniques was created and simulated for the SRM drive. The block diagram of the scheme is described in Figure 1. There are two fundamental processing stages in the control system. The first NN approximates the utility function by training the weights of NN using the least-squares (LS) method. To minimize the estimated cost function, the input signal is updated in the second NN processing block. Several data sets must be selected and estimated to train the cost function in the first NN.

To implement the proposed technique, three phases of 12/8 SRM were invested in and modeled. The nominal current of the motor was 6 A, and the resistance per phase was 2 Ω. The inductance curve fluctuated between 16 mH for maximum aligned inductance and 6 mH for minimum unaligned inductance. The rated wattage was 530 W, with a DC voltage of 100 V.

The cutoff frequency of the controller was set at 12 kHz. The developed control system's procedure should initiate with the stabilizing control policy, according to the policy iteration approach. To show the controller's functionality, the augmented state was set to $X_0 = \begin{bmatrix} -10 & 10 \end{bmatrix}^T$. In the utility function, Q and R are predefined matrices of appropriate size, with values of 100 and 0.001, respectively. The discount factor used to decrease future costs was set at $\gamma = 0.8$. A train of rectangular shape signals with an ultimate peak value of 4A is generated by the reference signal generator. The second NN examines 10 data objects every cycle to train its value and optimize the utility function using the least-squares technique.

The parameters of the second NN approach to their ideal values after 10 epochs to minimize the torque ripples and achieve excellent performance for the motor.

The optimal first NN values reached the ideal number values which could reduce the torque ripples at

$$S_{ui}^T = [100 - 100] \tag{25}$$

To test the suggested controller in this research, the speed of the SRM was kept constant and set at 60 RPM. The voltage provided to the motor was capped at 100 V because most of the real DC hardware's sources are rated to this limit.

Figure 3 shows the comparison between delta modulation and the proposed method. It can be seen that RL architecture using dual NNs could efficiently minimize the torque ripples. The total torque waveform per phase is demonstrated in Figure 4. In this figure, after the weights of NNs settle to their optimal number, the controller successfully minimize the torque ripples. Figure 5 clarifies how the NN parameters settle to their ideal numbers after the NNs are fully trained.

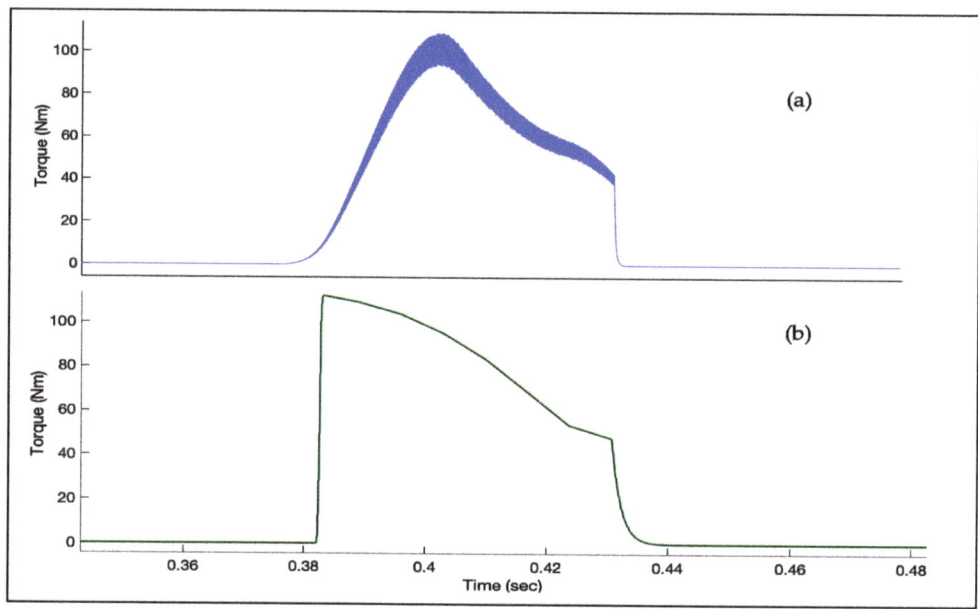

Figure 3. The phase torque of (**a**) the delta modulation method; (**b**) the proposed method.

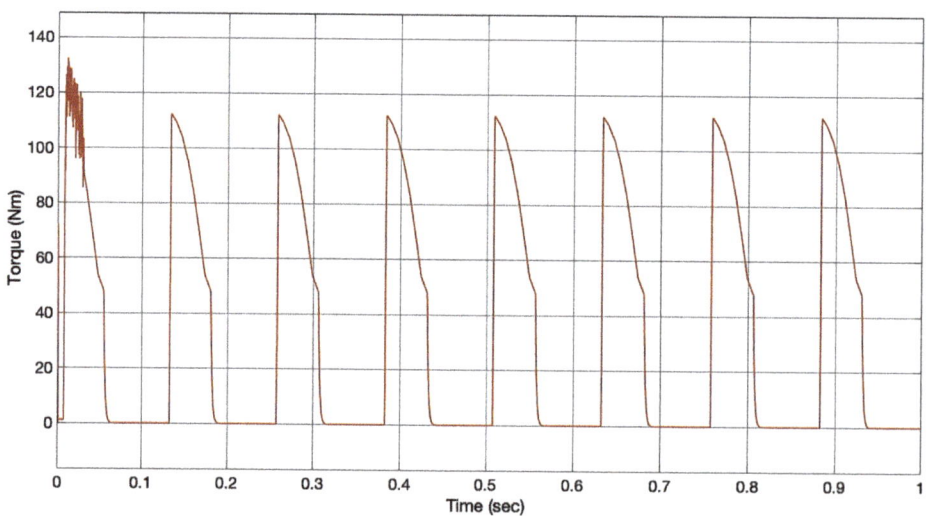

Figure 4. The total torque per phase at constant speed.

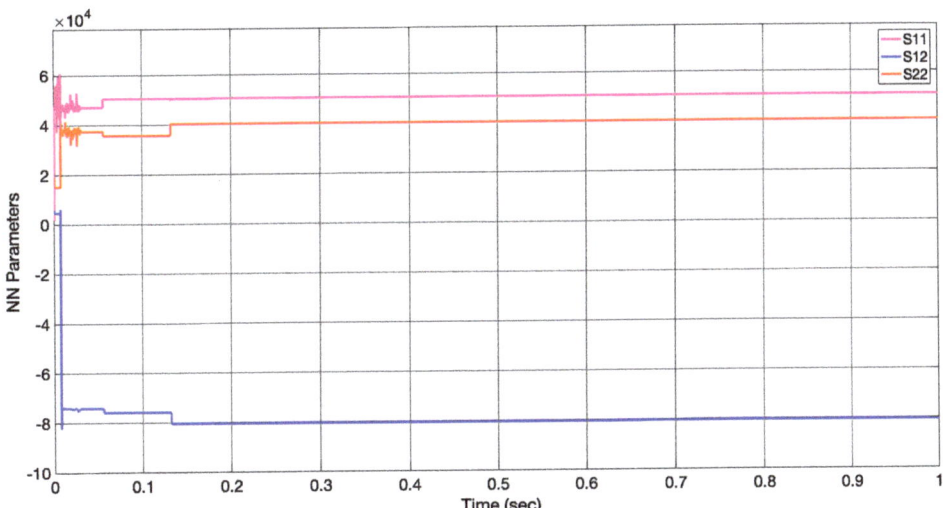

Figure 5. The convergence of the NN parameters to their optimal solutions.

In this article, the proposed dual NN architecture parameters are only the discount factor and the learning rate. These parameters are determined based on the trial-and-error technique. For weighting the Q and R matrices, the Q/R ratio is crucial for training NNs. The linear quadratic tracker will fail to follow the reference if the weight R has a high value due to the large cost in the control input. Additionally, if R = 0 or if the Q/R ratio is extremely high, the controller will follow the reference in the first step because of the extremely high applied control input. Hence, we chose the weights to be Q = 100 and R = 0.001 as they were the best selection based on the design technique.

4. Conclusions

This paper has presented a new strategy to minimize the torque ripples using the architecture of dual-stream NNs using Reinforcement Learning for the switched reluctance motor. A new enhanced architecture for SRM has been created, which will aid in the construction of the model's optimum tracking control. To assess the machine's control performance, a quadratic value function for tracking and reducing the torque pulsations on the motor was constructed. To do so, dual-stream NN estimation algorithms were adopted to estimate the value function and to generate the optimal control policy. The parameters of the first NN were trained online in real time using the least-squares method until convergence. Additionally, the gradient descent logic was applied to tune the second NN. The simulation results indicated that the suggested strategy was successful at adjusting the motor's torque and reducing its oscillations without adding additional procedures to cope with the nonlinearity of the model.

Funding: This research received no external funding.

Data Availability Statement: Not applicable.

Acknowledgments: The researchers would like to thank the Deanship of Scientific Research, Qas-sim University for funding the publication of this project.

Conflicts of Interest: The author declares no conflict of interest.

References

1. Chichate, K.R.; Gore, S.R.; Zadey, A. Modelling and Simulation of Switched Reluctance Motor for Speed Control Applications. In Proceedings of the 2020 2nd International Conference on Innovative Mechanisms for Industry Applications (ICIMIA), Bangalore, India, 5–7 March 2020; pp. 637–640.
2. Hao, Z.; Yu, Q.; Cao, X.; Deng, X.; Shen, X. An Improved Direct Torque Control for a Single-Winding Bearingless Switched Reluctance Motor. *IEEE Trans. Energy Convers.* **2020**, *35*, 1381–1393. [CrossRef]
3. Valdivia, V.; Todd, R.; Bryan, F.J.; Barrado, A.; Lázaro, A.; Forsyth, A.J. Behavioral Modeling of a Switched Reluctance Generator for Aircraft Power Systems. *IEEE Trans. Ind. Electron.* **2013**, *61*, 2690–2699. [CrossRef]
4. Gan, C.; Wu, J.; Sun, Q.; Kong, W.; Li, H.; Hu, Y. A Review on Machine Topologies and Control Techniques for Low-Noise Switched Reluctance Motors in Electric Vehicle Applications. *IEEE Access* **2018**, *6*, 31430–31443. [CrossRef]
5. Sun, X.; Wan, B.; Lei, G.; Tian, X.; Guo, Y.; Zhu, J. Multiobjective and Multiphysics Design Optimization of a Switched Reluctance Motor for Electric Vehicle Applications. *IEEE Trans. Energy Convers.* **2021**, *36*, 3294–3304. [CrossRef]
6. Kiyota, K.; Kakishima, T.; Chiba, A. Comparison of Test Result and Design Stage Prediction of Switched Reluctance Motor Competitive With 60-KW Rare-Earth PM Motor. *IEEE Trans. Ind. Electron.* **2014**, *61*, 5712–5721. [CrossRef]
7. Bilgin, B.; Howey, B.; Callegaro, A.D.; Liang, J.; Kordic, M.; Taylor, J.; Emadi, A. Making the Case for Switched Reluctance Motors for Propulsion Applications. *IEEE Trans. Veh. Technol.* **2020**, *69*, 7172–7186. [CrossRef]
8. Ghaffarpour, A.; Mirsalim, M. Split-Tooth Double-Rotor Permanent Magnet Switched Reluctance Motor. *IEEE Trans. Transp. Electrif.* **2022**, *8*, 2400–2411. [CrossRef]
9. Yan, W.; Chen, H.; Liao, S.; Liu, Y.; Cheng, H. Design of a Low-Ripple Double-Modular-Stator Switched Reluctance Machine for Electric Vehicle Applications. *IEEE Trans. Transp. Electrif.* **2021**, *7*, 1349–1358. [CrossRef]
10. Anwar, M.N.; Husain, I. Radial Force Calculation and Acoustic Noise Prediction in Switched Reluctance Machines. *IEEE Trans. Ind. Appl.* **2000**, *36*, 1589–1597. [CrossRef]
11. Husain, T.; Sozer, Y.; Husain, I. DC-Assisted Bipolar Switched Reluctance Machine. *IEEE Trans. Ind. Appl.* **2017**, *53*, 2098–2109. [CrossRef]
12. Rajendran, A.; Karthik, B. Design and Analysis of Fuzzy and PI Controllers for Switched Reluctance Motor Drive. *Mater. Today Proc.* **2021**, *37*, 1608–1612. [CrossRef]
13. Abraham, R.; Ashok, S. Data-Driven Optimization of Torque Distribution Function for Torque Ripple Minimization of Switched Reluctance Motor. In Proceedings of the 2020 International Conference for Emerging Technology (INCET), Belgaum, India, 5–7 June 2020; pp. 1–6.
14. Ellabban, O.; Abu-Rub, H. Torque Control Strategies for a High Performance Switched Reluctance Motor Drive System. In Proceedings of the 2013 7th IEEE GCC Conference and Exhibition (GCC), Doha, Qatar, 17–20 November 2013; pp. 257–262.
15. Rahman, K.M.; Schulz, S.E. High Performance Fully Digital Switched Reluctance Motor Controller for Vehicle Propulsion. In Proceedings of the Conference Record of the 2001 IEEE Industry Applications Conference. 36th IAS Annual Meeting (Cat. No.01CH37248), Chicago, IL, USA, 30 September–4 October 2001; Volume 1, pp. 18–25.
16. Gallegos-Lopez, G.; Kjaer, P.C.; Miller, T.J.E. A New Sensorless Method for Switched Reluctance Motor Drives. *IEEE Trans. Ind. Appl.* **1998**, *34*, 832–840. [CrossRef]
17. Scalcon, F.P.; Fang, G.; Vieira, R.P.; Grundling, H.A.; Emadi, A. Discrete-Time Super-Twisting Sliding Mode Current Controller with Fixed Switching Frequency for Switched Reluctance Motors. *IEEE Trans. Power Electron.* **2022**, *37*, 3321–3333. [CrossRef]
18. Alharkan, H.; Shamsi, P.; Saadatmand, S.; Ferdowsi, M. Q-Learning Scheduling for Tracking Current Control of Switched Reluctance Motor Drives. In Proceedings of the 2020 IEEE Power and Energy Conference at Illinois (PECI), Champaign, IL, USA, 27–28 February 2020; pp. 1–6.
19. Mehta, S.; Pramod, P.; Husain, I. Analysis of Dynamic Current Control Techniques for Switched Reluctance Motor Drives for High Performance Applications. In Proceedings of the 2019 IEEE Transportation Electrification Conference and Expo (ITEC), Detroit, MI, USA, 19–21 June 2019; pp. 1–7.
20. Yan, N.; Cao, X.; Deng, Z. Direct Torque Control for Switched Reluctance Motor to Obtain High Torque–Ampere Ratio. *IEEE Trans. Ind. Electron.* **2019**, *66*, 5144–5152. [CrossRef]
21. Gobbi, R.; Ramar, K. Optimisation Techniques for a Hysteresis Current Controller to Minimize Torque Ripple in Switched Reluctance Motors. *IET Electr. Power Appl.* **2009**, *3*, 453–460. [CrossRef]
22. Husain, I.; Ehsani, M. Torque Ripple Minimization in Switched Reluctance Motor Drives by PWM Current Control. *IEEE Trans. Power Electron.* **1996**, *11*, 83–88. [CrossRef]
23. Matwankar, C.S.; Pramanick, S.; Singh, B. Position Sensorless Torque Ripple Control of Switched Reluctance Motor Drive Using B-Spline Neural Network. In Proceedings of the IECON 2021—47th Annual Conference of the IEEE Industrial Electronics Society, Toronto, ON, Canada, 13–16 October 2021; pp. 1–5.
24. Schramm, D.S.; Williams, B.W.; Green, T.C. Torque Ripple Reduction of Switched Reluctance Motors by Phase Current Optimal Profiling. In Proceedings of the PESC'92 Record. 23rd Annual IEEE Power Electronics Specialists Conference, Toledo, Spain, 29 June–3 July 1992; pp. 857–860.
25. Lin, Z.; Reay, D.S.; Williams, B.W.; He, X. Torque Ripple Reduction in Switched Reluctance Motor Drives Using B-Spline Neural Networks. *IEEE Trans. Ind. Appl.* **2006**, *42*, 1445–1453. [CrossRef]

26. Rahman, K.M.; Gopalakrishnan, S.; Fahimi, B.; Velayutham Rajarathnam, A.; Ehsani, M. Optimized Torque Control of Switched Reluctance Motor at All Operational Regimes Using Neural Network. *IEEE Trans. Ind. Appl.* **2001**, *37*, 904–913. [CrossRef]
27. Liu, D.; Lewis, F.L.; Wei, Q. Editorial Special Issue on Adaptive Dynamic Programming and Reinforcement Learning. *IEEE Trans Syst. Man. Cybern. Syst.* **2020**, *50*, 3944–3947. [CrossRef]
28. Reinforcement Learning and Feedback Control: Using Natural Decision Methods to Design Optimal Adaptive Controllers. *IEEE Control Syst.* **2012**, *32*, 76–105. [CrossRef]
29. Dierks, T.; Jagannathan, S. Optimal Tracking Control of Affine Nonlinear Discrete-Time Systems with Unknown Internal Dynamics. In Proceedings of the 48h IEEE Conference on Decision and Control (CDC) Held Jointly with 2009 28th Chinese Control Conference, Shanghai, China, 15–18 December 2009; pp. 6750–6755.
30. Lewis, F.L.; Vrabie, D. Reinforcement Learning and Adaptive Dynamic Programming for Feedback Control. *IEEE Circuits Syst. Mag.* **2009**, *9*, 32–50. [CrossRef]
31. Kiumarsi, B.; Lewis, F.L. Actor–Critic-Based Optimal Tracking for Partially Unknown Nonlinear Discrete-Time Systems. *IEEE Trans. Neural Netw. Learn Syst.* **2015**, *26*, 140–151. [CrossRef] [PubMed]
32. Zhu, G.; Li, X.; Sun, R.; Yang, Y.; Zhang, P. Policy Iteration for Optimal Control of Discrete-Time Time-Varying Nonlinear Systems. *IEEE/CAA J. Autom. Sin.* **2023**, *10*, 781–791. [CrossRef]
33. Yang, Y.; Modares, H.; Vamvoudakis, K.G.; He, W.; Xu, C.-Z.; Wunsch, D.C. Hamiltonian-Driven Adaptive Dynamic Programming with Approximation Errors. *IEEE Trans. Cybern.* **2022**, *52*, 13762–13773. [CrossRef]
34. Ge, L.; Ralev, I.; Klein-Hessling, A.; Song, S.; De Doncker, R.W. A Simple Reluctance Calibration Strategy to Obtain the Flux-Linkage Characteristics of Switched Reluctance Machines. *IEEE Trans. Power Electron.* **2020**, *35*, 2787–2798. [CrossRef]
35. Hur, J.; Kang, G.H.; Lee, J.Y.; Hong, J.P.; Lee, B.K. Design and Optimization of High Torque, Low Ripple Switched Reluctance Motor with Flux Barrier for Direct Drive. In Proceedings of the Conference Record of the 2004 IEEE Industry Applications Conference. 39th IAS Annual Meeting, Seattle, WA, USA, 3–7 October 2004; Volume 1, pp. 401–408.

Disclaimer/Publisher's Note: The statements, opinions and data contained in all publications are solely those of the individual author(s) and contributor(s) and not of MDPI and/or the editor(s). MDPI and/or the editor(s) disclaim responsibility for any injury to people or property resulting from any ideas, methods, instructions or products referred to in the content.

Article

Effect of Ripple Control on Induction Motors

Piotr Gnaciński [1,*], Marcin Pepliński [1], Adam Muc [2], Damian Hallmann [1] and Piotr Jankowski [1]

1. Department of Ship Electrical Power Engineering, Faculty of Marine Electrical Engineering, Gdynia Maritime University, Morska St. 83, 81-225 Gdynia, Poland; m.peplinski@we.umg.edu.pl (M.P.); d.hallmann@we.umg.edu.pl (D.H.); p.jankowski@we.umg.edu.pl (P.J.)
2. Department of Ship Automation, Faculty of Marine Electrical Engineering, Gdynia Maritime University, Morska St. 83, 81-225 Gdynia, Poland; a.muc@we.umg.edu.pl
* Correspondence: p.gnacinski@we.umg.edu.pl; Tel.: +48-58-5586-382

Abstract: One method for the remote management of electrical equipment is ripple control (RC), based on the injection of voltage interharmonics into the power network to transmit information. The disadvantage of this method is its negative impact on energy consumers, such as light sources, speakers, and devices counting zero crossings. This study investigates the effect of RC on low-voltage induction motors through the use of experimental and finite element methods. The results show that the provisions concerning RC included in the European Standard EN 50160 Voltage Characteristics of Electricity Supplied by Public Distribution Network are imprecise, failing to protect induction motors against excessive vibration.

Keywords: induction motors; interharmonics; mains communication voltage; power quality; ripple control; vibration

1. Introduction

In many countries [1], operators of distribution systems (DSs) use power lines to transmit communication signals. One possible remote management method of DS operation [2] is based on the superimposition of interharmonics on the voltage waveform [1–11]—components of frequency not being an integer multiple of the fundamental frequency. The novelized version of the standard [12] (2019) calls the injected signals "mains communication voltage" (MCV) and specifies the frequency range as 0.1–100 kHz. In the case of interharmonics with a frequency less than 3 kHz, the method is commonly dubbed "ripple control" (RC) [1–11].

The RC signal was originally produced by motor–generator sets, which were later replaced by static frequency converters [9]. The signal is in the form of telegram code [4,9], for example, of duration ~100 s [5] and value 1–5% of the nominal grid voltage [1]. Of note, this percentage can increase because of resonance phenomena in the power system [4,7,8]. The signal is typically injected into a medium-voltage network and transmitted to a low-voltage grid via power transformers [6,9,11]. In the low-voltage network, it is used to manage customers' electric meters and various energy receivers [1–11]. Furthermore, it can be applied for load peak reduction in the network [6,10]. If the power demand reaches a programmable threshold, some loads, for example, hot water boilers, heat pumps, or swimming pool pumps, can be switched off [6]. In practice, individual receivers can be configured to recognize specific codes [6].

A new challenge for RC is the effective governing of residential photovoltaic systems (PVs) and electric vehicle chargers and batteries [4,5,7,10]. For example, RC allows the use of sun-tracking systems for PVs to adjust the generated power to the actual grid demand [10]. Controlling PVs with RC is much more cost effective than with the Internet [5]. In summary, RC is considered an efficient, remunerative, and inherently cyber-secure method of managing various electrical equipment [4,5].

One drawback of RC is the negative impact on some energy receivers. It is reported to cause light flickers, audible noise from speakers and ceiling fans, and incorrect working of devices counting zero crossings [2,7,9]. Moreover, voltage interharmonics, applied in this method, are considered harmful power quality disturbances (PQDs). Their occurrence results in the poor operation of rotating machines, light sources, transformers, power electronic appliances, and control systems [13–15]. Among the various equipment, rotating machines are particularly sensitive to interharmonics (based on [14–25]). They cause speed fluctuations, increases in power losses, torque pulsations, and lateral and torsional vibrations, posing a risk of drivetrain damage [14–24]. The most exposed to failure are some medium-voltage equipment, such as large synchronous generators, multi-megawatt drivetrains with synchronous motors, and turbomachinery (based on [15,24]), which is likely the reason the possible interharmonic limit values in the standard [26] are dedicated to non-generation installations.

Interharmonic contamination usually originates from the working of wind power stations and other renewable sources of energy, cycloconverters, various power electronic equipment, and time-varying loads, including those from AC motors driving a pulsating anti-torque [3,24,27–30]. Especially significant sources of interharmonics are double-frequency conversion systems, like high-voltage DC links and inverters [29,30]. That is, voltage fluctuations across the capacitor in a DC link (or fluctuations of current flowing through the inductance in a DC link) are transmitted to both the AC input and the AC output of the double-conversion system [29–31], which may result in high interharmonic contamination [29]. For example, [29] reported various co-occurring voltage interharmonics, with values as high as 1.17%. These interharmonics were caused by the working of high-power inverters.

To achieve appropriate voltage quality, power quality standards specify limited permissible levels of various PQDs. However, the limits generally do not contain interharmonics. In IEEE-519: Standard for Harmonic Control in Electric Power Systems [26], proposals for two alternative limit curves for non-generation installations are discussed. One curve generally limits interharmonic subgroups of frequencies less than 1 kHz to 0.3% and those having frequencies within 1–2.5 kHz to 0.5%. According to the other limit curve, the permissible value of interharmonics of frequencies less than 2.5 kHz is 0.5%. The exceptions are interharmonics of frequencies close to harmonic frequencies, especially the fundamental one. The limit of voltage interharmonics of frequencies of ~50–70 Hz (in a 60 Hz system) should be based on the IEC flickermeter indication. The standard [26] warns that, in some cases, no intentional emission of voltage interharmonics can interfere with RC signals and underlines that "compatibility of voltage interharmonics with ripple control is necessary (...) and requires country-based limits".

Further, the European Standard EN 50160 Voltage Characteristics of Electricity Supplied by Public Distribution Network [12] contains the following comment: "The level of interharmonics is increasing due to the development of the application of frequency converters and similar control equipment. Levels are under consideration, pending more experience." Nevertheless, the standard [12] provides permissible values of voltage interharmonics used for the MCV. The highest limit is for the frequency of 0.1–0.4 kHz—according to [12], "for 99% of a day the 3 s mean value of signal voltages shall be less or equal to" 9%. For the higher frequencies of the MCV, the limits are much lower—at 100 kHz, the permissible value is about 1%.

Previous research works [14–23,25] do not cover induction motors (IMs) under the interharmonic values and frequencies admitted in [12] for the MCV. Many works [17–22,25] deal with IMs in cyclic voltage fluctuations, which are considered the superposition of interharmonics and subharmonics (i.e., components of frequency less than the fundamental values) [3,13,22]. Notably, the results of these investigations [17–22,25] cannot be directly applied to assessing the effect of RC on IMs. The impact of a single interharmonic tone on IMs was analyzed in [14,15,22,23,25]. For instance, the authors of [25] presented currents and rotational speed fluctuations for interharmonics of frequencies not exceeding

100 Hz. Other works [14,15,22,23] focused on currents, power losses, torque pulsations, vibrations for interharmonic values of 1%, and frequencies below 200 Hz. However, for these interharmonic and frequency values (the lowest frequencies used in RC), a rather moderate vibration was observed [14]. In summary, based on the current state of knowledge, assessing the effect of RC on IMs is not possible.

Therefore, the objectives of this paper were formulated. This work aims to point out that the limits of MCVs included in EN 50160 [12] are too tolerant and do not prevent IMs from malfunctioning. The second aim is to extend the authors' previous works [14,15,22] and present the investigation results for interharmonic frequencies and values up to 400 Hz and 9%, respectively. The considerations included in this study are limited to low-voltage equipment and non-generation installations.

2. Methodology

The effect of RC on IMs was investigated using numerical and empirical methods. The computations were performed with the two-dimensional finite element method (FEM) for a cage induction motor TSg100L-4B (rated power of 3 kW), referred to as motor1. Its chosen parameters are provided in Table 1. The model of the investigated motor was identified based on measurement results [32,33] and design data. Firstly, an electromagnetic circuit model was worked out using the RMxprt module and motor data, including ratings, the magnetization characteristic of iron, and geometric dimensions. Furthermore, based on the circuit model, a preliminary FEM model was elaborated. The original mesh proposed by the RMxprt module consisted of about 5000 elements, and the air gap was divided into two regions. Finally, some modifications were made to the field model. To improve the solution convergence, the number of finite elements was increased, and the air gap was divided into three regions. The tau-type mesh used for this study consisted of ~22,000 triangle elements—the stator core was divided into ~6200 elements, while the rotor core comprised ~3700 elements. The maximal length of the stator core elements was about 0.68 mm and that of the rotor core elements was ~0.27 mm. For comparison, the inner stator diameter was 94 mm. For the numerical analysis, the MAXWELL-ANSYS environment (ANSYS Electronics Desktop version 2022R2.4, Canonsburg, USA) and a transient-type solver were employed. Of note, some calculation parameters were found on the grounds of the analysis of solution convergence. The impact of vibrations and deformations was omitted during computations. The experimental validation of the field model is included in [14,32,33].

Table 1. Chosen parameters of the investigated motors.

Motor	Type	Rated Power (kW)	Rated Speed (rpm)	Rated Voltage (V)	Rated Current (A)
motor1	TSg100L-4B	3	1420	380	6.9
motor2	1LE1003-1BB22-2AA4	4	1460	400	7.9

The measurement setup comprised an AC programmable power source, a cage induction motor, a system for vibration measurements, and a computer-based power quality analyzer. The applied power source comprised two units—a Chroma 61512 (master) and a Chroma A615103 (slave) connected in parallel—totaling a rated power of 36 kVA. Additionally, it was equipped with some protection appliances, such as a reverse current protective unit, Chroma A615106. The power source could produce a voltage with programmable PQDs, such as subharmonics and interharmonics (SaIs) of frequencies from 0.01 to 2400 Hz, harmonics, voltage and frequency fluctuations, and phase or amplitude voltage unbalance.

The investigated motor 1LE1003-1BB22-2AA4 (rated power of 4 kW, referred to as motor2) was coupled with an unloaded DC machine (PZMb 54a, working as a generator). The motor2 nameplate parameters are provided in Table 1. Of note, the presence of the DC generator resulted in a small anti-torque (caused by mechanical losses of the generator) and an increase in the moment of inertia of the powertrain (which significantly affects torque pulsations under SaIs [22,33,34]). Additionally, the presence of the coupling may also exert an impact on vibration. According to the authors' experience (e.g., [15,22]), the vibrations may differ considerably in the cases of an uncoupled motor and a motor coupled with any machine (for instance, with an unloaded DC generator).

For vibration measurement, a Bruel & Kjaer (B&K) system was employed, which included a four-channel data acquisition module (B&K model 3676-B-040), a three-axis magnetically mounted accelerometer (B&K model 4529-B, with a frequency range of 0.3–12,800 Hz, sensitivity of 10 mV/ms^{-2}, maximum shock level peak of 5100 g, and weight of 14.5 g), a calibrator (B&K model 4294), and a computer with installed B&K Connect 2022, version 26.1.0.251 installed. Since the motor casing was made of die-cast aluminum, the accelerometer was attached to dedicated steel stands screwed into the motor (Figure 1). Before each measurement session, the accelerometer was calibrated. After the measurements were taken, the recorded accelerometer indications were filtered through a low-pass filter and recalculated into the broad-band vibration velocity [35,36] using the B&K Connect software. The vibration velocity was determined as per the main provisions of ISO Standard 20816-1 Mechanical vibration—measurement and evaluation of machine vibration—part 1: General guidelines [36].

Figure 1. Motor2 and the accelerometer (indicated with the red arrow).

The voltage and current waveforms were recorded using a digital oscilloscope Tektronix TBS 2000 B equipped with additional measurement transducers. The interharmonic content in the supply voltage and motor current was computed offline, employing fast Fourier transform and software customized by the authors.

A simplified diagram of the measurement setup is presented in Figure 2 (based on [14]).

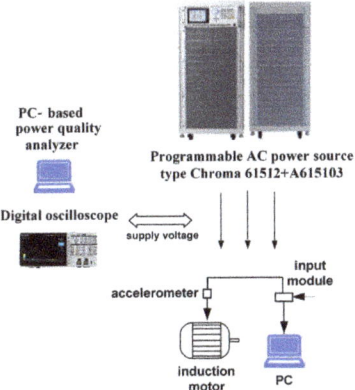

Figure 2. Simplified diagram of the measurement setup.

3. Results

3.1. Preliminary Remarks

During this study, the supply voltage was assumed to contain a single positive-sequence interharmonic of constant value. Numerical computations were performed for the omitted load inertia of the driven appliance (the justification is included in [14,34]). As the highest vibration of IMs caused by SaIs was observed for no load [15], all research results concern this state. Of note, some motors temporarily work with much less output power than rated [37,38] or even no-load conditions, for example, under the standard duty type S6 15% [39]. The torques, currents, and their frequency components are presented in relation to their rated values.

3.2. Currents

The primary source of the excessive vibration of IMs supplied with voltage with SaIs is torque pulsations caused by the flow of current SaIs (based on [15,40]).

A sample current waveform and its spectrum are shown in Figures 3 and 4, respectively, for motor1, in which the interharmonic frequency f_{ih} is 121 Hz and the value u_{ih} is 9%. Aside from the fundamental harmonic, the most significant frequency component is the interharmonic frequency f_{ih} = 121 Hz and value of 29.3% of the rated current I_{rat}. Additionally, the spectrum contains subharmonic components, which may produce voltage subharmonics in a power system. Notably, subharmonics of apparently inconsiderable values may harm the rotating machinery and transformers [13,15].

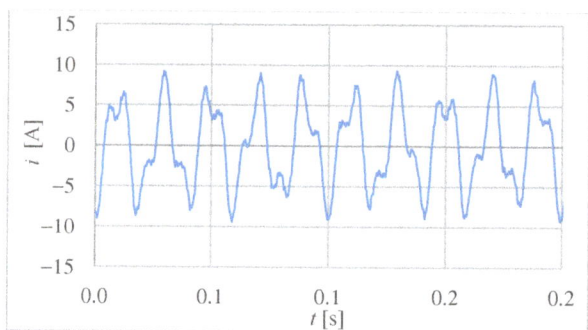

Figure 3. Measured current waveform of motor1 under no load, supplied with voltage containing interharmonics of value u_{ih} = 9% and frequency f_{ih} = 121 Hz.

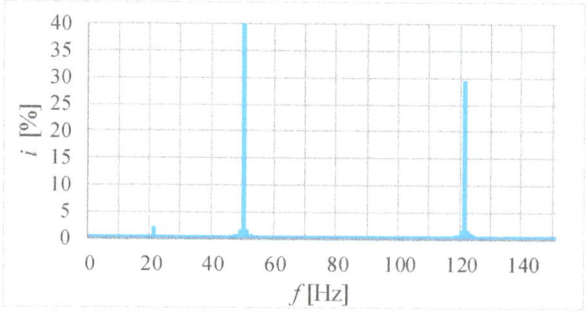

Figure 4. Spectrum of the current waveform presented in Figure 3.

The measured characteristic of the current interharmonics versus their frequency is provided in Figure 5 for motor2. The characteristic generally decreased but with small local extrema around the frequency $f_{ih} \approx 200$ Hz, which may be due to resonance phenomena.

The authors also carried out numerical investigations for motor1. Likewise, the computed characteristic of current subharmonics (Figure 6) decreased as the frequency f_{ih} increased. The general shape of the characteristics is due to the leakage inductance suppressing current interharmonics more significantly for the higher frequency f_{ih}.

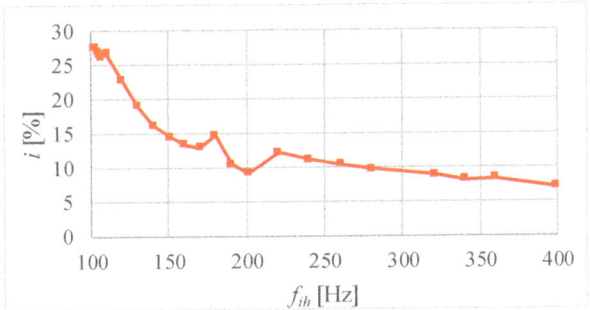

Figure 5. Measured current interharmonics versus their frequency for motor2 under no load. Current interharmonics are related to the rated motor current.

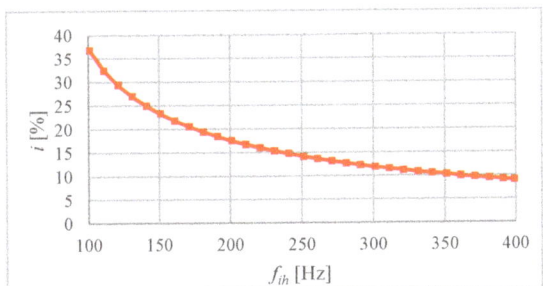

Figure 6. Computed current interharmonics versus their frequency for motor1 under no load. Current interharmonics are related to the rated motor current.

The analogical characteristics for the same motors at u_{ih} = 1% and frequencies of 50 to 100 Hz (motor1) or 200 Hz (motor2) are given in [14]. The main difference between these characteristics and those in Figures 5 and 6 is the global maxima below 100 Hz caused by the rigid-body resonance of the rotating mass.

In summary, for the investigated motors, the characteristics of current interharmonics generally decreased as the frequency f_{ih} increased and did not show global maxima in the considered frequency range.

3.3. Electromagnetic Torque Pulsations

Positive-sequence interharmonics cause a pulsating torque component (PTC) of the frequency f_p based on [25] the following:

$$f_p = f_{ih} - f_1 \qquad (1)$$

where f_1 is the fundamental frequency.

Figures 7 and 8 show the computed waveform and its spectrum of the electromagnetic torque for motor1, respectively, supplied with a voltage having an interharmonic frequency f_{ih} of 121 Hz and a value u_{ih} of 9%. The PTC frequency f_p = 71 Hz reached 39.7% of rated torque (T_{rat}). In contrast, the constant component (resulting from the first current harmonic) was approximately 1% of T_{rat}, typical for low-power, four-pole IMs under no load.

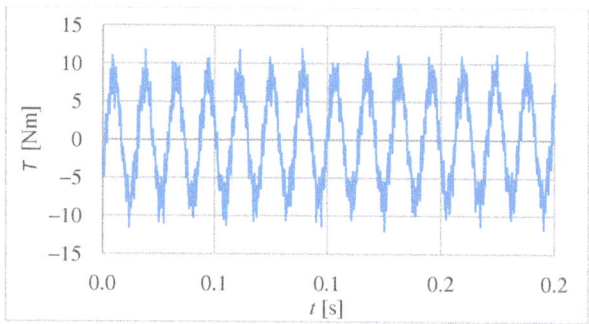

Figure 7. Torque waveform of motor1 under no load, supplied with the voltage containing the interharmonic value u_{ih} = 9% and frequency of f_{ih} = 121 Hz.

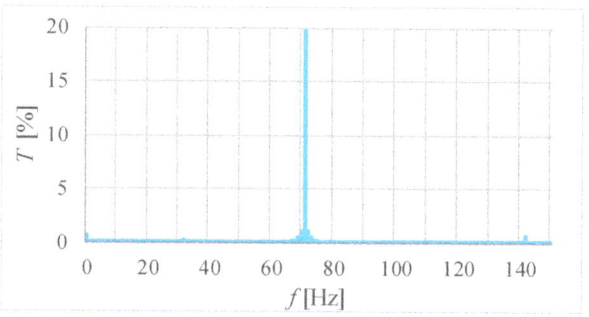

Figure 8. Spectrum of the torque waveform presented in Figure 7.

Figure 9 presents the PTCs of the frequency f_p versus the interharmonic frequency f_{ih}. For f_{ih} = 101 Hz, the PTC value was approximately four times that for f_{ih} = 399 Hz and reached ~50% of T_{rat}. Of note, this value is close to the maximal PTC observed for IMs supplied with voltage containing a single subharmonic value u_{sh} = 1% [21], resulting in extraordinarily high vibration [15,22].

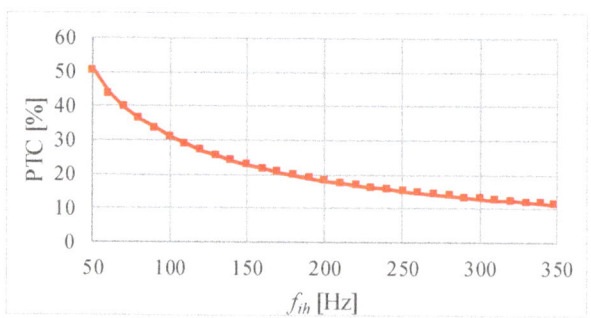

Figure 9. PTC versus the interharmonic frequency f_{ih} for u_{ih} = 9% and motor1. PTC is rated to the rated motor torque.

In some cases, the interharmonic value of 9% disturbed the starting process of motor2 (starting with the reduced supply voltage). This issue will be deeply analyzed in a separate paper.

In summary, RC caused a significant PTC, leading to excessive IM vibrations.

3.4. Vibration

For the assessment of vibration severity, the recommendations included in the standards [35,36] were employed. They specify four evaluation zones, denoted as Zone A, Zone B, Zone C, and Zone D. Zone C corresponds to vibrations admitted for a limited time, while the vibrations within Zone D "are normally considered to be of sufficient severity to cause damage to the machine". As the threshold values of each evaluation zone are not univocally specified in the current standard [36], they were assumed based on its former version [35]. Per [35], for low-power electric motors, a broad-band vibration velocity [35,36] between 1.8 and 4.5 mm/s corresponds to Zone C, and a vibration velocity greater than 4.5 mm/s corresponds to Zone D.

Figure 10 presents the characteristics of the broad-band vibration velocity versus the frequency of the voltage interharmonics for u_{ih} = 9% and motor2. The measured vibration velocity reached 5.025 mm/s, exceeding the boundaries of Zone D for the frequency $f_{ih} \leq 105$ Hz. Furthermore, for a frequency f_{ih} of 106 to 170 Hz, the vibration velocity fell into Zone C. Of note, the most severe vibration occurred for frequencies f_{ih} corresponding to the highest PTCs (see the previous subsection). Nevertheless, "the magnitude of the ... vibration directly depends on the mechanical behavior of the motor structure and the possibility of a resonance condition ... on the structure of an entire unit or on the motor components, such as a stator core or frame" [40]. Consequently, for other drivetrains, the highest vibration may appear for other frequency f_{ih} values. Furthermore, the shape of the characteristic under consideration can be explained by both the behavior of the mechanical structure and the effect of leakage inductance (see Sections 3.2 and 3.3).

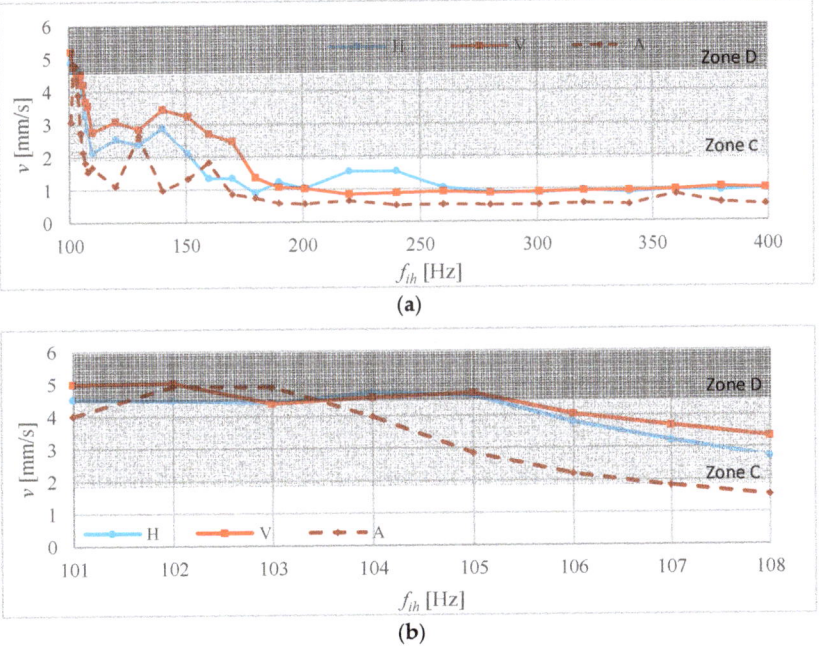

Figure 10. Measured broad-band vibration velocity in the horizontal (H), vertical (V), and axial (A) directions versus the voltage interharmonic frequency for motor2 and interharmonic value u_{ih} = 9%. Figure (**b**) is an enlarged fragment of Figure (**a**).

Figure 11 shows the characteristics of the broad-band vibration velocity versus the interharmonic value u_{ih} for the frequency f_{ih} = 101 Hz and motor2. The plots show significant non-linearity, probably due to the coupling reaction. For $u_{sh} \leq 7\%$, the vibration velocity gradually increased to 2.62 mm/s, exceeding the threshold value of Zone C for $u_{sh} \approx 5\%$. In turn, between $u_{ih} = 7\%$ and $u_{ih} = 8\%$, it rapidly increased to 4.97 mm/s and fell into Zone D.

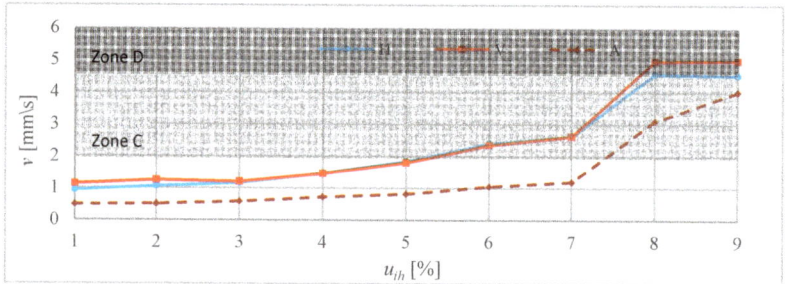

Figure 11. Measured broad-band vibration velocity in the horizontal (H), vertical (V), and axial (A) directions versus the voltage interharmonic value u_{ih} for motor2 and interharmonic frequency $f_{ih} = 101$ Hz.

The characteristics presented in Figure 11 were measured using the frequency f_{ih} corresponding to the highest vibration velocity (see Figure 10). Contrastingly, Figures 12 and 13 present analogical characteristics for frequencies at which motor2 showed comparatively low vibration. The appropriate experimental investigations were performed for exemplary RC signal frequencies [1,5,11]: $f_{ih} = 175$ Hz (Figure 12) and 208.3 Hz (Figure 13). The maximal vibration velocity did not exceed 1.672 mm/s and fell into Zone B.

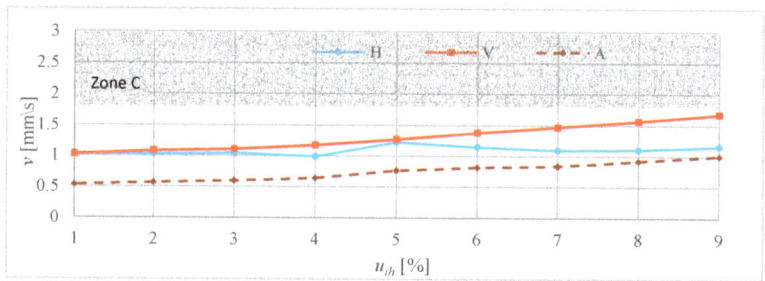

Figure 12. Measured broad-band vibration velocity in the horizontal (H), vertical (V), and axial (A) directions versus the voltage interharmonic value u_{ih} for motor2 and interharmonic frequency $f_{ih} = 175$ Hz.

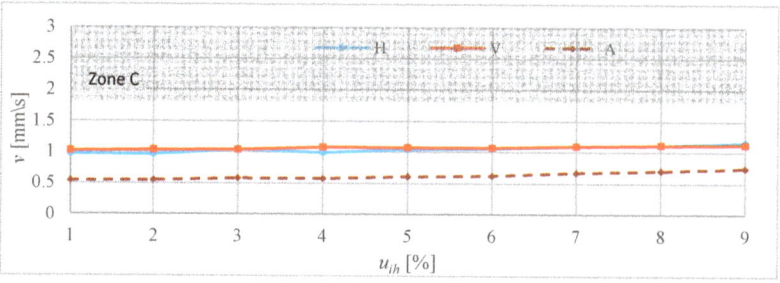

Figure 13. Measured broad-band vibration velocity in the horizontal (H), vertical (V), and axial (A) directions versus the voltage interharmonic value u_{ih} for motor2 and interharmonic frequency $f_{ih} = 208.3$ Hz.

In summary, the presented results of investigations indicate that the application of MCVs at the values permitted in the standard [12] may result in IM failures due to excessive vibration.

4. Discussion

A dynamic growth in the number of PV installations and electric cars presents a new challenge for RC. The distribution network is expected to be increasingly contaminated with RC signals, which should be considered as a specific case of voltage waveform distortions. One receiver especially susceptible to voltage waveform distortions (including voltage harmonics and SaIs) is an induction motor. Voltage waveform distortions cause various harmful phenomena, such as an increase in power losses, overheating, a local saturation of the magnetic circuit, torque ripples, and excessive lateral and torsional vibration [3,13–15,17–25,31–34,40–42], resulting even in powertrain destruction [31].

To prevent energy receivers from malfunctioning, power quality standards impose limitations on various PQDs. Many European countries apply the standard EN 50160 [12], which specifies the limits of RC signals. According to [12], within the frequency range of 0.1–0.4 kHz, "for 99% of a day the 3 s mean value of signal voltages shall be less or equal to" 9%. In practice, the standard does not limit the value of RC signals whose total duration is less than 1% of the day; in practice, any signal values can be found acceptable in light of [12]. Additionally, the 9% limit in [12] is inappropriate. Voltage interharmonics within this limit cause significant torque pulsations, leading to excessive vibrations. For the investigated motors, their levels fell into evaluation Zone D [35,36], in which they "are normally considered to be of sufficient severity to cause damage to the machine" [35,36].

Currently, only the practice used by DS operators protects IMs from destructive vibration, in which the value of RC signals is usually in the range of 1–5% [1], much less than that permitted by [12]. Of note, the value of RC signals can be significantly amplified because of resonance phenomena [4,7,8], even by a factor of three [8]. Such resonances in a power system were observed at frequencies of 1 kHz [4,8]. At the same time, the vibration of motor2 fell into Zone D for the interharmonic frequency $f_{ih} \leq 105$ Hz and the value $u_{ih} \geq 8\%$. In practice, such an RC signal is rather unlikely. Nevertheless, these standards should enable the electrical equipment to operate reliably and durably rather than the practice used by DS operators.

Furthermore, the PTC frequency may correspond to the natural torsional frequency of the elastic-body mode [14]. In drivetrains with IMs, the elastic-mode resonance [16,24,31,43,44] may lead to the amplification of PTCs by a factor exceeding 100 [31] and, consequently, a coupling or shaft failure [31,43,44]. Notably, the resonance may cause drivetrain destruction after a comparatively short time, for example, during repetitive starts [43]. The effect of the elastic-mode resonance on IMs will be the subject of future investigations.

Given the above considerations, the provisions in question [12] are unacceptable. They do not protect IMs from the potentially harmful impact of RC, especially with lateral and torsional vibration. Revising the standard [12] requires in-depth investigations of the undesirable phenomena caused by RC.

5. Conclusions

The provisions concerning RC laid in EN 50160 [12] are imprecise and too tolerant. According to [12], any level of RC signals can be considered acceptable, provided that their total duration is less than 1% of the day. For RC signals of longer total durations, the maximal permitted value is as high as 9%. This research shows that, even for interharmonics less than the limit, IM vibration may fall within evaluation Zone D, risking machine damage [35,36]. Presently, only practices used by DS operators prevent IMs from excessive vibration. The standard [12] should be modified taking into account the impact of RC on energy consumers, the real values of RC signals injected into DSs [1], a possible magnification of the signal for some frequencies due to resonance phenomena [4,7,8], and the possible interference of RC signals with voltage interharmonics occurring in the power system.

Author Contributions: Conceptualization, P.G., M.P. and D.H.; methodology, M.P., D.H. and P.J.; formal analysis, M.P., A.M. and D.H.; investigation, M.P., A.M. and D.H.; writing—original draft preparation, P.G.; supervision, P.G. All authors have read and agreed to the published version of the manuscript.

Funding: This project is financially supported under the framework of a program of the Ministry of Science and Higher Education (Poland) as "Regional Excellence Initiative" in the years 2019–2023, project number 006/RID/2018/19, amount of funding PLN 11 870 000.

Data Availability Statement: Data are contained within the article.

Conflicts of Interest: The authors declare no conflict of interest. The funders had no role in the design of this study; in the collection, analyses, or interpretation of data; in the writing of the manuscript; or in the decision to publish the results.

References

1. Garma, T.; Šesnić, S. Measurement and modeling of the propagation of the ripple control signal through the distribution network. *Int. J. Electr. Power Energy Syst.* **2014**, *63*, 674–680. [CrossRef]
2. Battacharyya, S.; Cobben, S.; Toonen, J. Impacts of ripple control signals at low voltage customer's installations. In Proceedings of the 22nd International Conference on Electricity Distribution, Stockholm, Sweden, 10–13 June 2013.
3. Bollen, M.H.J.; Gu, I.Y.H. Origin of power quality variations. In *Signal Processing of Power Quality Disturbances*; Wiley: New York, NY, USA, 2006; pp. 41–162.
4. Boutsiadis, E.; Tsiamitros, D.; Stimoniaris, D. Ripple signaling control for ancillary services in distribution networks. *Turk J. Electr. Power Energy Syst.* **2022**, *2*, 31–45. [CrossRef]
5. Boutsiadis, E.; Tsiamitros, D.; Stimoniaris, D. Distributed generation control via ripple signaling for establishment of ancillary services in distribution networks. In Proceedings of the 2021 13th International Conference on Electrical and Electronics Engineering (ELECO), Bursa, Turkey, 25–27 November 2021; pp. 18–23.
6. Dzung, D.; Berganza, I.; Sendin, A. Evolution of powerline communications for smart distribution: From ripple control to OFDM. In Proceedings of the 2011 IEEE International Symposium on Power Line Communications and Its Applications, Udine, Italy, 3–6 April 2011; pp. 474–478.
7. Muttaqi, K.M.; Rahman, O.; Sutanto, D.; Lipu, M.H.; Abdolrasol, M.G.; Hannan, M.A. High-frequency ripple injection signals for the effective utilization of residential EV storage in future power grids with rooftop PV system. *IEEE Trans. Ind. Appl.* **2022**, *58*, 6655–6665. [CrossRef]
8. Perera, B.S.P.; Nguyen, K.; Gosbell, V.J.; Browne, N.; Elphick, S.; Stones, J. Ripple signal amplification in distribution systems: A case study. In Proceedings of the 10th International Conference on Harmonics and Quality of Power, Rio de Janeiro, Brazil, 6–9 October 2002; Volume 1, pp. 93–98.
9. Rahman, O.; Elphick, S.; Muttaqi, K.M.; David, J. Investigation of LED lighting performance in the presence of ripple injection load control signals. *IEEE Trans. Ind. Appl.* **2019**, *55*, 5436–5444. [CrossRef]
10. Tsiakalos, A.; Tsiamitros, D.; Tsiakalos, A.; Stimoniaris, D.; Ozdemir, A.; Roumeliotis, M.; Asimopoulos, N. Development of an innovative grid ancillary service for PV installations: Methodology, communication issues and experimental results. *Sustain. Energy Technol. Assess.* **2021**, *44*, 101081. [CrossRef]
11. Yang, Y.; Dennetière, S. Modeling of the behavior of power electronic equipment to grid ripple control signal. *Electr. Power Syst. Res.* **2009**, *79*, 443–448. [CrossRef]
12. *EN Standard 50160, 2010/A2:2019*; Voltage Characteristics of Electricity Supplied by Public Distribution Network. CELENEC: Brussels, Belgium, 2019.
13. Gallo, D.; Landi, C.; Langella, R.A.; Testa, A. Limits for low frequency interharmonic voltages: Can they be based on the flickermeter use. In Proceedings of the 2005 IEEE Russia Power Tech, St. Petersburg, Russia, 27–30 June 2005; pp. 1–7.
14. Gnaciński, P.; Hallmann, D.; Klimczak, P.; Muc, A.; Pepliński, M. Effects of voltage interharmonics on cage induction motors. *Energies* **2021**, *14*, 1218. [CrossRef]
15. Gnacinski, P.; Peplinski, M.; Murawski, L.; Szelezinski, A. Vibration of induction machine supplied with voltage containing subharmonics and interharmonics. *IEEE Trans. Energy Convers.* **2019**, *34*, 1928–1937. [CrossRef]
16. Bongini, L.; Mastromauro, R.A. Subsynchronous torsional interactions and start-up issues in oil & gas plants: A real case study. In Proceedings of the AEIT International Annual Conference (AEIT), Firenze, Italy, 18–20 September 2019; pp. 1–6.
17. Ghaseminezhad, M.; Doroudi, A.; Hosseinian, S.H.; Jalilian, A. Analysis of voltage fluctuation impact on induction motors by an innovative equivalent circuit considering the speed changes. *IET Gener. Transm. Distrib.* **2017**, *11*, 512–519. [CrossRef]
18. Ghaseminezhad, M.; Doroudi, A.; Hosseinian, S.H.; Jalilian, A. An investigation of induction motor saturation under voltage fluctuation conditions. *J. Magn.* **2017**, *22*, 306–314. [CrossRef]
19. Ghaseminezhad, M.; Doroudi, A.; Hosseinian, S.H.; Jalilian, A. Analytical field study on induction motors under fluctuated voltages. *Iran. J. Electr. Electron. Eng.* **2021**, *17*, 1620.

20. Ghaseminezhad, M.; Doroudi, A.; Hosseinian, S.H.; Jalilian, A. Investigation of increased ohmic and core losses in induction motors under voltage fluctuation conditions. *Iran. J. Sci. Technol. Trans. Electr. Eng.* **2018**, *43*, 1–10. [CrossRef]
21. Ghaseminezhad, M.; Doroudi, A.; Hosseinian, S.H.; Jalilian, A. High torque and excessive vibration on the induction motors under special voltage fluctuation conditions. *COMPEL—Int. J. Comput. Math. Electr. Electron. Eng.* **2021**, *40*, 822–836. [CrossRef]
22. Gnaciński, P.; Hallmann, D.; Muc, A.; Klimczak, P.; Pepliński, M. Induction motor supplied with voltage containing symmetrical subharmonics and interharmonics. *Energies* **2022**, *15*, 7712. [CrossRef]
23. Gnaciński, P.; Pepliński, M.; Hallmann, D. Currents and Power Losses of Induction Machine Under Voltage Interharmonics. In Proceedings of the 21st European Conference on Power Electronics and Applications (EPE '19 ECCE Europe), Genova, Italy, 3–5 September 2019.
24. Schramm, S.; Sihler, C.; Song-Manguelle, J.; Rotondo, P. Damping torsional interharmonic effects of large drives. *IEEE Trans. Power Electron.* **2010**, *25*, 1090–1098. [CrossRef]
25. Tennakoon, S.; Perera, S.; Robinson, D. Flicker attenuation—Part I: Response of three-phase induction motors to regular voltage fluctuations. *IEEE Trans. Power Deliv.* **2008**, *23*, 1207–1214. [CrossRef]
26. *IEEE Standard 519-2022*; IEEE Recommended Practice and Requirements for Harmonic Control in Electric Power Systems. IEEE: New York, NY, USA, 2022.
27. Arkkio, A.; Cederström, S.; Awan, H.A.A.; Saarakkala, S.E.; Holopainen, T.P. Additional losses of electrical machines under torsional vibration. *IEEE Trans. Energy Convers.* **2018**, *33*, 245–251. [CrossRef]
28. Avdeev, B.A.; Vyngra, A.V.; Chernyi, S.G.; Zhilenkov, A.A.; Sokolov, S.S. Evaluation and procedure for estimation of interharmonics on the example of non-sinusoidal current of an induction motor with variable periodic load. *IEEE Access* **2021**, *9*, 158412–158419. [CrossRef]
29. Nassif, A.B. Assessing the impact of harmonics and interharmonics of top and mudpump variable frequency drives in drilling rigs. *IEEE Trans. Ind. Appl.* **2019**, *55*, 5574–5583. [CrossRef]
30. Testa, A.; Akram, M.F.; Burch, R.; Carpinelli, G.; Chang, G.; Dinavahi, V.; Hatziadoniu, C.; Grady, W.M.; Gunther, E.; Halpin, M.; et al. Interharmonics: Theory and modeling. *IEEE Trans. Power Deliv.* **2007**, *22*, 2335–2348. [CrossRef]
31. Tripp, H.; Kim, D.; Whitney, R. *A Comprehensive Cause Analysis of a Coupling Failure Induced by Torsional Oscillations in a Variable Speed Motor*; Texas A&M University, Turbomachinery Laboratories: College Station, TX, USA, 1993; pp. 17–24.
32. Gnaciński, P.; Pepliński, M.; Hallmann, D.; Jankowski, P. Induction cage machine thermal transients under lowered voltage quality. *IET Electr. Power Appl.* **2019**, *13*, 479–486. [CrossRef]
33. Gnaciński, P.; Pepliński, M.; Hallmann, D.; Jankowski, P. The effects of voltage subharmonics on cage induction machine. *Int. J. Electr. Power Energy Syst.* **2019**, *111*, 125–131. [CrossRef]
34. Gnaciński, P.; Klimczak, P. High-Power induction motors supplied with voltage containing subharmonics. *Energies* **2020**, *13*, 5894. [CrossRef]
35. *ISO Standard 10816-1*; Mechanical Vibration—Evaluation of Machine Vibration by Measurements on Non-Rotating Parts—Part 1: General guidelines. ISO: Genova, Switzerland, 1995.
36. *ISO Standard 20816-1*; Mechanical Vibration—Measurement and Evaluation of Machine Vibration—Part 1: General Guidelines. ISO: Genova, Switzerland, 2016.
37. Singh, R.R.; Raj, C.T.; Palka, R.; Indragandhi, V.; Wardach, M.; Paplicki, P. Energy optimal intelligent switching mechanism for induction motors with time varying load. In Proceedings of the IOP Conference Series: Materials Science and Engineering, Chennai, India, 9 June 2020; p. 906.
38. Singh, R.R.; Chelliah, T.R. Enforcement of cost-effective energy conservation on single-fed asynchronous machine using a novel switching strategy. *Energy* **2017**, *126*, 179–191. [CrossRef]
39. *IEC Standard 60034-1*; Rotating Electrical Machines. Part 1: Rating and Performance. IEC: Genova, Switzerland, 2004.
40. Tsypkin, M. The origin of the electromagnetic vibration of induction motors operating in modern industry: Practical experience—Analysis and diagnostics. *IEEE Trans. Ind. Appl.* **2017**, *53*, 1669–1676. [CrossRef]
41. Gonzalez-Abreu, A.D.; Osornio-Rios, R.A.; Jaen-Cuellar, A.Y.; Delgado-Prieto, M.; Antonino-Daviu, J.A.; Karlis, A. Advances in power quality analysis techniques for electrical machines and drives: A review. *Energies* **2022**, *15*, 1909. [CrossRef]
42. Sieklucki, G.; Sobieraj, S.; Gromba, J.; Necula, R.E. Analysis and approximation of THD and torque ripple of induction motor for SVPWM control of VSI. *Energies* **2023**, *16*, 4628. [CrossRef]
43. Feese, T.; Ryan, M. *Torsional Vibration Problem with Motor/ID Fan System Due to PWM Variable Frequency Drive*; Texas A&M University, Turbomachinery Laboratories: College Station, TX, USA, 2008.
44. Perez, R.X. *Design, Modeling and Reliability in Rotating Machinery*; Wiley: Hoboken, NJ, USA, 2022.

Disclaimer/Publisher's Note: The statements, opinions and data contained in all publications are solely those of the individual author(s) and contributor(s) and not of MDPI and/or the editor(s). MDPI and/or the editor(s) disclaim responsibility for any injury to people or property resulting from any ideas, methods, instructions or products referred to in the content.

Article

Improving Torque Analysis and Design Using the Air-Gap Field Modulation Principle for Permanent-Magnet Hub Machines

Yuhua Sun [1,2], Nicola Bianchi [1,*], Jinghua Ji [2] and Wenxiang Zhao [2,*]

[1] Department of Industrial Engineering, University of Padova, 35131 Padova, Italy; syh@stmail.ujs.edu.cn
[2] School of Electrical and Information Engineering, Jiangsu University, Zhenjiang 212013, China
* Correspondence: nicola.bianchi@unipd.it (N.B.); zwx@ujs.edu.cn (W.Z.)

Abstract: The Double Permanent Magnet Vernier (DPMV) machine is well known for its high torque density and magnet utilization ratio. This paper aims to investigate the torque generation mechanism and its improved design in DPMV machines for hub propulsion based on the field modulation principle. Firstly, the topology of the proposed DPMV machine is introduced, and a commercial PM machine is used as a benchmark. Secondly, the rotor PM, stator PM, and armature magnetic fields are derived and analyzed considering the modulation effect, respectively. Meanwhile, the contribution of each harmonic to average torque is pointed out. It can be concluded that the 7th-, 12th-, 19th- and 24th-order flux density harmonics are the main source of average torque. Thanks to the multi-working harmonic characteristics, the average torque of DPMV machines has significantly increased by 31.8% compared to the counterpart commercial PM machine, while also reducing the PM weight by 75%. Thirdly, the auxiliary barrier structure and dual three-phase winding configuration are proposed from the perspective of optimizing the phase and amplitude of working harmonics, respectively. The improvements in average torque are 9.9% and 5.4%, correspondingly.

Keywords: hub machine; dual permanent magnet vernier (DPMV); air-gap field modulation; torque

Citation: Sun, Y.; Bianchi, N.; Ji, J.; Zhao, W. Improving Torque Analysis and Design Using the Air-Gap Field Modulation Principle for Permanent-Magnet Hub Machines. *Energies* **2023**, *16*, 6214. https://doi.org/10.3390/en16176214

Academic Editor: Armando Pires

Received: 20 July 2023
Revised: 8 August 2023
Accepted: 24 August 2023
Published: 27 August 2023

Copyright: © 2023 by the authors. Licensee MDPI, Basel, Switzerland. This article is an open access article distributed under the terms and conditions of the Creative Commons Attribution (CC BY) license (https://creativecommons.org/licenses/by/4.0/).

1. Introduction

Due to increasing concerns about energy security and environmental impact, traditional vehicles with internal combustion engines are likely to be phased out in the future [1–3]. The electrification of transportation has become a key development trend, and such a revolution in mobility extends to light electric vehicles such as electric scooters and bicycles [4]. Permanent Magnet (PM) hub machines have attracted much attention due to their advantages of high efficiency, reliability, and compact structure [5]. With the increasing travel demand, the high torque density requirements of hub machines are becoming more stringent [6].

Significant works on torque improvement have been presented. Among them, increasing the machine size and PM usage are effective ways to improve torque. However, these methods also lead to unacceptable increases in weight and cost [7]. In [8], the stator structure with unequal teeth was proposed to enhance fundamental harmonic components, thereby offering useful performance benefits in terms of a higher torque capability and reduced torque ripple. However, this structure is only suitable for single-layer winding structures. In addition to optimizing stator structure, Halbach [9] and hybrid rotor [10] structures were adopted to increase torque capacity. The former structure leads to manufacturing difficulties, while the latter structure cannot meet the high torque density requirements in the speed range. Further, the current harmonic injection can also be used to increase torque capability, although it causes additional losses [11]. To sum up, the above methods all have their limitations, and the torque improvement effect is not significant. The single working harmonic characteristic of conventional PM machines restricts the potential for further torque improvement.

The improved torque density of Permanent Magnet Vernier (PMV) machines has garnered significant attention in electric wheel applications due to their multi-working harmonic characteristics [12–14]. The PMV machines can be divided into two types depending on the location of PM, namely Stator-PM (PMS) and Rotor-PM (PMR) styles [14]. Further, [15,16] proposed a novel PMV machine with double stator and double rotor, respectively. These machines achieve higher energy transmission and power conversion than the single stator or rotor counterparts. However, the mentioned PMV machine creates complex structures and increased difficulty in processing and assembly. By comparison, the Double Permanent Magnet Vernier (DPMV) machine was proposed and analyzed in [17], featuring the presence of PM on both the stator and rotor. Due to the bidirectional field modulation effect, air-gap flux density harmonics of the DPMV machines are more abundant than conventional PM machines. The torque capability of the DPMV machine is compared to conventional PM and PMV machines in [18,19], respectively. The results indicate that the DPMV machine can effectively improve the torque capability without increasing machine dimensions. The Consequent Pole (CP) rotor structure was proposed to replace conventional rotor structures such as surface mounted and spoke array structures [20,21]. In this case, the PM is magnetized in the North Pole direction, and the salient iron core serves as the South Pole. In [22], a 12-slot/10-pole PM machine with a CP structure achieves 92% output torque via 65% magnet usage of its counterpart with a surface-mounted structure. This shows that the CP structure in PMV machines can greatly improve the PM utilization rate. The purpose of this paper is to theoretically analyze the harmonic components of DPMV machine with a CP structure, verifying its multi-working harmonic characteristics and advantages in average torque improvement and PM usage reduction. The main novelty of our research is that the two new designs are proposed to further improve the average torque of the DPMV machine from different perspectives, e.g., auxiliary barrier structure and dual three-phase winding configuration.

This paper deals with the torque generation mechanism and its improvement design in the DPMV machine for hub propulsion. This paper is structured as follows. In Section 2, the topology and air-gap field modulation principle of the DPMV machine is presented. The conventional PM machine is used as a benchmark. In Section 3, the PMR, PMS, and armature magnetic fields are investigated in detail, and the emerging harmonics caused by modulation effect are recognized. Then, the torque generation of the DPMV machine is investigated, and the contribution of each harmonic to average torque is pointed out. Based on the above analyses, two new designs to improve the average torque of DPMV machines are proposed in Section 4. The improvement principle was elaborated from the perspective of optimizing the phase and amplitude of working harmonics. Finally, conclusions are presented in Section 5.

2. Topology and Modulation Principle Analysis

It is well known that the PMV machine is operated on the basis of the air-gap field modulation principle. The armature magnetic field with small pole pairs P_{AR} is modulated by the stator modulation poles P_S that correspond to the stator teeth, obtaining the harmonic components that can interact with the PMR field with high pole pairs P_{PMR}. The relationship between PMR pole pairs, stator modulation poles P_S, and armature winding pole-pairs should be satisfied as follows [14]:

$$P_{PMR} = P_S \pm P_{AR} \qquad (1)$$

To further improve torque by taking advantage of the field modulation effect, the PM is also placed on the stator modulation pole. Similarly, the armature magnetic field with small pole pairs is modulated by the rotor modulation poles P_{PMR} to obtain the harmonic components that can interact with the PMS magnetic field with high pole pairs P_{PMR}. Namely, it can be written as

$$yP_{PMS} = P_{PMR} \pm P_{AR} \qquad (2)$$

where y is positive integer.

A commercial PM hub machine in [4] for e-bike is selected as the benchmark and shown in Figure 1a, in which the 12-slot/10-pole combination and interior PM (IPM) type are adopted. In this section, the red, green, and blue windings always correspond to phase A, phase B, and phase C, respectively. The arrows in PM always represent the direction of magnetization. For comparison, the stator slot Q of the DPMV machine is 12 as well and adopts a split tooth structure, as shown in Figure 1b. The number of stator modulation poles P_S is 24. Then, the pole pair of armature winding remains consistent with that of the commercial hub machine, e.g., $P_{AR} = 5$. Based on (1), the number of PMR pole pairs P_{PMR} should be 19. It is worth noting that both the PMR and PMS of the proposed DPMV machine adopt the CP structure. The salient rotor teeth can also serve as modulation poles, which will be elaborated in the following section. Table 1 lists the main specifications of the two machines. They have identical volume, slot filling factor, and material. The PM weight and electromagnetic load of the proposed DPMV machine are only 75% and 87% of that of the IPM machine, respectively.

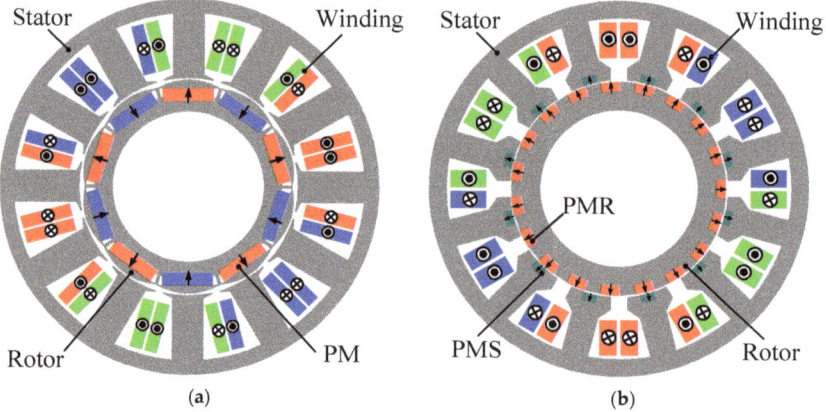

Figure 1. Cross-section of the PM machine: (**a**) commercial IPM machine; (**b**) DPMV machine.

Table 1. Main parameters of the IPM and DPMV machines.

Items	Symbol	IPM	DPMV
Pole number of PMR	P_{PMR}	10	19
Pole number of PMS	P_{PMS}	/	12
Number of stator slot	Q	12	12
Stator outer diameter (mm)	D_o	90	90
Stator inner diameter (mm)	D_s	52	52
Axial length (mm)	L_{sk}	30	30
Air-gap length (mm)	g	0.5	0.5
Stator slot area (mm^2)	S_{slot}	154	137
Turn number per coil	N_c	23	20
Thickness of PMR (mm)	h_r	3	2.2
Thickness of PMS (mm)	h_s	/	2
Pole-arc ratio of PMR	k_r	0.81	0.54
Pole-arc ratio of PMS	k_s	/	0.27
Total PM weight (g)	/	81	61
PM material	/	N40UH	N40UH

3. Torque Analyses with Multi-Working Harmonics

In this section, the PMR, PMS, and armature air-gap magnetic fields of the proposed DPMV machine are investigated independently. Their interaction and torque generation principle will be presented. Additionally, to obtain the analytical model of air-gap flux density, the derivation in this section is based on the following assumptions [17]:

(1) The tangential components of the air-gap magnetic field are neglected for simplicity;
(2) The leakage flux is ignored; therefore, the waveform of air-gap primitive MMF is considered as square waves. In addition, the end effect is also neglected, so the air-gap MMF is regarded as the same in the axial direction;
(3) The permeability of stator and rotor iron is infinite, so the iron reluctance is neglected.

The general methodology of this section is as follows: Firstly, both PMR and PMS are magnetized in the North Pole direction, and the salient iron core serves as the South Pole. Therefore, the primitive air-gap PM flux density waveform within the PM range is a positive square wave, while it is a negative square wave within the core range. Similarly, the primitive armature winding flux density is the superposition of a series of square waves considering the coil polarity. Secondly, the permeance functions accounting for winding, PMS, and PMR slotting effect can be obtained by using the path of the flux lines in the corresponding opening region. The flux line always flow through a smaller reluctance path. Thirdly, the harmonic characteristics of each magnetic field are acquired by using FFT, including the spatial order, amplitude, mechanical speed, and rotation direction. Finally, the frozen permeability method is adapted to separate the torque generated due to the interaction of different magnetic fields, recognizing the contribution of each harmonics to average torque. Moreover, the torque waveforms of DPMV and counterpart IPM machines are compared using the software Ansys Electronics Desktop.

3.1. PMR Flux Density

The primitive air-gap PMR flux density without modulation by the stator is shown in Figure 2. B_1 and B_1' are defined as the magnitudes of PMR and iron poles, respectively, which can be written as follows:

$$\begin{cases} B_1 = \dfrac{B_r}{1+\dfrac{g\mu_r}{h_r(1-k_r)}} \\ B_1' = \dfrac{k_r}{1-k_r} B_1 \end{cases} \quad (3)$$

where B_r is the remanence flux density of PM, and μ_r is the PM relative differential permeability. Further, the Fourier series expansion of the primitive PMR flux density B_1 can be deduced as follows:

$$\begin{cases} B_1(\theta_m, t) = \sum\limits_{j=1,2,3...}^{\infty} B_j \cos\{jP_{PMR}(\theta_m - \Omega_m t - \theta_0)\} \\ B_j = \dfrac{2B_r h_r \sin(j\pi k_r)}{j\pi\{(1-k_r)h_r + g\mu_r\}} \end{cases} \quad (4)$$

where Ω_m is the mechanical angular speed, t is time, θ_m is the angular position in stator reference, and θ_0 is the initial phase ($\theta_0 = 0$ in this section).

The influence of winding and PMS slots on the PMR magnetic field can be accounted by introducing a stator permeance function, as shown in Figure 3. Here, the brownish red line represents the permeance curve caused by PMS slot, and blue line represents the permeance curve caused by winding slot. The permeance function produced by winding slot Λ_{Slot} and PMR slot Λ_{PMS} can be expressed as follows:

$$\begin{cases} \Lambda_{Slot}(\theta_m) = A_0 + \sum\limits_{n=1,2,3} A_n \cos(nQ\theta_m) \\ \Lambda_{PMS}(\theta_m) = C_0 + \sum\limits_{n=1,2,3} C_n \cos\{nP_{PMS}(\theta_m - \tfrac{\pi}{12})\} \end{cases} \quad (5)$$

where A_0 and A_n are Fourier coefficients of the winding slot permeance function, and C_0 and C_n are Fourier factors of the PMS slot permeance function. The focus of this section is to highlight the modulation effects of topology structure on magnetic fields. Thus, the detailed expression of the above Fourier coefficients will not be discussed. Based on (5), the total stator permeance function is

$$\Lambda_s(\theta_m) = \Lambda_{\text{Slot}}(\theta_m) \cdot \Lambda_{\text{PMS}}(\theta_m) \approx \Lambda_{s0} + \sum_{n=1,2,\ldots}^{\infty} \Lambda_{sn} \cos(nP_S\theta_m) \tag{6}$$

where $P_S = Q + P_{\text{PMS}}$, and the Λ_{s0} and Λ_{sn} are the Fourier coefficients of the total stator permeance function. Thus, the modulated PMR air-gap flux density B_{PMR} can be expressed as follows:

$$B_{\text{PMR}}(\theta_m, t) = \sum_{j=1,2,3\ldots}^{\infty} B_j \Lambda_{s0} \cos[jP_{\text{PMR}}(\theta_m - \Omega_m t)] + \frac{1}{2}\left\{ \sum_{j=1,2\ldots}^{\infty} \sum_{n=1,2\ldots}^{\infty} B_j \Lambda_{sn} \cos[(jP_{\text{PMR}} + nP_S)\theta_m - jP_{\text{PMR}}\Omega_m t] + \sum_{j=1,2\ldots}^{\infty} \sum_{n=1,2\ldots}^{\infty} B_j \Lambda_{sn} \cos[(jP_{\text{PMR}} - nP_S)\theta_m - jP_{\text{PMR}}\Omega_m t] \right\} \tag{7}$$

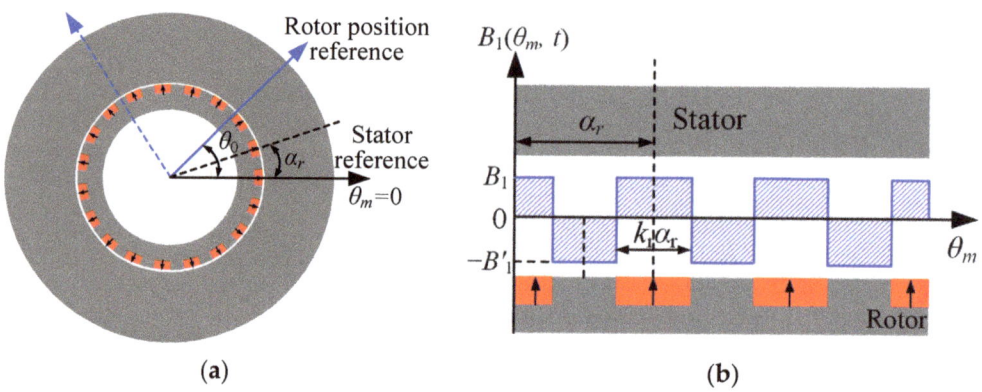

Figure 2. Air-gap PMR flux density without stator modulation. (**a**) Model. (**b**) Waveform ($\theta_0 = 0$, $t = 0$).

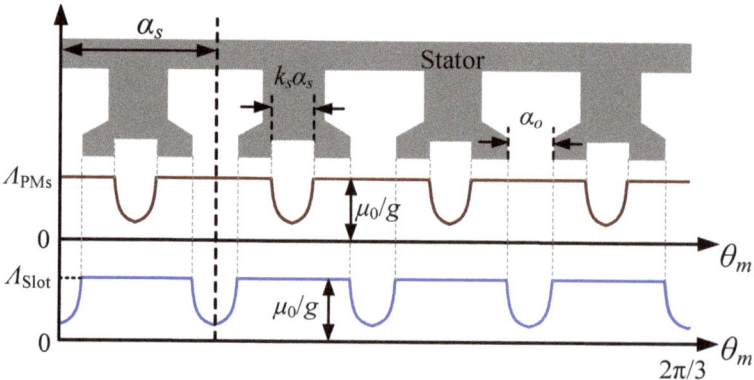

Figure 3. Air-gap permeance function accounting for winding and PMS slotting effect.

The last two items of (7) represent the modulation effect of stator structure on the PMR magnetic field. The harmonic components with $jP_{\text{PMR}} \pm nP_S$ are generated, and the related rotation speed is $jP_{\text{PMR}}\Omega_m/(jP_{\text{PMR}} \pm nP_S)$, as shown in Figure 4.

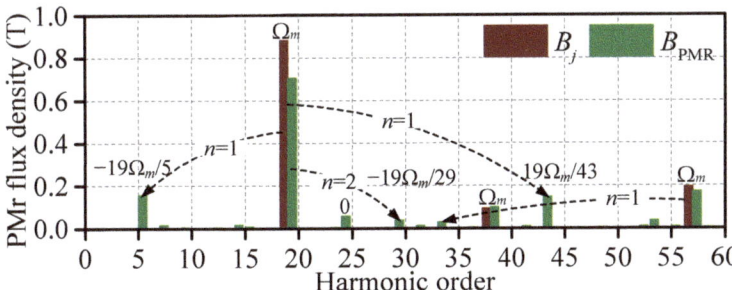

Figure 4. Spectrum comparison of PMR flux density before and after stator modulation by FEM.

3.2. PMS Flux Density

The primitive air-gap PMS flux density B_2 without PMR and winding slots modulation is shown in Figure 5. The Fourier series expansion of the primitive PMS flux density can be deduced as follows:

$$\begin{cases} B_2(\theta_m) = \sum\limits_{v=1,2,3...}^{\infty} B_v \cos(v P_{\text{PMS}} \theta_m) \\ B_v = \frac{2 B_r h_s \sin(v \pi k_s)}{v \pi [(1-k_s) h_s + g \mu_r]} \end{cases} \tag{8}$$

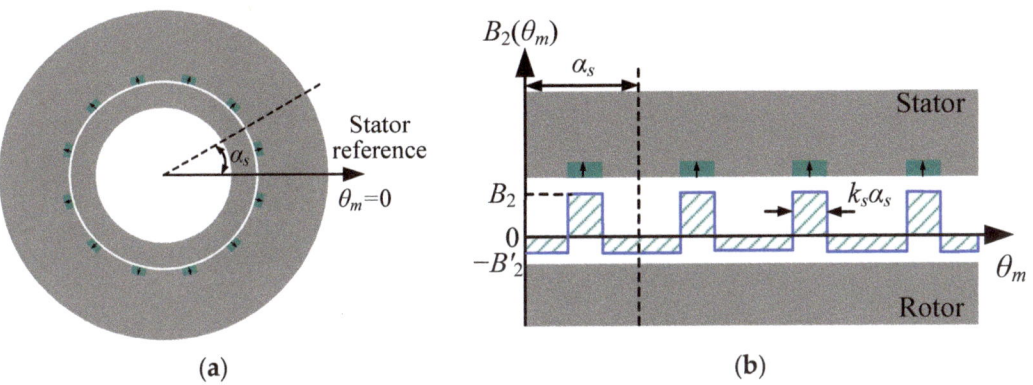

Figure 5. Air-gap PMS flux density without winding and PMR slots modulation. (**a**) Model. (**b**) Waveform.

Then, the permeance function accounting for PMR slotting effect can be written as follows:

$$\Lambda_{\text{PMR}}(\theta_m, t) = \Lambda_{r0} + \sum_{n=1,2,3} \Lambda_{rn} \cos[n P_{\text{PMR}}(\theta_m - \Omega_m t)] \tag{9}$$

The modulated PMS air-gap flux density B_{PMS} can be expressed as follows:

$$\begin{aligned} B_{\text{PMS}}(\theta_m, t) &= [B_2(\theta_m) \cdot \Lambda_{\text{Slot}}(\theta_m)] \cdot \Lambda_{\text{PMR}}(\theta_m, t) \\ &= \left[\sum_{v=1,2,3...}^{\infty} B'_v \cos(v P_{\text{PMS}} \theta_m) \right] \cdot \left[\Lambda_{r0} + \sum_{n=1,2,3} \Lambda_{rn} \cos[n P_{\text{PMR}}(\theta_m - \Omega_m t)] \right] \\ &= \sum_{j=1,2,3...}^{\infty} B'_v \Lambda_{r0} \cos(v P_{\text{PMS}} \theta_m) + \\ &\quad \frac{1}{2} \left\{ \sum_{v=1,2...}^{\infty} \sum_{n=1,2...}^{\infty} B'_v \Lambda_{rn} \cos[(v P_{\text{PMS}} + n P_{\text{PMR}}) \theta_m - n P_{\text{PMR}} \Omega_m t] + \sum_{j=1,2...}^{\infty} \sum_{n=1,2...}^{\infty} B'_v \Lambda_{rn} \cos[(v P_{\text{PMS}} - n P_{\text{PMR}}) \theta_m + n P_{\text{PMR}} \Omega_m t] \right\} \end{aligned} \tag{10}$$

It can be seen that the winding slot has no influence on the harmonic order of the PMS magnetic field, but only changes the harmonic amplitude. Therefore, the B'_v is used to denote the amplitude of vth-order harmonics after winding slot modulation. Finally, the new harmonic components with $vP_{PMS} \pm nP_{PMR}$ are produced by the PMR slot modulation. Correspondingly, the related rotation speed is $\pm nP_{PMR}\Omega_m/(vP_{PMS} \pm nP_{PMR})$, as shown in Figure 6.

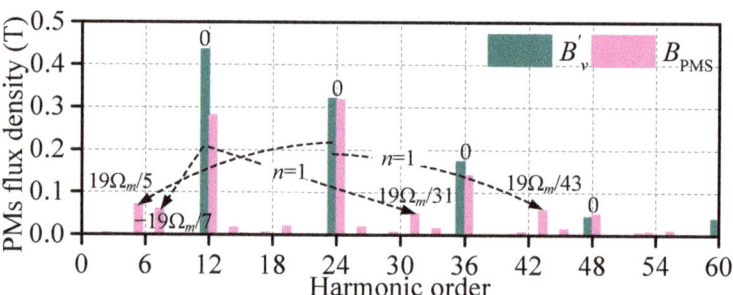

Figure 6. Spectrum comparison of PMS flux density before and after rotor modulation via FEM.

3.3. Armature Flux Density

Figure 7 shows the primitive air-gap armature MMF model and waveform, in which the initial MMF of each single phase is equivalent to an ideal square wave. Firstly, the winding function $N(\theta_m)$ of each phase can be expressed as follows [23]:

$$\begin{cases} N_A(\theta_m) = \sum_{h=1,3,5...} N_h \cdot \cos(h\theta_m + \gamma_h) \\ N_B(\theta_m) = \sum_{h=1,3,5...} N_h \cdot \cos\{h(\theta_m + \frac{2\pi}{3}) + \gamma_h\} \\ N_C(\theta_m) = \sum_{h=1,3,5...} N_h \cdot \cos\{h(\theta_m - \frac{2\pi}{3}) + \gamma_h\} \end{cases} \quad (11)$$

where h is the spatial harmonic order, and γ_h is the initial angle. Based on the winding distribution shown in Figure 7, $\gamma_h = -180°$ (h = 1, 5, 9, 13, etc.), $\gamma_h = 0°$ (h = 3, 7, 11, 15, etc.), N_h is the Fourier expansion factor, and $N_h = 2N_c k_{wh}/\pi h$, and k_{wh} is the winding factor of hth-order harmonics. Then, the MMF expression is obtained by multiplying the winding function by the current, yielding the following:

$$\begin{aligned} F(\theta_m, t) &= N_A(\theta_m)i_A(t) + N_B(\theta_m)i_B(t) + N_C(\theta_m)i_C(t) \\ &= \frac{3N_h I_{max}}{2} \left\{ \sum_{h=6l-1} \sin(h\theta_m + P_{PMR}\Omega_m t + \gamma_h) - \sum_{h=6l+1} \sin(h\theta_m - P_{PMR}\Omega_m t + \gamma_h) \right\} \end{aligned} \quad (12)$$

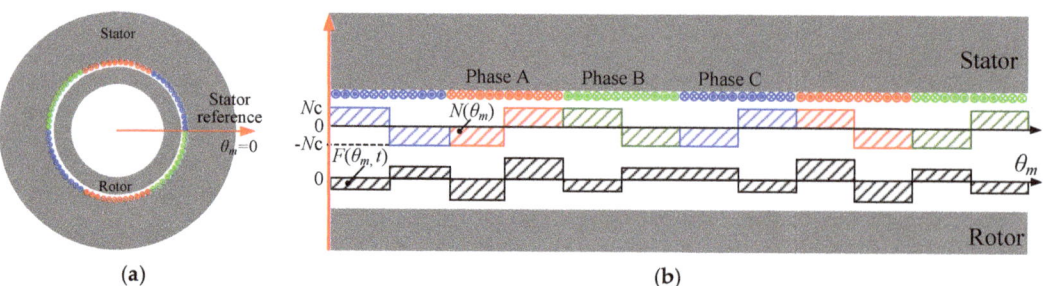

Figure 7. Air-gap armature MMF without winding, PMS, and PMR slots modulation. (**a**) Model. (**b**) Waveform (t = 0).

l is either 0 or a positive integer. I_{max} is the amplitude of phase current. The armature air-gap flux density B_{AR} considering rotor and stator modulation can be expressed as follows:

$$
\begin{aligned}
B_{AR}(\theta_m,t) &= \tfrac{\mu_0}{g}\cdot F(\theta_m,t)\cdot \Lambda_s(\theta_m)\cdot \Lambda_{PMR}(\theta_m,t)\\
&= \left[\sum_{h=6l-1} B_h\sin(h\theta_m + P_{PMR}\Omega_m t + \gamma'_h) - \sum_{h=6l+1} B_h\sin(h\theta_m - P_{PMR}\Omega_m t + \gamma'_h)\right]\cdot\left\{\Lambda_{r0} + \sum_{n=1,2,3}\Lambda_{rn}\cos[nP_{PMR}(\theta_m-\Omega_m t)]\right\}\\
&= \sum_{h=6l-1}\Lambda_{r0}B_h\sin(h\theta_m + P_{PMR}\Omega_m t + \gamma'_h) + \sum_{h=6l-1}\sum_{n=1,2,3}\tfrac{B_h\Lambda_{rn}}{2}\left\{\begin{array}{l}\sin[(h+nP_{PMR})\theta_m + (1-n)P_{PMR}\Omega_m t + \gamma'_h]\\ +\sin[(h-nP_{PMR})\theta_m + (1+n)P_{PMR}\Omega_m t + \gamma'_h]\end{array}\right\}\\
&\quad - \sum_{h=6l+1}\Lambda_{r0}B_h\sin(h\theta_m - P_{PMR}\Omega_m t + \gamma'_h) - \sum_{h=6l+1}\sum_{n=1,2,3}\tfrac{B_h\Lambda_{rn}}{2}\left\{\begin{array}{l}\sin[(h+nP_{PMR})\theta_m - (1+n)P_{PMR}\Omega_m t + \gamma'_h]\\ +\sin[(h-nP_{PMR})\theta_m - (1-n)P_{PMR}\Omega_m t + \gamma'_h]\end{array}\right\}
\end{aligned}
\tag{13}
$$

The modulation effect of winding and PMS slot on the armature magnetic field only changes the amplitude and phase, and does not result in new harmonic orders generation. The B_h and γ'_h represent the amplitude and phase of hth-order harmonics after winding and PMS slots modulation, respectively. The new harmonic orders with $h\pm nP_{PMR}$ emerged after rotor modulation, as shown in Figure 8.

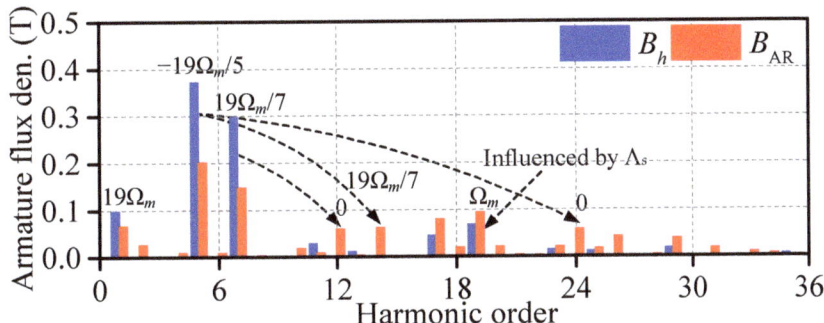

Figure 8. Spectrum comparison of PMS flux density before and after rotor modulation via FEM.

3.4. Torque Generation Principle

Based on the above analyses, the air-gap flux density harmonic order and corresponding mechanical speed of three magnetic fields considering bilateral modulation can be obtained, and they are presented in Table 2. The P, N, and S represent positive, negative, and stationary rotation directions, respectively. Conventionally, the average torque is produced when the harmonic components of different magnetic fields have the same order and speed [24]. As for the DPMV machine, there are two possible cases:

(1) The two magnetic fields have the same order and mechanical speed, and they can interact with each other directly and produce average torque;
(2) The two magnetic fields have different orders and mechanical speeds. However, there are flux modulation poles between them. The average torque can still be generated if two magnetic fields meet the following relationship:

$$
\begin{cases}
|jP_{PMR}\pm n_1 P_S| = |h\pm n_2 P_{PMR}|\\
\frac{jP_{PMR}\Omega_m}{jP_{PMR}\pm n_1 P_S} = \frac{(1\pm n)jP_{PMR}\Omega_m}{h\pm n_2 P_{PMR}}\ \text{or}\ \frac{-(1\mp n)jP_{PMR}\Omega_m}{h\pm n_2 P_{PMR}}
\end{cases}
\tag{14}
$$

$$
\begin{cases}
|vP_{PMS}\pm n_3 P_{PMR}| = |h\pm n_2 P_{PMR}|\\
\frac{\pm n_3 P_{PMR}\Omega_m}{vP_{PMS}\pm n_3 P_{PMR}} = \frac{(1\pm n)jP_{PMR}\Omega_m}{h\pm n_2 P_{PMR}}\ \text{or}\ \frac{-(1\mp n)jP_{PMR}\Omega_m}{h\pm n_2 P_{PMR}}
\end{cases}
\tag{15}
$$

For clarity, Figure 9 is used to describe different working points. Here, point a implies that the DPMV machine is jointly excited by three magnetic fields. Points b, c, and d indicate that the DPMV machine is only excited by PMR, PMS, and armature magnetic fields alone, respectively. There is almost no harmonic component between the PMR and PMS magnetic fields that satisfies (14) or (15), so the average torque at operating point e is approximately 0. The total average torque T_a of the DPMV machine is the superposition of the interaction between the PMR and armature magnetic fields, as well as the interaction between the PMS and armature magnetic fields. The contribution of each harmonic to torque can be expressed as follows:

$$\begin{aligned}
T_a(t) &= T_f(t) + T_g(t) \\
&= \frac{\pi r^2 L_{sk}}{\mu_0} \left\{ \underbrace{\int_0^{2\pi} B_{f\_ra}(\theta_m,t) B_{f\_ta}(\theta_m,t) d\theta_m}_{\text{Interaction between } B_{PMr} \text{ and } B_{Ar}} + \underbrace{\int_0^{2\pi} B_{g\_ra}(\theta_m,t) B_{g\_ta}(\theta_m,t) d\theta_m}_{\text{Interaction between } B_{PMs} \text{ and } B_{Ar}} \right\} \\
&= \sum_k \frac{\pi r^2 L_{sk}}{\mu_0} B_{ra\_k} B_{ta\_k} \cos[\theta_{ra\_k} - \theta_{ta\_k}]
\end{aligned} \tag{16}$$

where r is the air-gap radius, B_{ra} and B_{ta} represent the air-gap radial and tangential flux densities at corresponding working points, respectively, and k represents the harmonic order that satisfies (14) or (15).

Table 2. Air-gap flux density harmonics of different magnetic fields.

	Harmonic Order	Mechanical Speed	Rotate Direction
PMR magnetic field	jP_{PMR}	Ω_m	P
	$jP_{PMR} + nP_S$	$jP_{PMR}\Omega_m/(jP_{PMR} + nP_S)$	P
	$jP_{PMR} - nP_S$	$jP_{PMR}\Omega_m/(jP_{PMR} - nP_S)$	$(jP_{PMR} - nP_S > 0)$ P $(jP_{PMR} - nP_S < 0)$ N
PMS magnetic field	vP_{PMS}	0	S
	$vP_{PMS} + nP_{PMR}$	$nP_{PMR}\Omega_m/(vP_{PMS} + nP_{PMR})$	P
	$vP_{PMS} - nP_{PMR}$	$-nP_{PMR}\Omega_m/(vP_{PMs} - nP_{PMR})$	$(vP_{PMS} - nP_{PMR} > 0)$ N $(vP_{PMS} - nP_{PMR} < 0)$ P
Armature magnetic field ($h = 6l - 1$)	h	$-P_{PMR}\Omega_m/h$	N
	$h + nP_{PMR}$	$-(1-n)P_{PMR}\Omega_m/(h + nP_{PMR})$	$n \neq 1$ P $n = 1$ S
	$h - nP_{PMR}$	$-(1+n)P_{PMR}\Omega_m/(h - nP_{PMR})$	$(h - nP_{PMR} > 0)$ N $(h - nP_{PMR} < 0)$ P
Armature magnetic field ($h = 6l + 1$)	h	$-P_{PMR}\Omega_m/h$	P
	$h + nP_{PMR}$	$(1+n)P_{PMR}\Omega_m/(h + nP_{PMR})$	P
	$h - nP_{PMR}$	$(1-n)P_{PMR}\Omega_m/(h - nP_{PMR})$	$(h - nP_{PMR} > 0 \mid n \neq 1)$ N $n = 1$ S $(h - nP_{PMR} < 0 \mid n \neq 1)$ P

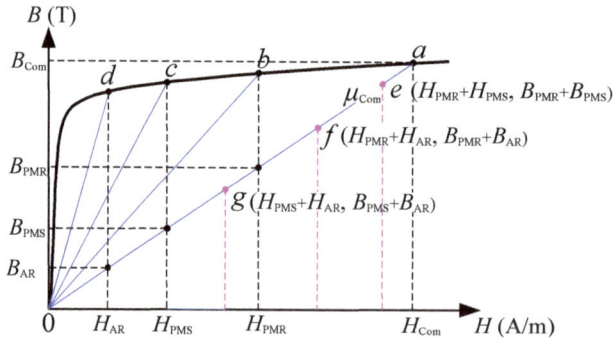

Figure 9. Different working points are represented by B-H curve.

Further, the contribution of each flux density harmonics to average torque is shown in Figure 10. It can be seen that the 19th-order harmonics of PMR and armature magnetic fields are the main source of average torque T_f. Similarly, the 7th-, 12th-, and 24th-order harmonics of PMS and armature magnetic fields are the main source of average torque T_g. By comparison, the working harmonic of commercial IPM machine is only 5th-order. This demonstrates the characteristics of multi-working harmonics in DPMV machine. Subsequently, the two torque components T_f and T_g are calculated with FEM considering frozen permeability, as shown in Figure 11a. Then, the torque waveforms of the proposed DPMV and commercial IPM machines are compared in Figure 11b. The average torque values of DPMV and IPM machines are 2.2 Nm and 2.9 Nm, respectively. The average torque of the DPMV machine is improved by 31.8% compared to the IPM machine. Moreover, the DPMV machine also has a torque ripple comparable to the IPM counterpart. Additionally, the variations in average torque with current amplitude is compared in Figure 12. Although the increment percent decreases as the current amplitude increases, the increment percent is always greater than 20% throughout the current range (0–30) A. This is mainly due to the higher harmonic components of the DPMV machine than the IPM machine. The above comparison results indicate that adopting the DPMV machine instead of the original IPM machine based on air-gap magnetic field modulation can effectively improve torque performance.

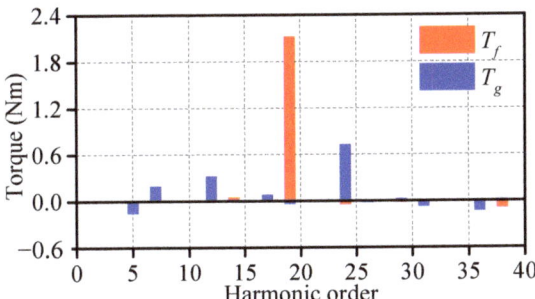

Figure 10. The contribution of each flux density harmonic to average torque.

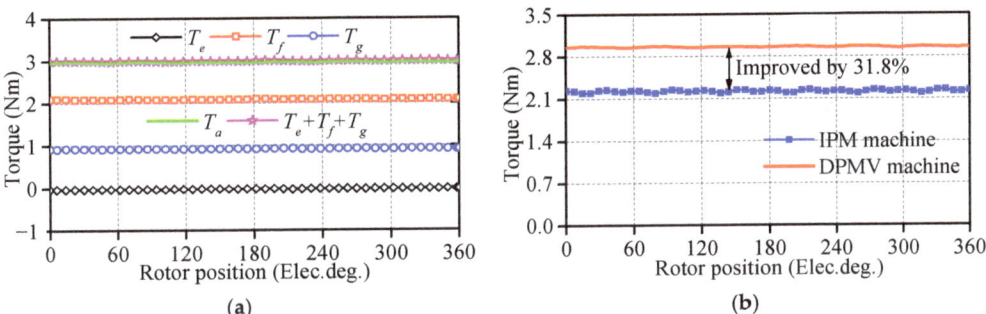

Figure 11. Torque waveforms. (**a**) Torque separation of DPMV machine. (**b**) Torque comparison between IPM and DPMV machines.

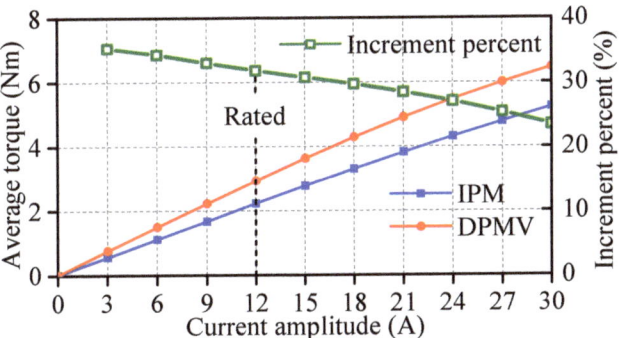

Figure 12. The variations in average torque with current amplitude.

4. New Design to Improve Torque

The working harmonics of the proposed DPMV machine are identified based on the magnetic field modulation, and the 19th-order harmonic is the largest contributor. In order to further improve the average torque of the DPMV machine, two main aspects can be taken from (16). On one hand, phase angle reconfiguration makes the phase difference between the radial and tangential of the 19th-order harmonics smaller. On the other hand, the 19th-harmonic amplitude increases. Correspondingly, the Auxiliary Barrier (AB) structure and Dual Three-Phase 30° (DTP-30°) winding are adopted in this section.

The detailed results of this section are all based on the commercial finite element software Ansys Electronics Desktop, in which the 2D simulated models with different structures are established. The air-gap flux density waveform represents its radial distribution at the air-gap centerline. Then, the amplitude and phase characteristics of spatial harmonics throughout the time region can be obtained using FFT. Finally, the torque waveform and its average value of different structures are compared.

4.1. Auxiliary Barrier Structure

Figure 13 shows the 1/3 model of the new stator structure with ABr, and other dimensions consistent with the original structure. The epoxy material is used at the AB to fix the PMS. The β_1 and β_2 is the angle of left and right ABs, respectively. The influence of AB on the air-gap flux density at the initial rotor position is shown in Figure 14. It can be seen that the waveform is shifted with the position of the AB. Then, Figure 15 shows the phase difference between the radial and tangential of the 19th-order harmonic throughout the position range. The cosine value of the phase difference between the radial and tangential components of the 19th-order harmonic increases from 0.23 to 0.25, and the amplitude of 19th-order harmonic remains unchanged basically. Undoubtedly, the average torque of the DPMV machine further increases with the cosine value [13].

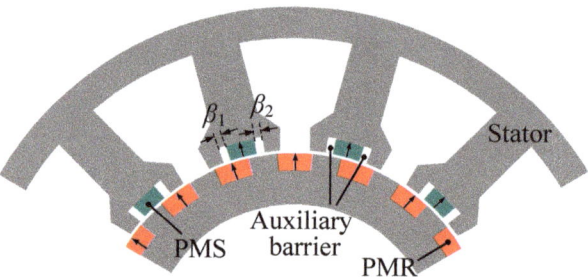

Figure 13. Schematic diagram of the stator with AB.

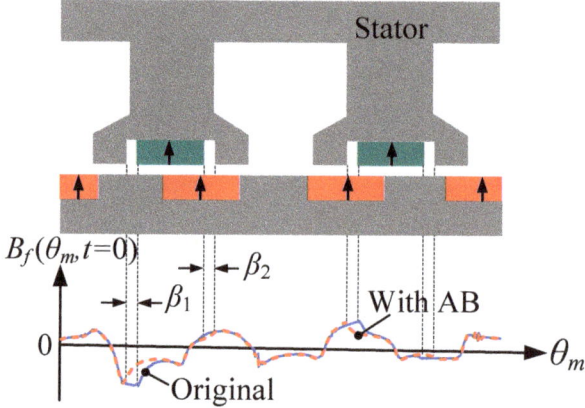

Figure 14. Influence of AB on the air-gap flux density at working point f.

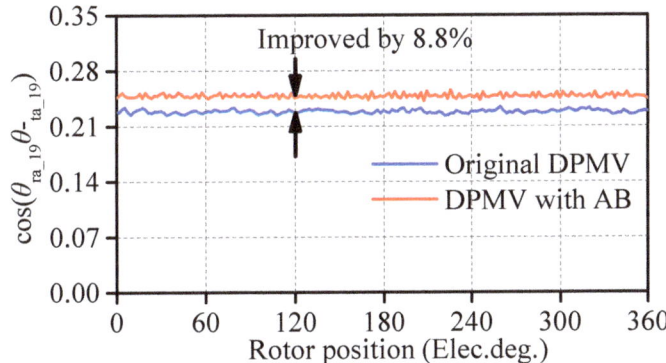

Figure 15. The cosine value of phase difference between radial and tangential components.

In addition, the effect of AB on average torque is also related to its dimensions. Figure 16 describes the variation in the total torque of the DPMV machine with angles β_1 and β_2. Consequently, the angles β_1 and β_2 both are determined to be 2°; in this case, the stator is still symmetrical. Finally, the total torque waveforms of original and new DPMV machines are compared in Figure 17. The total torque is increased from 2.9 Nm to 3.2 Nm without deteriorating torque ripple. This indicates that the proposed new structure with AB is feasible for improving torque density.

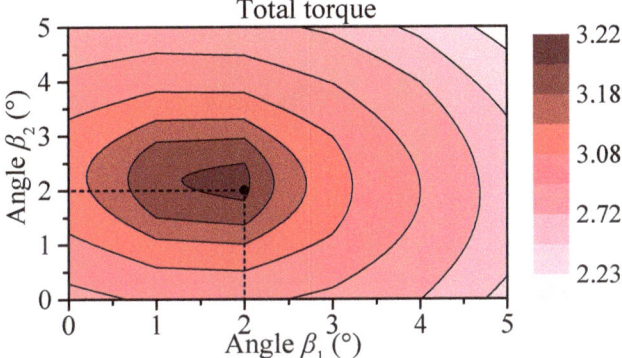

Figure 16. Total torque variation in the DPMV machine with angles β_1 and β_2.

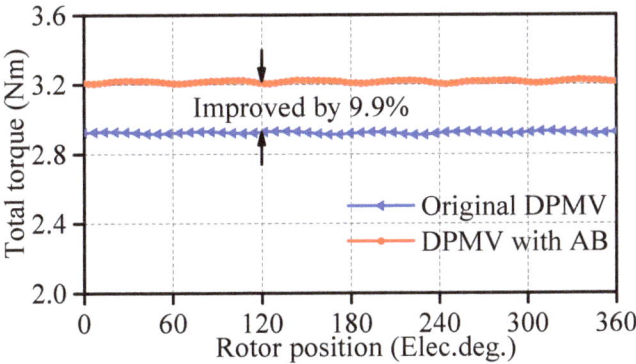

Figure 17. Total torque waveform comparison of the DPMV with and without AB.

4.2. Dual Three-Phase Winding

The DTP-30° winding configuration is conducive to increasing the winding factor and thus improving the average torque [25]. The winding factors of 5th-, 7th-, and 19th-order winding function harmonics are all 0.933 when the DPMV machine employs the original three-phase winding. By comparison, the winding factors of the above harmonics are all 0.966 when the DTP-30° winding configuration is employed. Figure 18 shows the DPMV machine with DTP-30° winding configuration, in which the ownership of winding corresponds to the color of the vector diagram. The winding configuration has no effect on the PM magnetic fields, and this section compares armature flux density with different winding configurations, as shown in Figure 19. Based on Figure 10, due to the increase in armature flux density of the 12th-, 19th- and 24th-order harmonics, the average torque is improved with employing DTP-30° winding. Figure 20 shows the total torque waveforms of DPMV with different winding configurations. The total torque is increased from 2.9 Nm to 3.0 Nm, and the torque ripple is superior as well. It should be pointed out that adopting the DTP-30° configuration results in complex control topology and increased control difficulty.

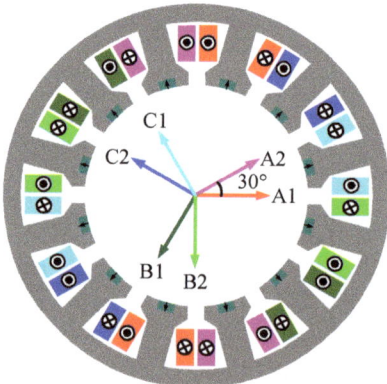

Figure 18. The DPMV with DTP-30° winding configuration.

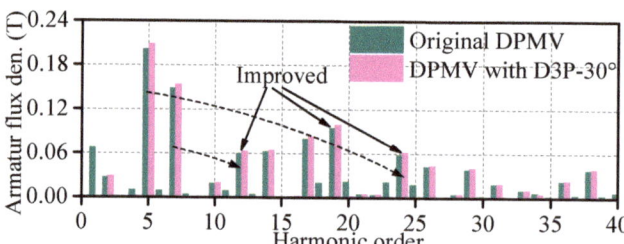

Figure 19. Spectrum comparison of armature flux density of the DPMV with different winding configurations at point d.

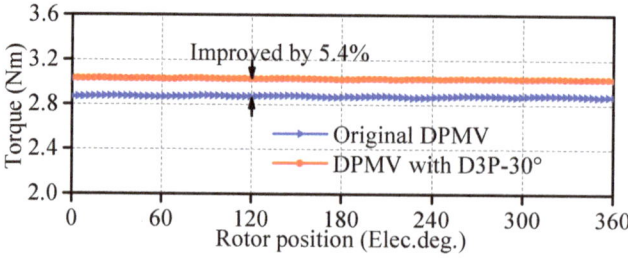

Figure 20. Total torque waveform comparison of the DPMV with different winding configurations.

5. Conclusions

This paper focuses on the torque analysis and improved design of the DPMV machine with the air-gap field modulation principle. The MMF permeance models of PMR, PMS, and armature magnetic fields have been established, and the modulation effect of topology structure has been analyzed in detail. Afterward, the torque generation mechanism of the DPMV machine has been investigated and the contribution of effective working harmonics to average torque has been identified with the frozen permeability method. The results show that the 7th-, 12th-, 19th- and 24th-order flux density harmonics are the main source of average torque, and especially the contribution of 19th-order harmonic exceeds 65%. Thanks to the multi-working harmonic characteristic, the proposed DPMV machine improves average torque by 31.8% with 75% PM weight of the IPM counterpart. The main contribution of this paper lies in proposing the auxiliary barrier structure and dual three-phase winding to improve the contribution of 19th-order harmonic to the average torque, respectively. While the auxiliary barrier structure is beneficial for increasing the angle difference between the radial and tangential components of the 19th-order harmonic, the dual three-phase winding can improve the amplitude of the 19th-order harmonic.

This paper solely focuses on the qualitative analyses of the torque generation mechanism of the DPMV machine. Therefore, the leakage flux, end effect, and iron reluctance are neglected. Future work will focus on the quantitative calculation of steady torque and torque ripple considering the nonlinear characteristics, and manufacturing a prototype for validation.

Author Contributions: Conceptualization, N.B. and Y.S.; methodology, W.Z. and J.J.; software, Y.S.; validation, J.J.; formal analysis, N.B., W.Z. and Y.S.; writing—original draft preparation, Y.S.; writing —review and editing, N.B. and W.Z.; visualization, N.B. and Y.S.; supervision, N.B. and W.Z.; project administration, N.B.; funding acquisition, N.B. All authors have read and agreed to the published version of the manuscript.

Funding: This research received no external funding.

Data Availability Statement: Not applicable.

Conflicts of Interest: The authors declare no conflict of interest.

References

1. Snoussi, J.; Elghali, S.B.; Benbouzid, M.; Mimouni, M.F. Optimal sizing of energy storage systems using frequency-separation-based energy management for fuel cell hybrid electric vehicles. *IEEE Trans. Veh. Technol.* **2018**, *67*, 9337–9346. [CrossRef]
2. Capata, R.; Calabria, A. High-performance electric/hybrid vehicle—Environmental, economic and technical assessments of electrical accumulators for sustainable mobility. *Energies* **2022**, *15*, 2134. [CrossRef]
3. Nasiri-Zarandi, R.; Karami-Shahnani, A.; Toulabi, M.S.; Tessarolo, A. Design and experimental performance assessment of an outer rotor PM-assisted SynRM for the electric bike propulsion. *IEEE Trans. Transport. Electrific.* **2023**, *9*, 727–736. [CrossRef]
4. Contò, C.; Bianchi, N. E-bike motor drive: A review of configurations and capabilities. *Energies* **2023**, *16*, 160. [CrossRef]
5. Asef, P.; Bargallo, R.; Lapthorn, A.; Tavernini, D.; Shao, L.; Sorniotti, A. Assessment of the energy consumption and drivability performance of an IPMSM-driven electric vehicle using different buried magnet arrangements. *Energies* **2021**, *14*, 1418. [CrossRef]
6. Abdelkefi, A.; Souissi, A.; Abdennadher, I.; Masmoudi, A. On the analysis and torque enhancement of flux-switching permanent magnet machines in electric power steering systems. *World Electr. Veh. J.* **2022**, *13*, 64. [CrossRef]
7. Fatemi, A.; Nehl, T.W.; Yang, X.; Hao, L.; Gopalakrishnan, S.; Omekanda, A.M.; Namuduri, C.S. Design optimization of an electric machine for a 48-V hybrid vehicle with comparison of rotor technologies and pole-slot combinations. *IEEE Trans. Ind. Appl.* **2020**, *56*, 4609–4622. [CrossRef]
8. Gu, Z.; Wang, K.; Zhu, Z.Q.; Wu, Z.; Liu, C.; Cao, R. Torque improvement in five-phase unequal tooth SPM machine by injecting third harmonic current. *IEEE Trans. Veh. Technol.* **2018**, *67*, 206–215. [CrossRef]
9. Lee, M.; Koo, B.; Nam, K. Analytic optimization of the Halbach array slotless motor considering stator yoke saturation. *IEEE Trans. Magn.* **2021**, *57*, 8200806. [CrossRef]
10. Zeng, X.; Quan, L.; Zhu, X.; Xu, L.; Liu, F. Investigation of an asymmetrical rotor hybrid permanent magnet motor for approaching maximum output torque. *IEEE Trans. Appl. Supercond.* **2019**, *29*, 0602704. [CrossRef]
11. Sun, Y.; Zhao, W.; Ji, J.; Zheng, J.; Cheng, Y. Torque improvement in dual m-phase permanent-magnet machines by phase shift for electric ship applications. *IEEE Trans. Veh. Technol.* **2020**, *69*, 9601–9612. [CrossRef]
12. Cheng, M.; Han, P.; Hua, W. General airgap field modulation theory for electrical machines. *IEEE Trans. Ind. Electron.* **2017**, *64*, 6063–6074. [CrossRef]
13. Tahanian, H.; Aliahmadi, M.; Faiz, J. Ferrite permanent magnets in electrical machines: Opportunities and challenges of a non-rare-earth alternative. *IEEE Trans. Magn.* **2020**, *56*, 900120. [CrossRef]

14. Guendouz, W.; Tounzi, A.; Rekioua, T. Design of quasi-halbach permanent-magnet vernier machine for direct-drive urban vehicle application. *Machines* **2023**, *11*, 136. [CrossRef]
15. Akuru, U.B.; Ullah, W.; Idoko, H.C.; Khan, F. Comparative performance evaluation and prototyping of double-stator wound-field flux modulation machines. In Proceedings of the 2022 International Conference on Electrical Machines (ICEM), Valencia, Spain, 5–8 September 2022; pp. 1893–1898.
16. Alavijeh, M.M.; Mirsalim, M. Design and optimization of a new dual-rotor Vernier machine for wind-turbine application. In Proceedings of the 2020 28th Iranian Conference on Electrical Engineering (ICEE), Tabriz, Iran, 4–6 August 2020; pp. 1–6.
17. Shi, Y.; Zhong, J.; Jian, L. Quantitative analysis of back-EMF of a dual-permanent-magnet-excited machine: Alert to flux density harmonics which make a negative contribution to back-EMF. *IEEE Access* **2021**, *9*, 94064–94077. [CrossRef]
18. Jang, D.; Chang, J. Investigation of doubly salient structure for permanent magnet vernier machines using flux modulation effects. *IEEE Trans. Energy Convers.* **2019**, *34*, 2019–2028. [CrossRef]
19. Jian, L.; Shi, Y.; Ching, T.W. Quantitative comparison of two typical field-modulated permanent magnet machines: Unidirectional field modulation effect versus bidirectional field modulation effect. In Proceedings of the 2019 22nd International Conference on Electrical Machines and Systems (ICEMS), Harbin, China, 11–14 August 2019; pp. 1–6.
20. Qi, J.; Zhu, Z.Q.; Yan, L.; Jewell, G.W.; Gan, C.; Ren, Y.; Brockway, S.; Hilton, C. Effect of pole shaping on torque characteristics of consequent pole PM machines. *IEEE Trans. Ind. Appl.* **2022**, *58*, 3511–3521. [CrossRef]
21. Li, J.; Wang, K. A novel spoke-type PM machine employing asymmetric modular consequent-pole rotor. *IEEE/ASME Trans. Mechatronics* **2019**, *24*, 2182–2192. [CrossRef]
22. Wang, K.; Li, J.; Zhu, S.; Liu, C. Novel hybrid-pole rotors for consequent-pole PM machines without unipolar leakage flux. *IEEE Trans. Ind. Electron.* **2019**, *66*, 6811–6823. [CrossRef]
23. Sun, Y.; Zhao, W.; Ji, J.; Zheng, J.; Song, X. Effect of phase shift on inductance and short-circuit current in dual three-phase 48-slot/22-pole permanent-magnet machines. *IEEE Trans. Ind. Electron.* **2022**, *69*, 1135–1145. [CrossRef]
24. Zhu, Z.Q.; Liu, Y. Analysis of air-gap field modulation and magnetic gearing effect in fractional-slot concentrated-winding permanent-magnet synchronous machines. *IEEE Trans. Ind. Electron.* **2018**, *65*, 3688–3698. [CrossRef]
25. Li, Y.; Zhu, Z.Q.; Wu, X.; Thomas, A.S.; Wu, Z. Comparative study of modular dual 3-phase permanent magnet machines with overlapping/non-overlapping windings. *IEEE Trans. Ind. Appl.* **2019**, *55*, 3566–3576. [CrossRef]

Disclaimer/Publisher's Note: The statements, opinions and data contained in all publications are solely those of the individual author(s) and contributor(s) and not of MDPI and/or the editor(s). MDPI and/or the editor(s) disclaim responsibility for any injury to people or property resulting from any ideas, methods, instructions or products referred to in the content.

Article

Insulation Condition Assessment in Inverter-Fed Motors Using the High-Frequency Common Mode Current: A Case Study

Mariam Saeed, Daniel Fernández, Juan Manuel Guerrero, Ignacio Díaz and Fernando Briz *

Department of Electrical, Computer & System Engineering, University of Oviedo, 33204 Gijón, Spain; saeedmariam@uniovi.es (M.S.); fernandezalodaniel@uniovi.es (D.F.); guerrero@uniovi.es (J.M.G.); idiaz@uniovi.es (I.D.)
* Correspondence: fbriz@uniovi.es

Abstract: The use of the common mode current for stator winding insulation condition assessment has been extensively studied. Two main approaches have been followed. The first models the electric behavior of ground-wall insulation as an equivalent *RC* circuit; these methods have been successfully applied to high-voltage high-power machines. The second uses the high frequency of the common mode current which results from the voltage pulses applied by the inverter. This approach has mainly been studied for the case of low-voltage, inverter-fed machines, and has not yet reached the level of maturity of the first. One fact noticed after a literature review is that in most cases, the faults being detected were induced by connecting external elements between winding and stator magnetic core. This paper presents a case study on the use of the high-frequency common mode current to monitor the stator insulation condition. Insulation degradation occurred progressively with the machine operating normally; no exogenous elements were added. Signal processing able to detect the degradation at early stages will be discussed.

Keywords: high-frequency common mode current; inverter-fed motors; insulation monitoring

Citation: Saeed, M.; Fernández, D.; Guerrero, J.M.; Díaz, I.; Briz, F. Insulation Condition Assessment in Inverter-Fed Motors Using the High-Frequency Common Mode Current: A Case Study. *Energies* 2024, 17, 470. https://doi.org/10.3390/en17020470

Academic Editors: Moussa Boukhnifer and Larbi Djilali

Received: 12 December 2023
Revised: 12 January 2024
Accepted: 15 January 2024
Published: 18 January 2024

Copyright: © 2024 by the authors. Licensee MDPI, Basel, Switzerland. This article is an open access article distributed under the terms and conditions of the Creative Commons Attribution (CC BY) license (https://creativecommons.org/licenses/by/4.0/).

1. Introduction

Variable speed drives are commonly used in multiple fields such as transportation, wind power, and industrial machinery, requiring high reliability. Stator insulation failure has been reported as the second most frequently occurring fault in induction machines [1–3]. The exposition of the machine windings to high rates of voltage change (dv/dt) due to switches commutation has been early reported to have adverse effects on the insulation [4], which are worsened with the use of new fast switching wide-bandgap devices [5–9].

The better-established insulation monitoring methods are offline, most of them being specific for high-voltage machines [2,3,10,11]. These include insulation resistance, high potential, capacitance, dissipation factor, and partial discharge among others. A main drawback is that the machine has to be removed from service; also, some of these tests are invasive.

A number of methods have also been proposed for insulation monitoring of inverter-fed low-voltage machines. Especially appealing are methods that use the phase current as a vast majority of modern drives include current sensors for control and protection purposes. These methods are often referred to as *Motor Current Signature Analysis (MCSA)* [12–16], but noting that many forms of signal processing are possible [17,18]. A concern with these methods is their capability to detect insulation faults at an early stage. Current sensors are selected according to the control needs, and might not comply with the bandwidth and sensitivity requirements to detect incipient faults. It is noted in this regard that for the experimental verification, it is a common practice to add external resistors to the test machine to emulate the fault [13–15,17–19]; there is no evidence that these kinds of artificial faults will produce similar effects to those due to the actual insulation degradation.

The use of the common mode current has been explored as a means to detect insulation degradation at early stages. While the method is well established for high-voltage, line-connected machines [2,3,20–22], its application to low-voltage, inverter-fed machines are less mature [7–10,23–29].

This paper presents a case study on the use of the high-frequency common mode current to monitor the stator insulation condition. Although this paper will focus on the case of an induction machine, the conclusions might be extended to other types of AC machines with similar stator designs as permanent magnet and synchronous reluctance machines. Aging was accelerated by performing a sequence of experiments in which the machine was forced to operate at temperatures above its insulation class. Insulation degradation occurred progressively, and without adding exogenous elements. Methods for the signal processing capable of detecting the degradation at early stages will be discussed.

The main contributions of the presented paper are: (1) insulation degradation is performed progressively, without artificially provoking the fault and without any exogenous elements (e.g., external resistors or capacitors) following, therefore, a process closer to that occurring in real-world conditions; (2) it has been confirmed that the high-frequency behavior of the zero sequence current is sensitive to insulation degradation even at incipient stages (when the DC insulation resistance is still very high); (3) it has been shown that the behavior of the zero sequence current during the degradation process differs significantly from the behavior observed using artificially induced faults; (4) consequently, it has been shown that signal processing methods and metrics developed based on results obtained using such artificially induced faults might fail in real-world implementations.

This paper is organized as follows: common mode current modeling is presented in Section 2; Section 3 describes test-bench and experiments; common mode current measurement is addressed in Section 4; motor degradation is discussed in Section 5; signal processing and results are presented in Section 6; finally, conclusions are summarized in Section 7.

2. Common Mode Current Modeling

Two main approaches have been proposed in the literature for the use of the common mode current for stator winding insulation assessment. The first requires a finite resistance between winding and magnetic core; the second is based on the analysis of *HF* common mode current resulting from the voltage pulses applied by the inverter. Both are discussed, with the second being the approach used in this paper.

2.1. Modeling Using an Equivalent RC Circuit

For ground-wall insulation, the model in Figure 1a has been widely used [2,3,7,10,20–24,29]. The *Dissipation Factor* (*DF*) or alternatively the *Power Factor* (*PF*) can be used (1). These methods are especially indicated for machines with voltage ratings of 6 kV and higher [21,22].

$$DF = \tan(\delta) = \frac{|I_r|}{|I_c|} \quad ; \quad PF = \cos(90 - \delta) = \frac{|I_r|}{|I_0|} \qquad (1)$$

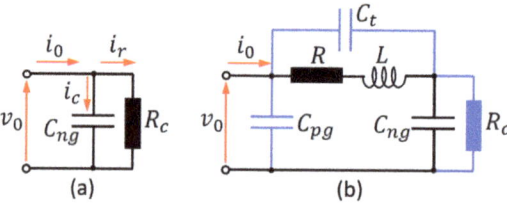

Figure 1. Equivalent circuits: (**a**) *RC* used mainly for ground-wall insulation analysis; (**b**) *RLC* used for *HF* response analysis. Components in blue can exist or not, depending on the approach.

Detection of ground-wall insulation faults of inverter-fed machines has been reported too. However, in the experiments provided in most of these works, the fault was induced artificially. Capacitors [7] or resistors [2,10,24] were inserted between the neutral of the machine and ground (frame). Capacitors connected to winding taps were used in [6,30] to emulate changes in the inter-turn capacitance, but noting that in these works the phase current transients instead of common mode current were analyzed. In all the cases, the test machine voltage was in the range of hundreds of volts.

A concern for these reports is to what extent conclusions obtained from the analysis of faults induced artificially can be extended to the real faults. A further concern is the sensitivity required by the current sensors used to measure the leakage current in low-voltage machines [7,20,29,30].

2.2. Common Mode Voltage Excitation

Common mode voltage pulses applied by the inverter provide a useful form of *HF* excitation for methods using the *HF* components of the common mode current. The common mode voltage applied by the inverter is defined as (2).

$$v_0 = \frac{v_a + v_b + v_c}{3} \quad (2)$$

The output voltage of a two-level inverter can take only two possible values with respect to the midpoint of the inverter: $-V_{dc}/2$ and $V_{dc}/2$. It is deduced from (2) that v_0 can take four possible values, $v_0 = \{-V_{dc}/2, -V_{dc}/6, V_{dc}/6, V_{dc}/2\}$. Use of (2) is simplified if the phase voltage commands are zero (3), as in this case all phases will switch simultaneously.

$$v_a^* = v_b^* = v_c^* = 0 \;;\; v_a = v_b = v_c = v_0 \quad (3)$$

For inverters using *PWM/SVM*, the three-phase voltages and the common mode voltage will be a square wave signal (50% duty) varying between $-V_{dc}/2$ and $+V_{dc}/2$ in this case [see voltage wave shape in Section 3 Figure 5a] at the switching frequency of the inverter. All the results shown in this paper will be obtained with this type of voltage excitation. One possible disadvantage of this approach is that the method could not be considered *online*, as it is unusual that the inverter operates with zero voltage command. However, this is not considered a drawback. On one hand, the type of fault being detected develops very slowly, with continuous monitoring not being required. In addition, operating with a voltage command equal to zero is easy to achieve after the turn-on of the drive. Furthermore, in this case, the machine would be at (or close to) ambient temperature, mitigating the influence of temperature discussed later.

2.3. Modeling Using an Equivalent Resonant Circuit

The use of the *HF* components of the common mode current for diagnostic purposes has been analyzed in [8,25–28]. Modeling of the oscillations of the common mode current using equivalent circuits with passive elements requires the presence of a resonant *LC* network. Several models of this type have been proposed in the literature. The simplest circuit consists of an *RLC* network [28]. This would correspond to the circuit in black in Figure 1b (i.e., $C_t = C_{pg} = 0$ and $R_c = \infty$). The corresponding transfer function is (4) in this case.

$$\frac{I_0}{V_0} = \frac{C_{ng}S}{LC_{ng}S^2 + RC_{ng}S + 1} \quad (4)$$

The ground-wall resistance R_c can be included in the model, the resulting transfer function being (5).

$$\frac{I_0}{V_0} = \frac{C_{ng}R_cS + 1}{LC_{ng}R_cS^2 + (L + C_{ng}RR_c)S + (R + R_c)} \quad (5)$$

This model has been used in [27], and with a slightly different arrangement in [25,26,31]. It is noted that [26,27,31] used distributed parameters, while in [25], lumped parameters are used. In all the experiments performed during this research, it was not possible to detect any dc zero currents between windings and magnetic core. However, the sensors and acquisition being used were not specific for the detection of small leakage currents (see Section 4); also, the voltage being applied was relatively small (hundred volts). Consequently, for the analysis presented in this paper, the model in (5) would not add any benefits compared to (4). However, the fact that it includes a zero was found useful for the identification-based analysis presented in Section 6.1.

The model in Figure 1b including capacitors C_t and C_{ng} has been used in [8,32]. It is noted that [8,32] used distributed parameters. The fact that the common mode current can flow through a purely capacitive path results in an improper transfer function, i.e., with more zeros than poles. Therefore, model identification discussed in Section 6.1 cannot be applied in this case. Consequently, this model will not be considered. Further discussion on the use of the models presented in this section for insulation assessment is presented in Section 6.1.

3. Test Bench and Experiments Description

One objective for the research presented in this paper was that insulation degradation followed a similar process to that occurring in real-world conditions. The use of thermal chambers to achieve accelerated aging has been reported [9,27,29,33]. In [29], the recommendations from IEEE Std 117-2015 [34] were followed. The test conditions applied to Class F motors in [27] were significantly more aggressive than those recommended in [34]. Maximum temperatures of 200 °C and 230 °C are reported in [9,33], respectively, but details on the exposure times are not provided.

Regardless of the benefits of using a thermal chamber, drawbacks must also be considered. Feeding motors operated in a thermal chamber can be extremely challenging. In addition, the thermal chamber results in temperature distribution within the motor, which can be rather different from the temperature distribution in real operating conditions. Although a climatic chamber is available for motor testing, its use was disregarded.

3.1. Test Machine and Three-Phase Inverter

The main test bench consists of two identical induction machines, denoted as *Device Under Test* (*DUT*) and *Auxiliary* (*AUX*) motors (see Figures 2 and 3). Main motor parameters are shown in Figure 3. Further details can be found in Section 5. The *DUT* fan was removed; consequently, it reached higher temperatures than *AUX*. Both motors are fed by two three-phase inverters connected back-to-back (see Figures 2 and 3), equipped with 600 V IGBTs. The dc-link voltage was limited to 400 V. Phase currents and voltages of both machines were measured using Hall-effect current and voltage sensors of 100 kHz bandwidth. Signals from these sensors were sampled and 750 kHz with 16-bit resolution. A 1024-line encoder is used to measure the speed. Five type-K thermocouple temperature sensors were installed: two inserted into the end-windings of *DUT* motor (see Section 4), two attached to *DUT* and *AUX* frames, and one for ambient temperature. Temperatures were sampled at 1 Hz. For some of the initial experiments, temperature sensors were not operational (0 °C in Figure 7).

The common mode current resonance frequency was found to occur at ≈3 MHz. Consequently, the current sensor bandwidths and sampling frequencies described above are inadequate to capture this signal. A second concern with the three-phase inverter was that maximum dc link voltage was limited to 400 V. Higher voltages are desirable to evaluate the influence of the dc voltage on the *HF* common mode current. To overcome these limitations, a *Full-Bridge* (*FB*) and dedicated sensorization were developed, they are described in Sections 3.2 and 4, respectively.

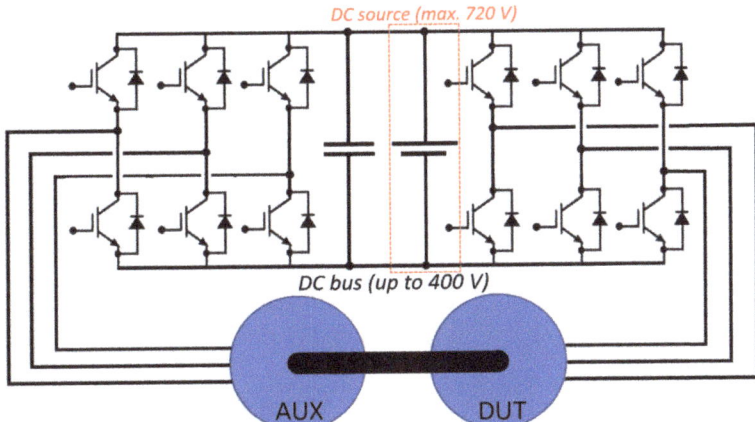

Figure 2. Schematic representation of the experimental test bench using two back-to-back inverters.

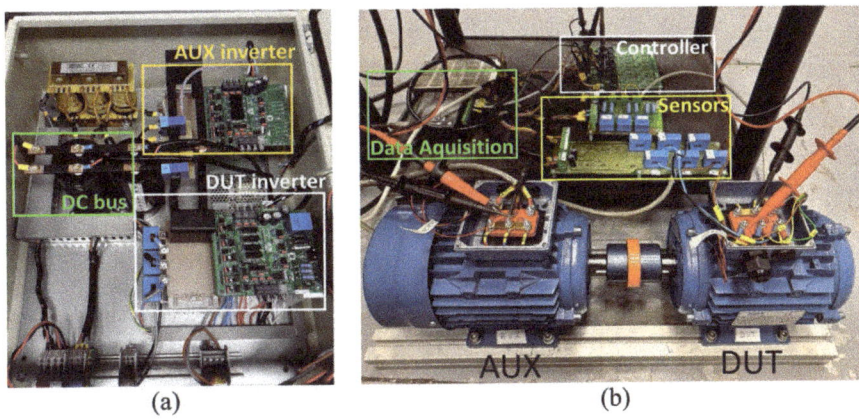

Figure 3. Test bench pictures: (**a**) back-to-back inverters and, (**b**) *DUT* and *AUX* motors. Rated values: $V = 440$ V; $I = 2.74$ A; $P = 1.1$ kW; $f = 50$ Hz; $\omega_r = 1390$ rpm; 4 poles.

3.2. Full-Bridge Converter

The schematic representation of the *FB* converter and *HF* sensorization developed to capture the *HF* common mode current is shown in Figure 4. The *FB* converter uses 1.2 kV *SiC MOSFET*, with a *dc*-link voltage up to 720 V. The voltage applied by the *FB* is the same as when the three-phase inverter is commanded zero voltage (3). However, *SiC MOSFET* produces faster commutations than *Si IGBTs*. Furthermore, a voltage step up to 1440 V is now possible, compared to the voltage step of 400 V achievable with the three-phase inverter.

Figure 5a,b show the voltage pulses applied by the *FB* and the resulting common mode current. Figure 5c,d show the same signals using a zoomed timescale. A conclusion from Figure 5c,d is that the magnitude of the voltage pulses has no visible effects on the common mode current transient. However, when the insulation fault developed, the magnitude of the voltage clearly affected the speed of degradation. This will be discussed later.

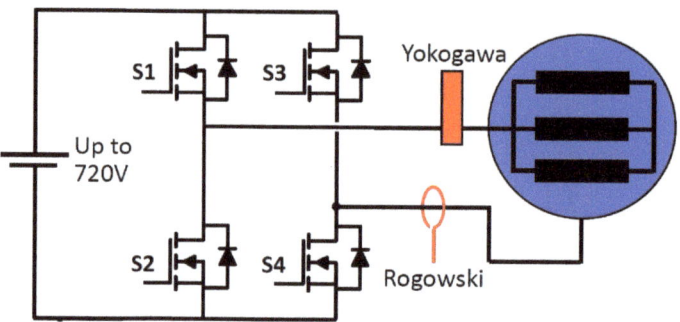

Figure 4. Full-bridge and current sensors used to measure the common mode current.

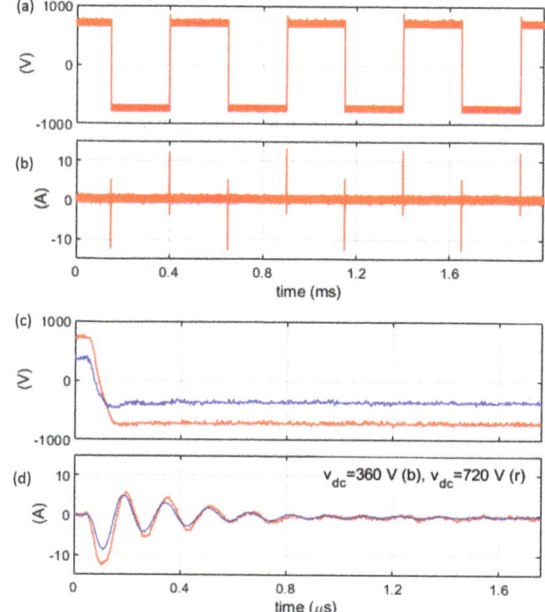

Figure 5. (a) Common mode voltage; (b) common mode current. *FB* switches at 2 kHz, $v_{dc} = 720$ V; (c,d) zoomed signals for the case of $v_{dc} = 720$ V (red) and $v_{dc} = 360$ V (blue), respectively.

3.3. Experiment Description

Experiments were performed at a maximum rate of one per day, always starting with the motors at ambient temperature. The *DUT* temperature increase is due exclusively to losses induced during its operation. To accelerate degradation, in some of the experiments *DUT* was forced to operate at temperatures above its class for small periods of time. As mentioned, *DUT* fan was removed. Consequently, *AUX* temperature is significantly lower, its insulation not being jeopardized.

Motors are fed using the back-to-back inverters in Figure 2. Phase currents, voltages, speed, and temperatures are measured and stored. The operating conditions of the motors (control and modulation strategy of inverter feeding *DUT* machine, speed, and torque) vary from experiment to experiment, not being relevant to the contents of this paper. Most of the time the machines operate at rated load as this produces higher losses.

The number of experiments before failure was 28. Figure 6 shows temperatures vs. time for two of them. Both took ≈110 min, with a maximum *DUT* stator winding temperature of ≈110 °C and ≈160 °C, respectively. Figure 7 shows the duration and maximum temperature for all the experiments carried out. Further discussion on Figure 7 is presented in Section 5.

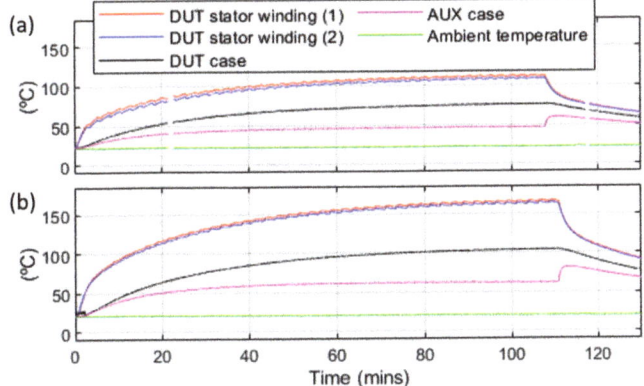

Figure 6. Temperatures for experiments (**a**) #27 and (**b**) #20 (see Figure 7).

DUT phase-to-phase and phase-to-frame insulation was measured using an insulation tester immediately before and after each experiment. The *HF* common mode current measurement using the *FB* was not performed for all the experiments, as disconnection/reconnection of power converters was impractical. However, it was measured before starting any new experiment once phase-to-phase insulation degradation was detected for the first time with the insulation tester (experiment #24 in Figure 7).

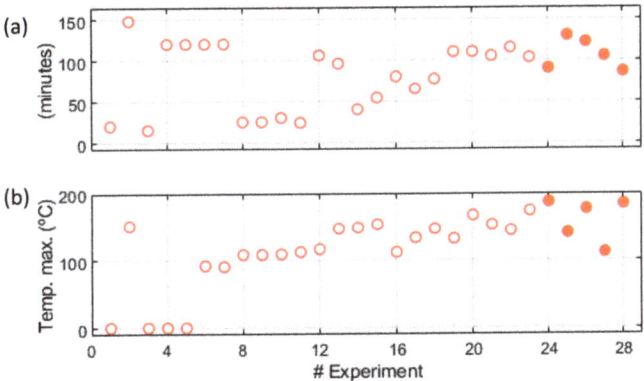

Figure 7. Summary of experiments: (**a**) length in minutes; (**b**) maximum temperature. Experiments marked with a solid dot indicate that a decrease in phase-to-phase insulation had been detected with the insulation tester.

4. Common Mode Current Measurement

A variety of sensors have been reported in the literature for common mode current analysis. The use of impedance analyzers with embedded sensors is reported in [25,28]; high sensitivity 2 MHz bandwidth differential current transformers were used in [3,20], but noting that in this case, the objective was to detect the leakage current through the winding-to-magnetic core resistance; shunt resistors were mentioned in [27]; use of Rogowski coils is reported in [35,36]; in [35] a 2 MHz magnetoresistive sensor was used; [7,8,29] used a 1 MHz bandwidth current transformer; a current transformer was also used in [26], but the bandwidth was not specified.

From the observed properties for the common mode current, it is concluded that bandwidths in the range of MHz are required. Two current sensors were evaluated: (1) High bandwidth, Hall-effect type instrumentation probe, and (2) Rogowski coil (See Figure 8). For signal acquisition, a conventional digital scope was used. Sensors configuration and connection are shown in Figures 4 and 8a. Measured signals are shown in Figure 9a. Both provide similar responses, the differences could come from the fact that one sensor was located at the cable feeding the stator and the other at the return cable.

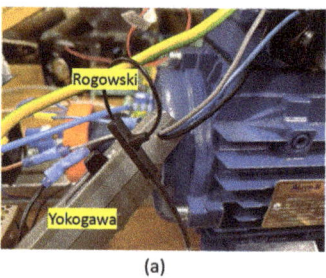

Figure 8. (a) Yokogawa Hall-effect and PEM Rogowski sensors. Both provide 50 MHz bandwidth and a maximum current of 30 A; (b) end-frame with the cables connected to two temperature sensors. Temperature sensors are attached to the end coils.

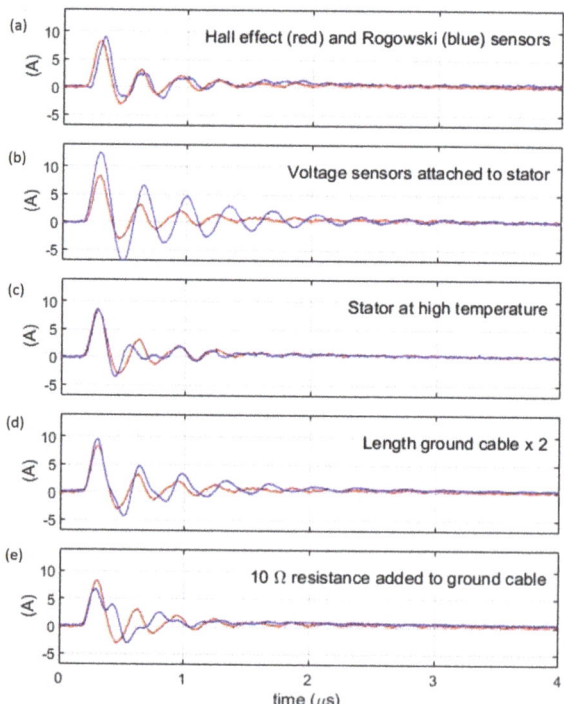

Figure 9. Measured common mode current: (a) Hall sensor vs. Rogowski coil at ambient temperature; (b) effect of a voltage sensor attached to the stator; (c) effect due to temperature; (d) effect due to ground cable length; (e) effect due to ground cable resistance. $V_{dc} = 360\ V$. Hall sensor was used for cases (b–e). The trace in red in (a) is shown in all the subplots for reference.

For a real implementation, there are two relevant aspects to consider. First, whether the sensor would be installed permanently, or only when a measurement is to be taken; second, the accuracy required both for sensors and associated electronics to achieve reliable measurements. Considering that the faults being tracked develop slowly and the sensors can be relatively expensive, having the sensors permanently installed might seem an inefficient solution. However, installing the sensor when a measurement is to be taken will give rise to additional concerns, e.g., if the sensor changes from measurement to measurement, or tolerances mounting the sensor. Having the sensor permanently installed would significantly reduce these concerns. In this case, sensor repetitiveness rather than sensor accuracy would be the parameter to consider. On the other hand, *Analog-to-Digital Converters* (*ADC*) with sampling rates in the range of at least tens of MHz would be required to sample the *HF* common mode current. *ADCs* providing such sampling rates with 16-bit resolution can be found at a reasonable cost. Having them permanently installed might not be therefore prohibitive in some applications.

If the sensor is to be installed only when a measurement is required, the use of open-core sensors would be advantageous. Consequently, the Rogowski coil seems a good option in this case, as it is a flexible, clip-around sensor. Since the information of interest is at *HF*, and in principle the low-frequency components of the common mode currents do not contain useful information, the use of a Rogowski coil without an integrator at its output might be viable [36]. It is noted, however, that this option has not been evaluated experimentally. Sensitivity of Rogowski coil to environmental conditions should also be considered [37].

HF Common Mode Current Sensitivity Analysis

A number of experiments were performed to understand the sensitivity of the *HF* common mode current to operating and implementation issues. The results are shown in Figure 9b–f. All the measurements were made using the Hall effect sensor. The trace in red in Figure 9a is shown in all the cases for reference. The remaining subplots in Figure 9 show the effect of: (b) connecting Hall-effect type voltage sensors to the stator; (c) increasing the stator temperature; (d) increasing the ground cable length ×2; (d) adding a resistor to the ground cable.

It is concluded from inspection of Figure 9 that the *HF* common mode current is sensitive to changes in the machine temperature, cable impedance, as well as to other elements that could be connected to the stator as voltage sensors. Consequently, to increase the reliability of the measurements, special attention should be paid to minimizing the changes of these parameters during measurements.

5. Motor Degradation and Post-Fault Analysis

The insulation level was measured before and after each experiment using an insulation tester. Phase-to-phase insulation and phase-to-frame insulation for experiments #1 to #23 was >10 GΩ, which was the limit value of the insulation tester. $U - W$ insulation decreased to be in the range between <10 GΩ and >5 GΩ after experiment #24. $U - W$ failure occurred during common mode current measurement following experiment #28. After this experiment, noise due to partial discharges was readily audible when a common mode voltage of 720 V was applied. Initially, the noise disappeared when the common mode voltage was 360 V. However, the fault evolved very quickly resulting in a net insulation failure between phase $U - W$. No ground-wall fault was detected. Phase V remained healthy as well. *DUT* motor was open after failure, with no deterioration in the stator winding being visible.

Insulation fault occurred between phases U and W. As the stator winding has a single layer with one phase per slot, the fault necessarily occurred in the end-winding. End-windings have been reported to be the hottest spot in electric machines [38]. Figure 10 shows the measured resistance between phases U and W using the four possible combinations and the estimated location of the fault. The phase resistances and phase-to-frame capacitance for the healthy case and after the last test are shown in Table 1. Capacitance for phase U and W after the fault is the same as they are short-circuited. Resistance and capacitance for the healthy phase increased slightly.

$R_{U1-W1} = 10.69\ \Omega$
$R_{U1-W2} = 5.89\ \Omega$
$R_{U2-W1} = 13.86\ \Omega$
$R_{U2-W2} = 9.04\ \Omega$

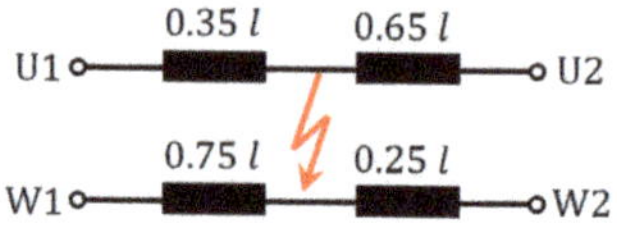

Figure 10. Measured resistances and estimated location of the fault between phases U and W in p.u. of winding length l.

Table 1. DUT dc stator resistance and capacitance.

	Healthy	Faulty		Healthy	Faulty
R_{U1-U2}	9.67 Ω	9.8 Ω	C_{U-g}	0.852 nF	1.43 nF
R_{V1-V2}	9.55 Ω	9.75 Ω	C_{V-g}	0.888 nF	0.977 nF
R_{W1-W2}	9.7 Ω	9.9 Ω	C_{W-g}	0.959 nF	1.43 nF

Figure 11 shows captures with the thermal imaging camera when a *dc* voltage is applied to the stator terminals indicated in the corresponding captions. The voltage was adjusted manually to obtain the nominal current. The images confirm that the insulation fault occurred in the end-winding, and not in the cables connecting the terminal box to the winding.

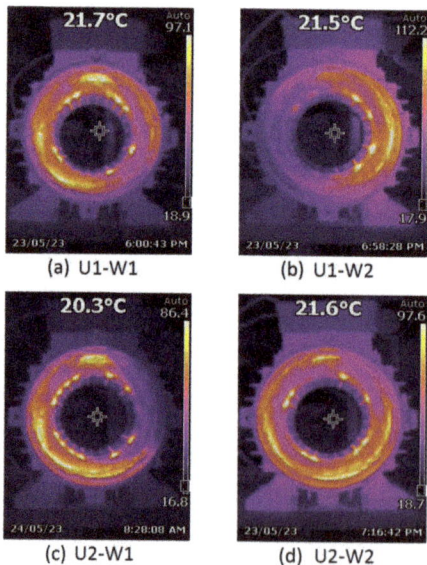

Figure 11. Captures of the thermal imaging camera after test #28 when the stator is fed from a *dc* voltage source between terminals indicated in the subcaptions. The stator has 24 slots, single-layer winding.

6. HF Common Mode Current Analysis and Processing

Figure 12a shows the common mode current for the case of a healthy machine, and for the case when an insulation decrease was detected for the first time with the insulation tester (experiment #24, see Figure 7). It is noted that although the figure shows a single pulse, the response is highly repetitive.

At first glance, the response for both cases looks very similar. A closer look reveals that the coincidence is most remarkable during the initial part of the transient. Figure 12b shows the common mode current after removing the initial part of the transient in Figure 12a. This signal is denoted as i_{0W}. It is observed that the oscillations of DUT once insulation degradation has started last longer (lower damping). Interestingly, this change in behavior of the common mode current remained almost unaltered for all the measurements of the common mode current performed previous to the start of experiments #25 to #28 in Figure 7. Furthermore, the behavior persisted even when the machine could not be fed from the inverter due to the lack of insulation between phases.

Different forms of signal processing of the common mode current are discussed in the following subsections.

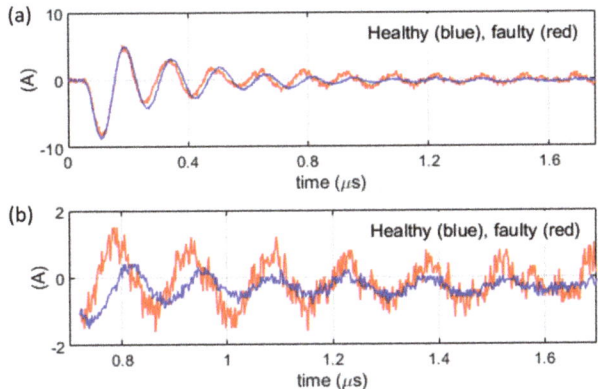

Figure 12. (a) Transient common mode current for the healthy and faulty cases. (b) Windowed common mode current i_{0W}; obtained by applying a window W to i_0. $W = 1$ for 0.75 μs $< t <$ 1.7 μs; $W = 0$ otherwise.

6.1. Insulation Assessment Based on Model Estimation

The peak value of the common mode current was used in [26,27] as a possible indicator of the insulation condition. However, from the results obtained in this work, the peak value alone was not found to be a reliable metric. The common mode voltage-to-current frequency response function was used in [8,25,32]. Changes either in the peak value or in the frequency response function would be connected with system pole migration in (4), (5) due to changes in machine parameters. It would, therefore, be expected that the migration of model parameters could provide useful information on the insulation condition.

Model identification using the measured common mode current was implemented using Matlab. The function `procest` with a model structure of two underdamped poles and a zero provided the best fit to estimation data, the system resulting from the identification being of the form (6).

$$\frac{I_0}{V_0} = k \frac{(\tau_z s + 1)}{(\tau_w s)^2 + 2\xi \tau_w s + 1} \quad (6)$$

Under the assumption that $R_c \gg R$, the following relationship can be found between (6) and (5).

$$\tau_w \approx \sqrt{LC_{ng}}; \; \xi \approx \frac{R}{2}\sqrt{\frac{C_{ng}}{L}}; \; \tau_z = C_{ng}R_c; \; k = \frac{1}{L} \quad (7)$$

To reduce the sensitivity to noise or other possible disturbances, parameter identification was performed by averaging the common mode current resulting in eight successive common mode voltage transitions, as shown in Figure 5.

To validate the estimated model, its response was obtained using Matlab function `lsim`, with the input being the measured voltage. Figure 13 shows the actual and simulated common mode current. A good agreement is observed in general.

Table 2 shows the result obtained averaging the eight estimations, $\omega_d = \omega_n \sqrt{1-\zeta^2}$ being the damped natural frequency. A difference between the healthy and faulty cases in Figure 12 is the reduction in the estimated damping for the case of the faulty machine (oscillations last longer). This is consistent with the behavior of the damping factor ζ observed in Table 2. It is deduced from (7) that a decrease in ζ can be due to a decrease in either R and/or C_{ng}, or to an increase on L. Unfortunately, the decrease observed is in the range of 5% to 10%, which might not be enough for reliable detection. A second difference observed in Figure 12 between the healthy and the faulty cases is an increase in the damped natural frequency ω_d. However, this is not confirmed by the results in Table 2, as the trend observed for the cases 360 V and 720 V are opposite.

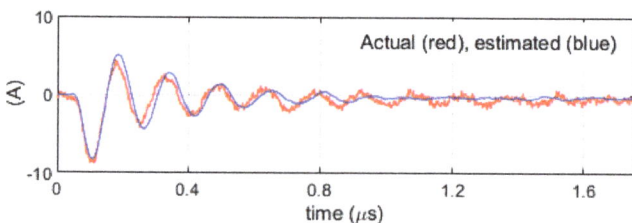

Figure 13. Actual and estimated common mode current using the parametric model in (6).

Table 2. Model identification: Experimental results.

DUT	Voltage (V)	ω_d (Hz)	ζ	$1/\tau_z$ (Hz)	k
Healthy	360	6,749,592	0.1167	6,820,495	380
Faulty	360	70,073,690	0.1042	7,089,280	340
Healthy	720	7,216,209	0.1103	7,261,304	417
Faulty	720	6,888,639	0.1049	6,941,588	359

6.2. Insulation Assessment Based on Frequency Analysis

It was already shown in Figure 12 that a remarkable difference between the healthy and faulty cases was the persistence of the oscillations for the second case. This suggests the use of frequency-based methods. For this purpose, it is advantageous to remove the initial part of the common mode current transient due to two reasons: first, no relevant differences are observed between the healthy and faulty cases during the initial part of the transient; second, frequency-based methods are not effective in analyzing transient phenomena. A similar approach to the one described following was proposed in [35].

Figure 14 shows the *FFT* of the common mode current after removing the initial part of the transient (see Figure 12b. It is noted that with the time window used in Figure 12b, the oscillation occurs at the 7th harmonic. The increase in the peak values between the faulty and healthy cases observed in Figure 14 is seen in all cases in the range of 220%, and can be considered therefore significantly more reliable than the changes in ζ observed in Section 6.1.

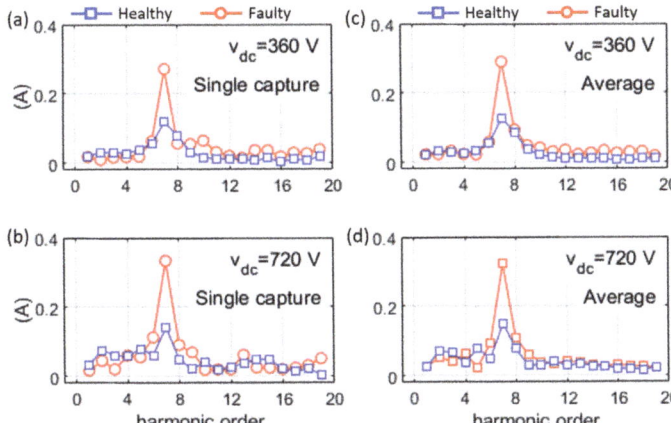

Figure 14. $|FFT(i_{0W})|$ [see Figure 12b]. Left (**a,b**): single capture, right (**c,d**): average of eight successive captures. Top (**a,c**): $V_{dc} = 360\ V$, bottom (**b,d**): $V_{dc} = 720\ V$. Blue/Red: healthy/faulty machine.

Regardless of the promising results, there are a few aspects to consider for the described signal processing. First, it is based on the assumption that the frequency of the oscillation does not change significantly between the healthy and faulty cases. Also, the length of the initial transient being removed, as well as the length of the common mode current window being processed, were adjusted manually. It would be desirable to automatize this process. For example, *PLL* could be used to estimate the frequency of the oscillation and adapt the *FFT*. However, more experimental data is needed to solve these uncertainties.

6.3. Other Forms of Signal Processing

The dependency on the oscillation frequency discussed in the previous section could be avoided if instead of focusing on a single component of the spectrum, the total energy of the common mode current is considered. The *RMS* value of the common mode current can be used for this purpose. Table 3 summarizes the results obtained using this approach. It is observed that if the initial part of the transient is not removed, $RMS(i_0)$ actually decreases for the faulty case with respect to the healthy case. This is due to the fact that most of the energy of the signal occurs on the initial part of the transient, which is not affected by insulation deterioration. This problem is solved by using $RMS(i_{0W})$. However, significant differences are observed for the case of $V_{dc} = 360$ and 720 V. Also, the sensitivity is significantly smaller compared to using the *FFT*.

Table 3. $RMS(i_0)$.

DUT	Voltage (V)	i_0 (A)	i_{0W} (A)	Voltage (V)	i_0 (A)	i_{0W} (A)
Healthy	360	1.807	0.436	720	2.67	0.792
Faulty	360	1.737	0.644	720	2.4722	0.897
$\Delta(i_0)\%$	360	-4%	47%	720	-7%	13%

It is finally noted that many other forms of signal processing, e.g., based on correlations or wavelet-based analysis [39], could be used. Independent of the approach being used, it is concluded from the results presented in this work that a single measurement of the *HF* common mode current is not enough to determine the insulation condition of a machine. Tracking the deviations with respect to when the machine was healthy is required. It is noted that the same applies to other methods for insulation degradation detection [21,22].

7. Conclusions

A case study of stator windings insulation condition assessment for inverter-fed machines using the HF common mode current has been presented in this paper. Implementation of the proposed method requires the use of a high bandwidth current sensor and acquisition system. The cost of these elements can be relevant for low-power, cheap induction motor drives. However, it could be fully justified in high-power (hundred kW), high efficiency, expensive induction motor drives, e.g., for railway traction. Implementation of the method does not require changes in the system layout and does not interfere with drive normal operation either. These elements could be installed permanently or only when a measurement is to be taken, both options having pros and cons.

Aging was accelerated by performing a sequence of experiments in which the machine was forced to operate at temperatures above its class. Phase-to-phase insulation failure finally occurred in the end-winding. Visual inspection did not reveal any anomaly.

Two main types of signal processing methods were used: model-based and FFT-based, with the second showing a significantly better detection capability when applied to the windowed common mode current. As the behavior of the common mode current is specific to each machine design, the signal processing should be tuned accordingly.

The sensitivity of the HF common mode current to operating conditions and system configuration was also evaluated. Winding temperatures can have significant effects. Consequently, measurements should be carried out with the machine at a similar temperature. Realizing the measurements at the start-up of the drive seems to be the most reliable option.

From the results shown in this paper, it is concluded that a single measurement of the HF common mode current is not enough to determine the insulation condition, trends over time should be tracked. It is noted that the same concern applies to other methods as DF and PF.

A limitation of the results presented in this paper is that the analysis was limited to a specific fault. However, this fault resulted from operating the machine repeatedly under extreme working conditions; no external elements were added. Consequently, the results shown are believed to be a true subset of the phenomena that could happen in practice.

Author Contributions: Conceptualization, M.S. and F.B.; methodology, M.S., J.M.G. and F.B.; software, M.S., J.M.G. and D.F.; validation, F.B., I.D. and M.S.; formal analysis, J.M.G. and F.B.; investigation, M.S. and J.M.G.; resources, F.B. and J.M.G.; data curation, F.B. and I.D.; writing—original draft preparation, F.B. and M.S.; writing—review and editing, J.M.G., I.D. and D.F.; visualization, M.S. and F.B.; supervision, F.B.; project administration, F.B.; funding acquisition, F.B. and J.M.G. All authors have read and agreed to the published version of the manuscript.

Funding: This work was supported in part by Ingeteam Power Technology S.A. and by the Government of the Principality of Asturias under project AYUD/2021/50988.

Data Availability Statement: Data are contained within the article.

Conflicts of Interest: The authors declare no conflicts of interest.

Abbreviations

The following abbreviations are used in this manuscript:

$MCSA$	Motor Current Signature Analysis
HF	High Frequency
DUT	Device Under Test
AUX	Auxiliary
FB	Full-Bridge
ADC	Analog-to-Digital Converters
DF	Dissipation Factor
PF	Power Factor

References

1. Surya, G.N.; Khan, Z.J.; Ballal, M.S.; Suryawanshi, H.M. A Simplified Frequency-Domain Detection of Stator Turn Fault in Squirrel-Cage Induction Motors Using an Observer Coil Technique. *IEEE Trans. Ind. Electron.* **2017**, *64*, 1495–1506. [CrossRef]
2. Lee, S.B.; Younsi, K.; Kliman, G. An online technique for monitoring the insulation condition of AC machine stator windings. *IEEE Trans. Energy Convers.* **2005**, *20*, 737–745. [CrossRef]
3. Zhang, P.; Younsi, K.; Neti, P. A Novel Online Stator Ground-Wall Insulation Monitoring Scheme for Inverter-Fed AC Motors. *IEEE Trans. Ind. Appl.* **2015**, *51*, 2201–2207. [CrossRef]
4. Stone, G.; Campbell, S.; Tetreault, S. Inverter-fed drives: Which motor stators are at risk? *IEEE Ind. Appl. Mag.* **2000**, *6*, 17–22. [CrossRef]
5. Xu, Y.; Yuan, X.; Ye, F.; Wang, Z.; Zhang, Y.; Diab, M.; Zhou, W. Impact of High Switching Speed and High Switching Frequency of Wide-Bandgap Motor Drives on Electric Machines. *IEEE Access* **2021**, *9*, 82866–82880. [CrossRef]
6. Zoeller, C.; Vogelsberger, M.A.; Wolbank, T.M.; Ertl, H. Impact of SiC semiconductors switching transition speed on insulation health state monitoring of traction machines. *IET Power Electron.* **2016**, *9*, 2769–2775. [CrossRef]
7. Zheng, D.; Zhang, P. An Online Groundwall and Phase-to-Phase Stator Insulation Monitoring Method for Inverter-Fed Machine. *IEEE Trans. Ind. Electron.* **2021**, *68*, 5303–5313. [CrossRef]
8. Zheng, D.; Lu, G.; Zhang, P. An Improved Online Stator Insulation Monitoring Method Based on Common-Mode Impedance Spectrum Considering the Effect of Aging Position. *IEEE Trans. Ind. Appl.* **2022**, *58*, 3558–3566. [CrossRef]
9. Alvarez-Gonzalez, F.; Hewitt, D.; Griffo, A.; Wang, J.; Diab, M.; Yuan, X. Design of Experiments for Stator Windings Insulation Degradation under High dv/dt and High Switching Frequency. In Proceedings of the IEEE Energy Conversion Congress and Exposition (ECCE), Detroit, MI, USA, 11–15 October 2020; pp. 789–795. [CrossRef]
10. Lee, S.B.; Yang, J.; Younsi, K.; Bharadwaj, R. An online groundwall and phase-to-phase insulation quality assessment technique for AC-machine stator windings. *IEEE Trans. Ind. Appl.* **2006**, *42*, 946–957. [CrossRef]
11. Cruz, J.d.S.; Fruett, F.; Lopes, R.d.R.; Takaki, F.L.; Tambascia, C.d.A.; Lima, E.R.d.; Giesbrecht, M. Partial Discharges Monitoring for Electric Machines Diagnosis: A Review. *Energies* **2022**, *15*, 7966. [CrossRef]
12. Jung, J.H.; Lee, J.J.; Kwon, B.H. Online Diagnosis of Induction Motors Using MCSA. *IEEE Trans. Ind. Electron.* **2006**, *53*, 1842–1852. [CrossRef]
13. Afrandideh, S.; Haghjoo, F.; Cruz, S.; Eshaghi Milasi, M. Detection of Turn-to-Turn Faults in the Stator and Rotor of Synchronous Machines During Startup. *IEEE Trans. Ind. Electron.* **2021**, *68*, 7485–7495. [CrossRef]
14. Gandhi, A.; Corrigan, T.; Parsa, L. Recent Advances in Modeling and Online Detection of Stator Interturn Faults in Electrical Motors. *IEEE Trans. Ind. Electron.* **2011**, *58*, 1564–1575. [CrossRef]
15. Briz, F.; Degner, M.W.; Garcia, P.; Diez, A.B. High-Frequency Carrier-Signal Voltage Selection for Stator Winding Fault Diagnosis in Inverter-Fed AC Machines. *IEEE Trans. Ind. Electron.* **2008**, *55*, 4181–4190. [CrossRef]
16. Pietrzak, P.; Wolkiewicz, M. Comparison of Selected Methods for the Stator Winding Condition Monitoring of a PMSM Using the Stator Phase Currents. *Energies* **2021**, *14*, 1630. [CrossRef]
17. Wolkiewicz, M.; Tarchała, G.; Orłowska, T.; Kowalski, C.T. Online Stator Interturn Short Circuits Monitoring in the DFOC Induction-Motor Drive. *IEEE Trans. Ind. Electron.* **2016**, *63*, 2517–2528. [CrossRef]
18. Bazan, G.H.; Scalassara, P.R.; Endo, W.; Goedtel, A.; Palácios, R.H.C.; Godoy, W.F. Stator Short-Circuit Diagnosis in Induction Motors Using Mutual Information and Intelligent Systems. *IEEE Trans. Ind. Electron.* **2019**, *66*, 3237–3246. [CrossRef]
19. Mazzoletti, M.A.; Bossio, G.R.; De Angelo, C.H.; Espinoza-Trejo, D.R. A Model-Based Strategy for Interturn Short-Circuit Fault Diagnosis in PMSM. *IEEE Trans. Ind. Electron.* **2017**, *64*, 7218–7228. [CrossRef]
20. Neti, P.; Younsi, K.; Shah, M.R. A novel high sensitivity differential current transformer for online health monitoring of industrial motor ground-wall insulation. In Proceedings of the 2013 IEEE Energy Conversion Congress and Exposition, Denver, CO, USA, 15–19 September 2013; pp. 2493–2499. [CrossRef]
21. *IEC 60034-27-3:2015*; Rotating Electrical Machines—Part 27-3: Dielectric Dissipation Factor Measurements on Stator Winding Insulation of Rotating Electrical Machines. IEC Standards: Geneve, Switzerland, 2015; pp. 1–34.
22. *IEEE Std 286-2000*; IEEE Recommended Practice for Measurement of Power Factor Tip-Up of Electric Machinery Stator Coil Insulation. IEEE Standards Association: Piscataway, NJ, USA, 2000. [CrossRef]
23. Pascoli, G.; Hribernik, W.; Ujvari, G. A practical investigation on the correlation between aging and the dissipation factor value of mica insulated generator windings. In Proceedings of the International Conference on Condition Monitoring and Diagnosis, Beijing, China, 21–24 April 2008; pp. 268–271. [CrossRef]
24. Yang, J.; Lee, S.B.; Yoo, J.; Lee, S.; Oh, Y.; Choi, C. A Stator Winding Insulation Condition Monitoring Technique for Inverter-Fed Machines. *IEEE Trans. Power Electron.* **2007**, *22*, 2026–2033. [CrossRef]
25. Neti, P.; Grubic, S. Online Broadband Insulation Spectroscopy of Induction Machines Using Signal Injection. *IEEE Trans. Ind. Appl.* **2017**, *53*, 1054–1062. [CrossRef]
26. Niu, F.; Wang, Y.; Huang, S.; Wu, L.; Huang, X.; Fang, Y.; Yang, T. An Online Groundwall Insulation Monitoring Method Based on Transient Characteristics of Leakage Current for Inverter-Fed Motors. *IEEE Trans. Power Electron.* **2022**, *37*, 9745–9753. [CrossRef]
27. Jensen, W.R.; Strangas, E.G.; Foster, S.N. A Method for Online Stator Insulation Prognosis for Inverter-Driven Machines. *IEEE Trans. Ind. Appl.* **2018**, *54*, 5897–5906. [CrossRef]

28. Cao, S.; Niu, F.; Huang, X.; Huang, S.; Wang, Y.; Li, K.; Fang, Y. Time-Frequency Characteristics Research of Common Mode Current in PWM Motor System. *IEEE Trans. Power Electron.* **2020**, *35*, 1450–1458. [CrossRef]
29. Tsyokhla, I.; Griffo, A.; Wang, J. Online Condition Monitoring for Diagnosis and Prognosis of Insulation Degradation of Inverter-Fed Machines. *IEEE Trans. Ind. Electron.* **2019**, *66*, 8126–8135. [CrossRef]
30. Nussbaumer, P.; Vogelsberger, M.A.; Wolbank, T.M. Induction Machine Insulation Health State Monitoring Based on Online Switching Transient Exploitation. *IEEE Trans. Ind. Electron.* **2015**, *62*, 1835–1845. [CrossRef]
31. Ryu, Y.; Park, B.R.; Han, K.J. Estimation of High-Frequency Parameters of AC Machine From Transmission Line Model. *IEEE Trans. Magn.* **2015**, *51*, 8101404. [CrossRef]
32. Zheng, D.; Zhang, P. A Novel Method of Monitoring and Locating Stator Winding Insulation Ageing for Inverter-fed Machine based on Switching Harmonics. In Proceedings of the IEEE Energy Conversion Congress and Exposition (ECCE), Detroit, MI, USA, 11–15 October 2020; pp. 4474–4479. [CrossRef]
33. Bolgova, V.; Lefebvre, S.; Hlioui, S.; Boucenna, N.; Costa, F.; Leonov, A. Development of testing methods for winding turn-to-turn insulation of low voltage motors fed by PWM converters. In Proceedings of the 19th European Conference on Power Electronics and Applications (EPE'17 ECCE Europe), Warsaw, Poland, 11–14 September 2017; pp. 1–10. [CrossRef]
34. *IEEE Std 117-2015 (Revision of IEEE Std 117-1974)*; IEEE Standard Test Procedure for Thermal Evaluation of Systems of Insulating Materials for Random-Wound AC Electric Machinery. IEEE Standards Association: Piscataway, NJ, USA, 2016; pp. 1–34. [CrossRef]
35. Zanuso, G.; Peretti, L. Evaluation of High-Frequency Current Ringing Measurements for Insulation Health Monitoring in Electrical Machines. *IEEE Trans. Energy Convers.* **2022**, *37*, 2637–2644. [CrossRef]
36. Poncelas, O.; Rosero, J.A.; Cusido, J.; Ortega, J.A.; Romeral, L. Motor Fault Detection Using a Rogowski Sensor Without an Integrator. *IEEE Trans. Ind. Electron.* **2009**, *56*, 4062–4070. [CrossRef]
37. Mingotti, A.; Costa, F.; Peretto, L.; Tinarelli, R. Accuracy Type Test for Rogowski Coils Subjected to Distorted Signals, Temperature, Humidity, and Position Variations. *Sensors* **2022**, *22*, 1397. [CrossRef]
38. Tovar-Barranco, A.; López-de Heredia, A.; Villar, I.; Briz, F. Modeling of End-Space Convection Heat-Transfer for Internal and External Rotor PMSMs With Fractional-Slot Concentrated Windings. *IEEE Trans. Ind. Electron.* **2021**, *68*, 1928–1937. [CrossRef]
39. Briz, F.; Degner, M.W.; Garcia, P.; Bragado, D. Broken Rotor Bar Detection in Line-Fed IM Using Complex Wavelet Analysis of Startup Transients. *IEEE Trans. Ind. Appl.* **2008**, *44*, 760–768. [CrossRef]

Disclaimer/Publisher's Note: The statements, opinions and data contained in all publications are solely those of the individual author(s) and contributor(s) and not of MDPI and/or the editor(s). MDPI and/or the editor(s) disclaim responsibility for any injury to people or property resulting from any ideas, methods, instructions or products referred to in the content.

Article

Neural Inverse Optimal Control of a Regenerative Braking System for Electric Vehicles

Jose A. Ruz-Hernandez [1,*], Larbi Djilali [1], Mario Antonio Ruz Canul [1], Moussa Boukhnifer [2] and Edgar N. Sanchez [3]

1 Faculty of Engineering, Universidad Autonoma del Carmen, Campeche 24180, Campeche, Mexico
2 Université de Lorraine, LCOMS, 57000 Metz, France
3 Department of Electrical Engineering, Cinvestav Guadalajara, Av. del Bosque 1145, Col. El Bajío, Zapopan 45019, Jalisco, Mexico
* Correspondence: jruz@pampano.unacar.mx

Abstract: This paper presents the development of a neural inverse optimal control (NIOC) for a regenerative braking system installed in electric vehicles (EVs), which is composed of a main energy system (MES) including a storage system and an auxiliary energy system (AES). This last one is composed of a supercapacitor and a buck–boost converter. The AES aims to recover the energy generated during braking that the MES is incapable of saving and using later during the speed increase. To build up the NIOC, a neural identifier has been trained with an extended Kalman filter (EKF) to estimate the real dynamics of the buck–boost converter. The NIOC is implemented to regulate the voltage and current dynamics in the AES. For testing the drive system of the EV, a DC motor is considered where the speed is controlled using a PID controller to regulate the tracking source in the regenerative braking. Simulation results illustrate the efficiency of the proposed control scheme to track time-varying references of the AES voltage and current dynamics measured at the buck–boost converter and to guarantee the charging and discharging operation modes of the supercapacitor. In addition, it is demonstrated that the proposed control scheme enhances the EV storage system's efficacy and performance when the regenerative braking system is working. Furthermore, the mean squared error is calculated to prove and compare the proposed control scheme with the mean squared error for a PID controller.

Keywords: electric vehicles; regenerative braking; inverse optimal control; buck–boost converter; neural identifier

Citation: Ruz-Hernandez, J.A.; Djilali, L.; Ruz Canul, M.A.; Boukhnifer, M.; Sanchez, E.N. Neural Inverse Optimal Control of a Regenerative Braking System for Electric Vehicles. *Energies* **2022**, *15*, 8975. https://doi.org/10.3390/en15238975

Academic Editor: Joao L. Afonso

Received: 1 October 2022
Accepted: 18 November 2022
Published: 28 November 2022

Publisher's Note: MDPI stays neutral with regard to jurisdictional claims in published maps and institutional affiliations.

Copyright: © 2022 by the authors. Licensee MDPI, Basel, Switzerland. This article is an open access article distributed under the terms and conditions of the Creative Commons Attribution (CC BY) license (https://creativecommons.org/licenses/by/4.0/).

1. Introduction

Electric vehicles (EVs) have demonstrated, in the present day, their importance in the solution of the environmental impact generated by conventional vehicles, such as air pollution and CO_2 emissions, and economic issues such as the gasoline prices [1]. The facility of using energy stored from the conversion of kinetic and potential energy into electrical energy only by changing the operation mode of an electrical motor to use it as a generator is one of the advantages of EVs that improve the driving performance and the life of the storage system [2]. One proposal to enhance EVs' driving range is the use of range extenders such as internal combustion engines, free-piston linear generator, fuel cells, micro gas turbines, and zinc–air batteries [3]. However, many disadvantages have been found such as the nonreduction of gas emission in some combinations, and the hard accessibility to some of the proposed range extenders. On the other hand, hybrid vehicles have been presented as an alternative to improve the performance of EVs by combining an internal combustion engine with an electric motor and reducing the emission of polluting gases [4]. Another proposed solution is the use of an external EV charger to administer energy to the battery bank. This type of EV is called a plug-in hybrid electric vehicle [5]. The operation of hybrid vehicles offers many advantages which also come with many challenges to ensure the switching between both installed supply systems because of the combination of many technologies; as a result, these complex hybrid controllers are required [6]. The use of fully EV technology combining the main energy system (MES) and an auxiliary energy system (AES) can reduce the challenges described above [7]. The AES contains battery banks, supercapacitors, and power electronic devices, which improves the efficiency of these systems because of the latest advancements in MOSFETs [8].

In recent years, the regenerative braking capability in EVs has been one of the most important characteristics because it helps to improve the operation and efficacy of the regenerative braking system in electric vehicles [9]. As a result, numerous ideas have been put forth to achieve greater performance when operating EVs in a variety of scenarios where the main dynamics in storage systems are controlled. Due to the increasing production and demand of EVs on a global scale, studies have demonstrated that regenerative braking is an excellent strategy for energy conservation because it can retain any energy lost during an electric vehicle's braking.

Lately, many control strategies have been developed in different regenerative braking architectures. A case study created in [10] considered a unilateral boost operation connected to a DC motor and simulated the switched operation of the converter produced, which was mainly comprised of IGBT bridges. A Lyapunov stability analysis was applied to ensure the system's stability, and a proposed switching control law was implemented to achieve robust control.

In [11], a model predictive control was employed to manage the torque distributions, optimizing the hydraulic braking and motor torque, maximizing the regenerative braking system, and enhancing the energy storage system. This application of regenerative braking was explored with Simulink's AMESim software and was utilized to model the proposed control strategy and analyze various driving scenarios. Additionally, a real-time test was executed showing positive results.

The improvement of the driving range and battery extended life cycle was demonstrated in [12] using a regenerative braking architecture consisting of a three-phase induction motor powered by a DC–DC buck–boost converter connected in parallel with a lithium-ion battery and a supercapacitor, where the current dynamics of the regenerative braking mode were controlled by a PI controller. Additionally, a three-phase inverter and the braking forces produced by the traction on the EV's wheels were approximated using an artificial neural network (ANN).

In [13], to recover the energy wasted during the deceleration, a regenerative braking system composed of an ultracapacitor pack and battery was designed, obtaining an improvement in the efficiency of the regenerative braking in comparison with a standalone battery system because of the additional ultracapacitor pack. In [14], a PI controller was implemented with the same design as mentioned above to regulate the buck–boost converter output voltage. Using the exponential reaching law and a parameter optimization, a fuzzy logic sliding mode controller was implemented in [15] to keep the optimal slip value for an antilock braking system in an EV. Comparing the fuzzy sliding mode control in [15] and the fuzzy one in [16] with an intelligent sliding mode controller employed to track the desired slip during braking implemented in [17], the energy recuperation was improved considerably without overcharging the battery. Recently, nonlinear control algorithms, such as the inverse optimal, feedback linearization, and sliding mode, have been implemented in electrical drives, win systems, and biomedical applications among others.

In [18], inverse optimal control (IOC) was implemented to regulate the voltage of a DC–DC converter and compared with a PID controller under the same conditions resulting in better performance with the IOC. In [19], the same control scheme was proposed to ensure the tracking of the desired trajectory of an induction motor and to avoid the instability generated by disturbances. In [20], inverse optimal control was used in a feedback stochastic nonlinear system and it was proved that the asymptotic stability was guaranteed for the probability of control systems. However, the controllers previously mentioned require previous knowledge of the system parameters since the analysis of the control algorithms is based on the mathematical models of the controlled system and these are not always easy to access in real operations. Additionally, their robustness and stability are not assured in the presence of disturbances [21].

The advances in technology create a need to solve problems presented in systems with complex, unknown dynamics, and highly coupled behavior. Engineers should make use of mathematical tools to solve these control problems. Neural networks are widely implemented to obtain a mathematical model approximating the unknown dynamics and use this information as the base to implement a conventional control algorithm. Different control problems have been resolved by using neural control such as in biomedical applications [22], microgrids [23], and in multiagent stabilization systems [24]. Nevertheless, this neural control is not widely implemented on regenerative braking systems for EVs [25].

This paper presents neural inverse optimal control (NIOC) for a regenerative braking system implemented in EVs. The proposed controller is used to regulate the current and voltage of the buck–boost converter related to the AES to recover the wasted energy during braking and enhance the MES's efficiency. The main contributions of the present paper are: (1) An online-identification-based recurrent high order neural network (RHONN) trained by an extended kalman filter (EKF) as a build-up to approximate the DC buck–boost behaviors. (2) Based on the obtained neural model, the inverse optimal control strategy is synthesized and implemented to track the buck–boost current and voltage desired dynamics. (3) Since the proposed controller is based on a neural identifier, robustness to parameter variations and disturbances is ensured. (4) To verify stability and robustness of the proposed control scheme, a comparison with the conventional PID controller is implemented. (5) By the implementation of the proposed controller for the AES, the storage of energy in the MES has more efficiency, and the loss of energy is largely reduced in comparison with a standalone MES.

The rest of the paper is organized as follows: In Section 2, the material and methods used in the article are described and the steps followed to structure this paper are briefly explained to get the major idea and process of this work. In Section 3, the regenerative braking problem is described. In addition, the buck and boost operation of the buck–boost converter is explained. In Section 4, mathematical preliminaries are introduced where the fundamentals of the corresponding equations used to develop the system identification and proposed control scheme for the regenerative braking system are presented. In Section 5, the buck–boost converter system modeling, and DC motor mathematical modeling are described. In Section 6, the neural controller design is presented. Additionally, the design of the reference generator and the DC motor control equations are presented. Section 7 illustrates the simulation results for the different steps implemented in the article where the validation of the neural controller with and without the regenerative braking system is shown. Furthermore, the robustness test is implemented where the results are compared with a PID controller and illustrated not only graphically but with results obtained from the mean squared error. Finally, Section 8 is the conclusion of the article where the obtained results are discussed and future work is proposed.

2. Materials and Methods

The method used to achieve the results obtained in this article follows the next steps:

- The goal of this article is to improve the regenerative braking system of an electric vehicle. The element of that system that allows the control of the current and voltage variables is the buck–boost converter. The validation and simulation of the proposed controller and regenerative braking system are implemented using the SimPower System toolbox of Matlab (Matlab, Simulink. de 1994–2022, ©The Math Works, Inc.).
- A mathematical model of the buck–boost converter [26] is used to develop the RHONN equations as in [27].
- After the RHONN equations are acquired, the extended Kalman filter is used to train the identifier and estimate the values of the dynamics in the buck–boost converter. The validation is illustrated in Figures 1 and 2.

- The trained RHONN allows the design of the neural controller. In our case, it is a neural inverse optimal controller.
- The validation of the proposed control scheme is to track the proposed time-varying trajectories without connecting the complete regenerative braking system. These results are illustrated in Figures 3–5.
- After the control scheme is validated, the design of a reference generator is developed. This reference generator provides the value in volts within which the buck–boost converter must operate during a driving operation. This signal is generated through the motor's DC dynamics, which are regulated using a PID controller.
- Once the whole regenerative braking system is connected (battery bank, supercapacitor and buck–boost converter, DC motor, etc.) the correct operation of the regenerative braking system is validated.
- From this validation the controlled variables, the better performance in the state of charge of the battery bank, and the correct operation of the supercapacitor charge and discharge operation modes are illustrated in Figures 6–11.
- Lastly, the robustness test is implemented by comparing the performance of the controller with a classic PID controller. In addition, not only the graphic results are demonstrated in Figures 11–13 but the mean squared error is calculated to validate the result obtained.

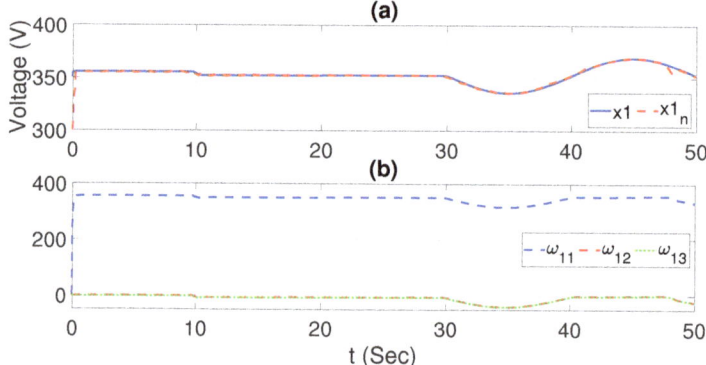

Figure 1. Voltage identification (**a**) and NN's weights (**b**).

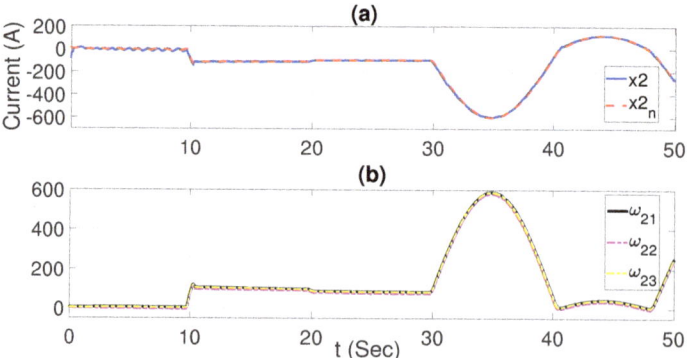

Figure 2. Current identification (**a**) and NN's weights (**b**).

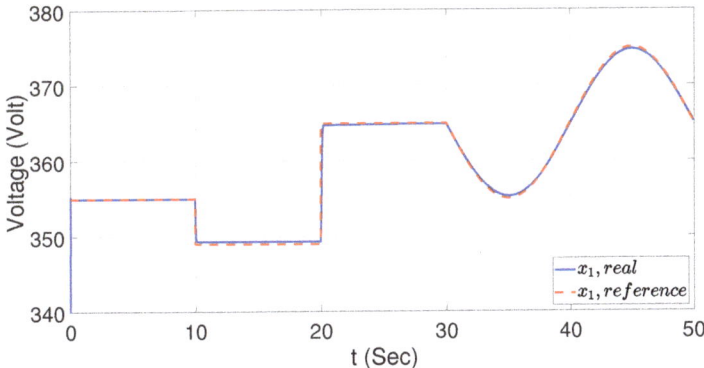

Figure 3. AES voltage trajectory tracking.

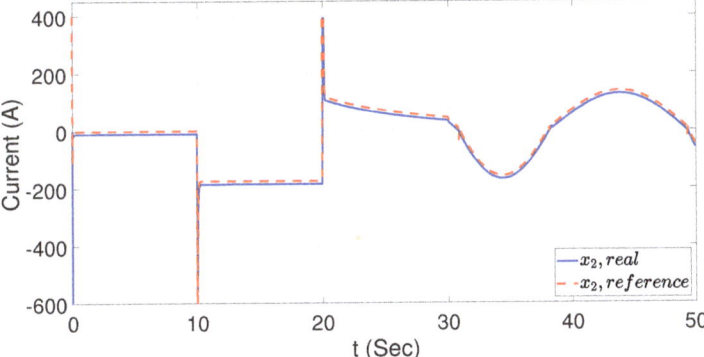

Figure 4. AES current trajectory tracking.

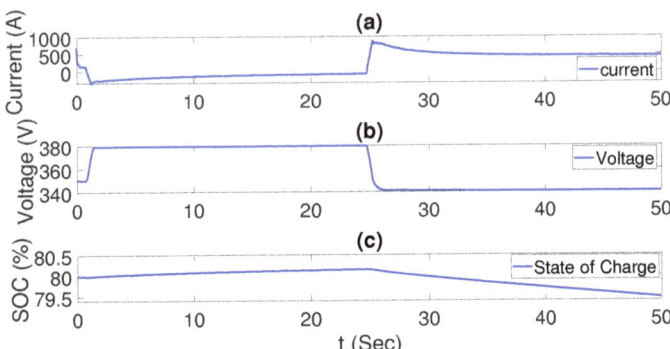

Figure 5. AES charging during trajectory tracking. (a) Illustrate the obtained current during the tracking operation, (b) the obtained voltage during the tracking operation and (c) the state of charge of the supercapacitor during the tracking operation. and discharging.

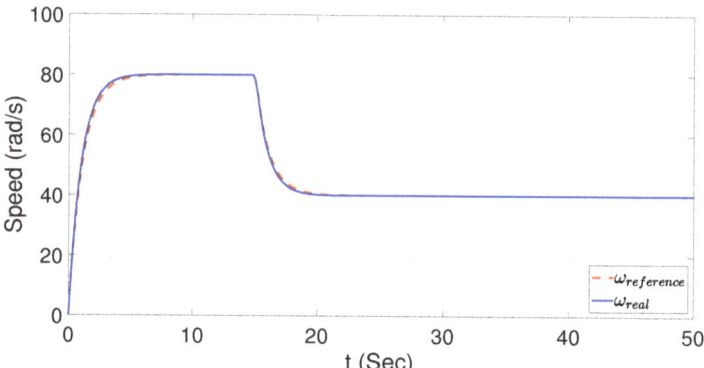

Figure 6. Motor speed control.

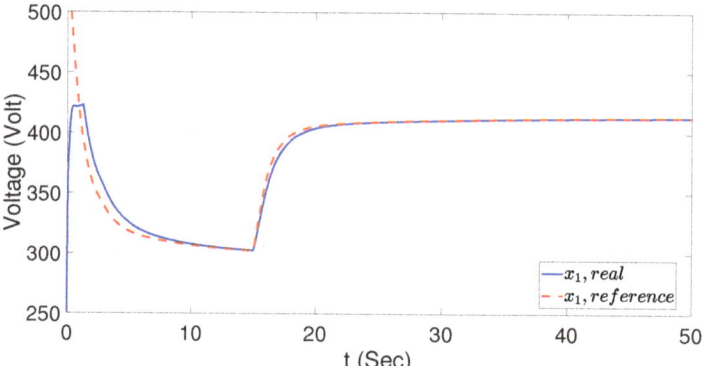

Figure 7. AES voltage control during regenerative braking.

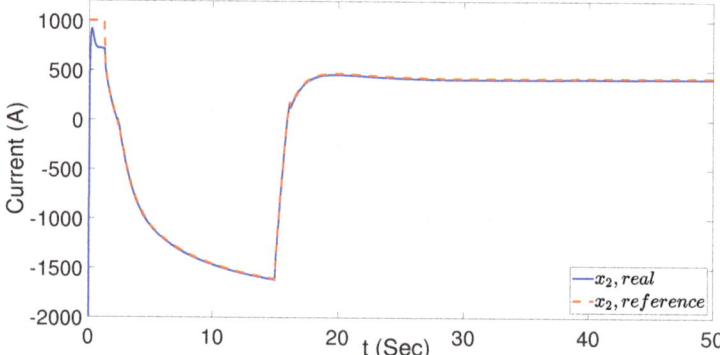

Figure 8. AES current control during regenerative braking.

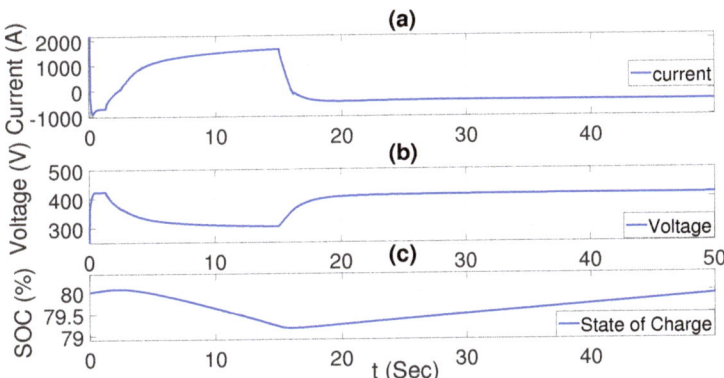

Figure 9. AES supercapacitor SOC. (**a**) Illustrate the obtained current during the vehicle operation, (**b**) the obtained voltage during the vehicle operation and (**c**) the state of charge of the supercapacitor during the operation.

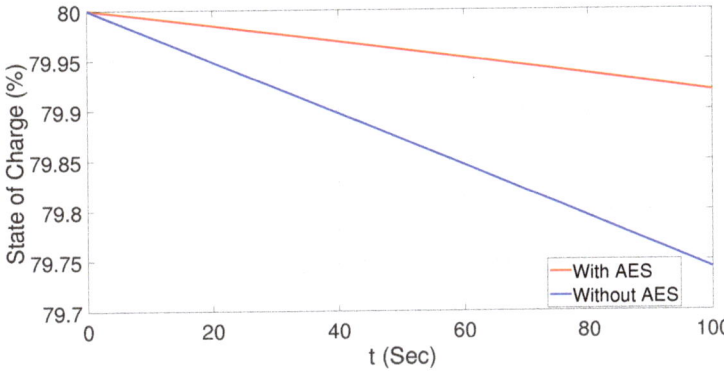

Figure 10. MES battery bank SOC comparison with and without AES.

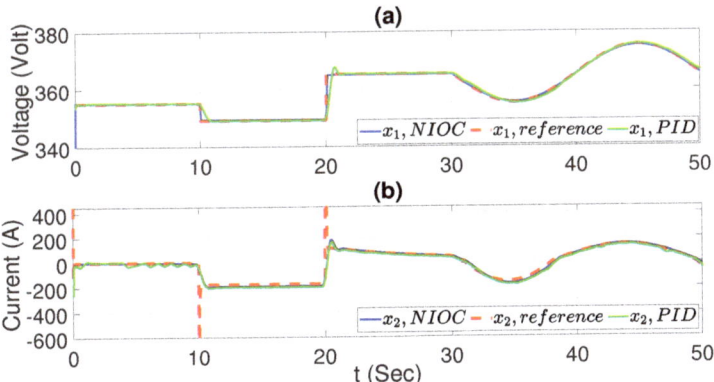

Figure 11. Influence of R changes on PI and NIOC: (**a**) voltage, (**b**) current.

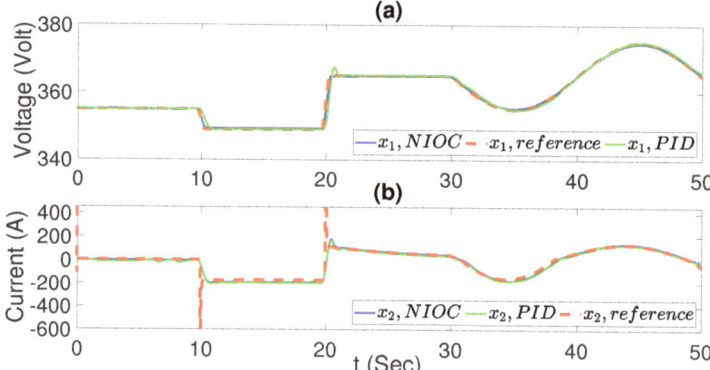

Figure 12. Influence of L changes on PI and NIOC: (**a**) voltage, (**b**) current.

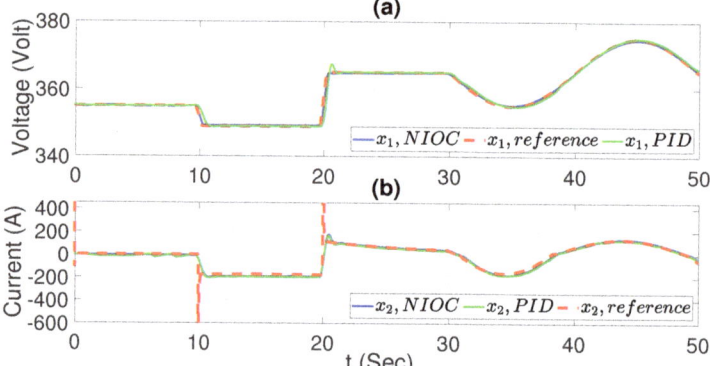

Figure 13. Influence of C changes on PI and NIOC: (**a**) voltage, (**b**) current.

3. Regenerative Braking Description

A regenerative braking system as depicted in Figure 14 allows the recovery of kinetic energy produced during braking and its utilization to improve the energy storage efficiency and extend the operating distance of the EV [2]. This system is composed of a supercapacitor and buck–boost converter, which are part of the AES. In addition, a battery bank is used to administer the energy to the electrical motor contained in the MES. The supercapacitor and the buck–boost converter are connected as illustrated in Figure 15, with the objective of increasing or decreasing the output voltage depending on the following operation modes.

Buck operation: In this mode, the output voltage is decreased regarding the input voltage. To achieve this, T1 is off and T2 is activated, then, the energy is transferred from the capacitor (V_c) to the supercapacitor voltage (V_{sc}). At the moment T2 is turned on, current flows from the capacitor C, generating current I_c to the supercapacitor. As a result, a fraction of this energy is charged into inductance L. On the other hand, when T2 is turned off, the current charged in L is discharged into V_c through diode D1, driving the current in the direction of capacitor C [14].

Boost operation: On the other hand, in this mode, the output voltage is increased. To do so, T2 is deactivated and T1 is activated to transfer energy from supercapacitor Vsc to battery bank Vc. When T1 is on, the energy is acquired from the capacitor, and stored in inductance L. Reversely, when T1 is OFF, the energy stored in the inductance is transferred into the capacitor through diode D2, and kept in the battery bank.

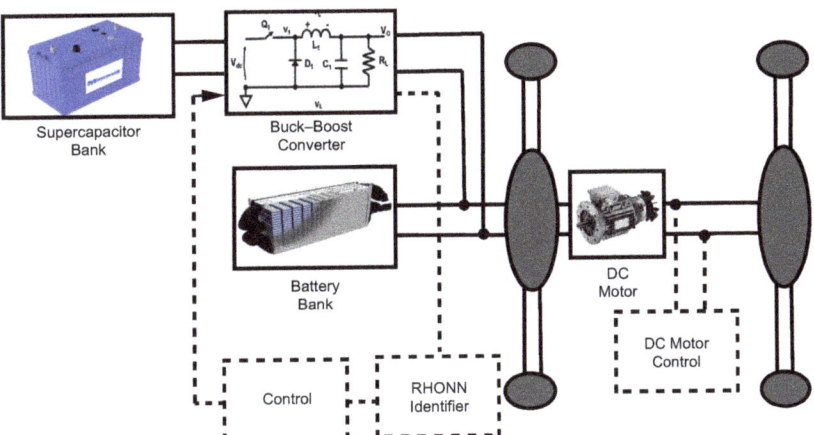

Figure 14. Regenerative braking system topology.

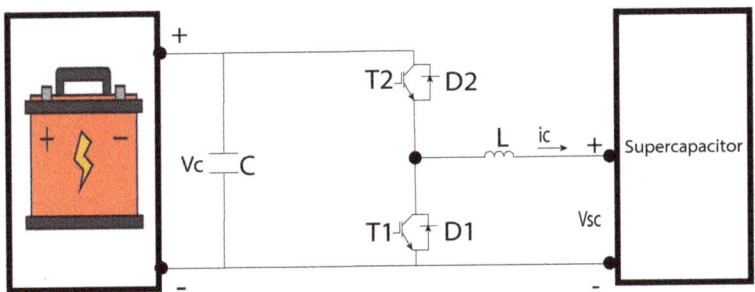

Figure 15. Buck–boost converter topology.

During the braking operation, the brake manages the electricity generated by the motor into the batteries or capacitors. The DC–DC converter operates in boost function during acceleration while it operates into buck function in deceleration, which makes it easier to charge up the supercapacitor.

4. Mathematical Preliminaries

4.1. Discrete-Time Inverse Optimal Control

Consider the following perturbed discrete-time nonlinear system [28]

$$x_{k+1} = f(x_k, k) + B(x_k)u(x_k, k) + d(x_k) \tag{1}$$

$$y_k = h(x_k) \tag{2}$$

with x_k the state variable, u_k the input vector, y_k the output vector to be controlled, $f(x_k)$, $B(x_k)$, and $h(x_k)$ smooth and bounded vectors. Considering y_k contains the full state vector, the objective is to force the controlled dynamics to track selected trajectories, then the tracking error is as follows

$$e_{k+1} = x_{k+1} - x_{ref,k+1} \tag{3}$$

with $x_{ref,k}$ the desired trajectory vector. The error dynamics at $k+1$ is expressed by

$$e_{k+1} = f(x_k, k) + B(x_k)u(x_k, k) + d(x_k) - x_{ref,k+1} \tag{4}$$

For the optimal problem solution, the cost function is minimized by solving the Hamilton–Jacobi–Bellman (HJB) partial differential equation (PDE). However, in some cases the solution of these classes of equations is difficult to obtain [29]. For the tracking trajectory, the cost function of system (4) is selected as

$$J(e_k) = \sum_{k=0}^{\infty} (l(e_k) + u - k^T R u_k) \tag{5}$$

where $J: \Re^n \to \Re^+$ is a performance measure, $l: \Re^n \to \Re^+$ is a positive semidefinite function, and $R: \Re^n \to \Re^{n \times m}$ is a positive real symmetric matrix. When the cost function J is optimal, it is noted as $J*$ and it is defined as Lyapunov function $V(e_k)$, which is time-invariant and should satisfy the discrete-time Bellman equation defined as follows

$$V(e_k) = \min_{u_k} l(e_k) + u_k^T R(e_k) u_k + V(e_{k+1}) \tag{6}$$

Hence, the discrete-time Hamiltonian equation is expressed as follows

$$H(e_k, u_k) = l(e_k) + u_k^T R(e_k) u_k + V(e_{k+1}) - V(e_k) \tag{7}$$

The optimal control law is obtained using $H(e_k, u_k) = 0$, and the gradient of (7)'s right-hand side is calculated with respect to u_k [28], then

$$u_k^* = -\frac{1}{2} R(e_k)^{-1} B(x_k)^T \frac{\partial V(e_{k+1})}{\partial e_{k+1}} \tag{8}$$

where $V(0) = 0$ is the boundary condition of $V(e_k)$ which should be satisfied and u_k* is the optimal control law. Using (8) in (6), the discrete-time HJB equation is

$$V(e_k) = \frac{1}{4} \frac{\partial V^T(e_{k+1})}{\partial e_{k+1}} R(e_k)^{-1} B(x_k)^T \frac{\partial V(e_{k+1})}{\partial e_{k+1}} + l(e_k) + V(e_{k+1}) \tag{9}$$

Determining the solution of the HJB PDE (9) for $V(e_k)$ is not trivial. To do so, the discrete-time inverse optimal control (IOC) technique and a Lyapunov function are used to synthesize the respective control law [29,30]. To state the above problem as an IOC one, the following definition is established.

Definition 1 ([28]). *For system (1), the control law in (8) is considered to be IOC (globally) stabilizing if:*

(1) *It ensures that (8) has (global) asymptotic stability for $e_k = 0$;*
(2) *It minimizes the cost function (5) for which $V(e_k)$ is positive definite function such that*

$$\overline{V} := V(e_{k+1}) - V(e_k) + u_k^* B(x_k) u_k^* \leq 0. \tag{10}$$

Thus, the IOC synthesis is based on $V(e_k)$ from the previous definition. Then,

Definition 2 ([28]). *Let us select $V(x_k)$, which is established to be a radially bounded positive definite function such that for each x_k there exist u_k and*

$$\Delta V(e_k, u_k) < 0 \tag{11}$$

where $V(e_k)$ is a discrete-time control Lyapunov function (CLF), which should be defined to satisfy conditions (1) and (2) of Definition 1. Thus, the CLF is selected as follows

$$V(e_k) = \frac{1}{2} e_k^T P(e_k) \tag{12}$$

with $P \in \Re^{n \times n}$ and $P = P^T > 0$. By selecting an appropriate matrix P, the control signal (8) guarantees the equilibrium point $e_k = 0$ of (4)'s stability. Additionally, the control law (8) with (12), which is considered as an inverse optimal control law for (1), optimizes the meaningful cost function in (5). Moreover, by using (8) in (12), the IOC law is established as follows:

$$u_k^* = \frac{1}{2}\left(R + \frac{1}{2}B(x_k)^T P B(x_k)\right)^{-1} B(x_k)^T P(f(x_k) - x_{ref,k+1}) \tag{13}$$

where P and R are positive definite matrices. Details about the NIOC synthesis is explained in [28]. To achieve adequate performance of the discrete-time IOC scheme, a priori knowledge of the model parameters is requested, which is not always fulfilled in real-time applications. In addition, since this control scheme is based on a mathematical model, robustness to parameters variations and disturbances cannot be ensured. To improve it, an RHONN identifier trained online with an EKF is proposed.

4.2. Discrete-Time Recurrent High-Order Neural Networks

In these last years, recurrent neural networks have been implemented to identify and approximate the mathematical models of complex systems [21]. The RHONN has demonstrated that is a good choice in nonlinear system identification, which consists of adjusting the parameters of an appropriately selected model according to an adaptive law. Using a series–parallel configuration, the estimated state variable of a nonlinear system using an RHONN identifier is given by [27]

$$\chi_{i,k+1} = \omega_i^T \phi_i(x_k) + \overline{\omega}_i^T \varphi_i(x_k, u_k) \tag{14}$$

where $\chi_{i,k+1}$ is the state of the ith neuron which identifies the ith component of x_k, $x_k = [x_{1,k}, ..., x_{n,k}]$ is the state vector, $\omega_{i,k} \in \Re^{L_i}$ are the adjustable synaptic weights of the NN, $\omega_{i,k}$ represent the adjustable weights, and $\overline{\omega}_{i,k}$ are the fixed weights, φ_i is a linear function of the state vector or vector input u_k depending to the system structure or external inputs to the RHONN model, and $u \in \Re^m$ $u = [u_{1,k}, u_{2,k}, \ldots, u_{m,k}^T]$ is the input vector to the network. The function $S(.)$ is a hyperbolic tangent function defined as

$$S(x_k) = \alpha_i \tanh(\beta_i x_k) \tag{15}$$

where x_k is the state variable; α and β are positive constants. Figure 16 illustrates the ith RHONN identifier scheme.

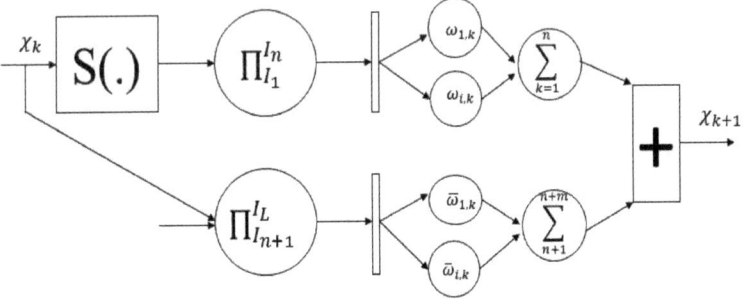

Figure 16. RHONN identifier scheme.

To train the proposed RHONN identifier, an EKF is used. The algorithm model is defined as follows

$$w_{i,k+1} = w_{i,k} + \eta_i K_{i,k} e_{i,k} \tag{16}$$
$$K_{i,k} = P_{i,k} H_{i,k} M_{i,k} \tag{17}$$
$$P_{i,k+1} = P_{i,k} - K_{i,k} H_{i,k}^T P_{i,k} + Q_{i,k} \tag{18}$$
$$e_{i,k} = x_{i,k} - \chi_{i,k} \quad i = 1, 2, \cdots, n \tag{19}$$
$$M_{i,k} = \left[R_{i,k} + H_{i,k}^T P_{i,k} H_{i,k} \right]^{-1} \tag{20}$$

with $e_i \in \mathbb{R}$ the identification error to be minimized, η_i a training algorithm design parameter, $K_{i,k} \in \mathbb{R}^{L_i \times m}$ the Kalman matrix, $Q_{i,k} \in \mathbb{R}^{L_i \times L_i}$ and $R_{i,k} \in \mathbb{R}^{m \times m}$ positive definite constant matrices, $P_i \in \mathbb{R}^{L_i \times L_i}$ an adjustable diagonal matrix, and $H_i \in \mathbb{R}^{L_i \times m}$ an adjustable matrix defined as the state derivative with respect to the neural identifier's adjustable weights. Details of the RHONN identifier and the respective EKF training algorithm, including a stability proof is explained in [21,31].

5. System Modeling and Neural Control
5.1. Buck–Boost Model

The used DC–DC converter in this application was composed of boost and buck converters. The first one is used under charge conditions while the second one is used under discharge conditions. The boost converter model is defined as [26]

$$x_{1,k} = (1 - \frac{ts}{RC}) x_{1,k} - \frac{ts}{C} x_{2,k} \tag{21}$$
$$x_{2,k} = x_{2,k} + \frac{ts}{L} U_{btt} u_c \tag{22}$$

The buck converter model is given by [26]

$$x_{1,k} = (1 - \frac{ts}{RC}) x_{1,k} + \frac{ts}{C} x_{2,k} \tag{23}$$
$$x_{2,k} = x_{2,k} + \frac{ts}{L} U_{btt} u_c \tag{24}$$

where $x_{1,k}$ is the converter output voltage, $x_{2,k}$ is the output current, U_{btt} is the battery voltage, u_c is the input vector, L is the inductance (H), R is the load resistance (Ω), C is the capacitor (F), and t_s is the sample time.

5.2. DC Motor

To illustrate the performance of the regenerative braking and for system completeness, a DC Motor was used as a drive system of the EV [14]. The DC machine's dynamics are governed by two attached first-order equations concerning the armature current and angular velocity as in [32]. The mathematical model is defined by [32]:

$$L \frac{di}{dt} = u - Ri - \lambda_0 \tag{25}$$
$$J \frac{d\omega}{dt} = k_t i - \tau_l \tag{26}$$

where i is the armature current (A), u is the terminal voltage (V), ω is the angular velocity (rad/s), J is the inertia of the motor rotor and load (kg m^2), R is the armature resistance (ω), L is the armature inductance (H), λ_0 is the back electromotive force (EMF) constant, k_t is the torque constant, and τ_l is the load torque.

6. Neural Controller Design

To approximate the used buck–boost power converter's dynamics, an RHONN identifier trained online by an EKF was employed, then based on the obtained model, the IOC was synthesized to manage the current flow and ensure the charging and discharging operating modes of the AES. Due to the similarity between the buck and boost converter models and the adaptive nature of the RHONN, a single neural identifier is proposed for both cases as follows

$$\begin{aligned}\hat{x}_{1,k} &= \omega_{1,1}(k)S(x_1) + \omega_{1,2}(k)S(x_2) \\ &\quad + w_{1,3}S(x_1)S(x_2) + \varpi_1 x_2 \end{aligned} \quad (27)$$

$$\begin{aligned}\hat{x}_{2,k} &= \omega_{2,1}(k)S(x_2) + \omega_{2,2}(k)S(x_1) \\ &\quad + w_{2,3}S(x_1)S(x_2) \end{aligned} \quad (28)$$

Using the compact form, (27) and (28) can be rewritten as follows

$$\hat{x}_{k+1} = \hat{F}(x_k) + \hat{B} u_k^* \quad (29)$$

$$\hat{y}_k = x_{2,k} \quad (30)$$

where $[\hat{x}_{1,k+1}, \hat{x}_{2,k+1}]^T$ are the estimated dynamics of $[x_{1,k}, x_{2,k}]^T$, u_k is the input signal, $\hat{y}_k$ is the output to be tracked, and $\hat{B}$ is the control matrix defined as $\hat{B} = diag[0, \varpi_2]$. For the controller design, the proposed controller was carried out for the current trajectory tracking. The current tracking error at $k + 1$ was obtained as

$$\begin{aligned}\hat{e}_{k+1} &= \omega_{2,1}(k)S(x_2) + \omega_{2,2}(k)S(x_1) \\ &\quad + w_{2,3}S(x_1)S(x_2)\varpi_2 u_k - x_{ref,k+1} \end{aligned} \quad (31)$$

Then, the equivalent NIOC was calculated using the same steps as in Section 4.1 as follows

$$u_k^* = \frac{1}{2}\left(R + \frac{1}{2}B(x_{2,k})^T P B(x_{2,k})\right)^{-1} B(x_{2,k})^T P e_{k+1} \quad (32)$$

where P and R are positive definite matrices.

Reference Generator Development

To define the buck–boost current desired value, a current reference generator was developed. The charge reference was defined as the energy contained in the supercapacitor as a function of the energy generated by the DC motor. Considering the work and energy theorem **"the work done between point A and point B on a particle results on the increase of its kinetic energy"** cited in [33], the following expression can be written

$$W_{A \to B} = \int_A^B P dt \quad (33)$$

Using the work and energy theorem on the DC motor, the energy can be estimated for an interval time $t \in [k\delta, (k+1)\delta)]$ as [14]

$$E(t) = \int_{k\delta}^{t} P_k d\varsigma + E_k \quad (34)$$

where the DC motor power can be estimated as $P = \tau_e \omega$. The energy in the supercapacitors during the charging mode can be estimated as

$$E_C^{ref-c}(t) = -k_p \int_{k\delta}^{t} sat_1(P) d\varsigma + E_{ck} \quad (35)$$

and during the discharging mode, it can be given as

$$E_C^{ref-d}(t) = k_p \int_{k\delta}^{t} sat_2(P)d\xi + E_{ck} \tag{36}$$

with $E_C^{ref-c}(t)$, $E_C^{ref-d}(t)$ and $0 < k_p < 1$ as the charge reference, discharge reference, and a constant representing the lost energy during transformation, respectively. sat_1 represent a saturation function in the range of $(-\infty, 0)$ and sat_2 the same function in the range of $(0, \infty)$. The energy reference for the supercapacitor is estimated as the sum result of both references.

$$E_C^{ref} = E_C^{ref-c} + E_C^{ref-d} \tag{37}$$

Then, the buck–boost voltage reference can be obtained using the last supercapacitor energy equation resolved for the voltage

$$V_{cr} = \sqrt{\frac{2E_C}{C}} \tag{38}$$

where V_{cr} is the reference voltage in the supercapacitor, E_c is the energy stored, and C is the capacitance.

To track the buck–boost voltage, the NIOC scheme was applied. Using (27), the voltage error at $k+1$, evc_{k+1} was calculated as follows

$$\begin{aligned} e_{Vcr,k+1} &= \omega_{1,1}(k)S(x_1) + \omega_{1,2}(k)S(x_2) \\ &+ w_{1,3}S(x_1)S(x_2) + \omega_1 x_2 - V_{cr} \end{aligned} \tag{39}$$

Then, the NIOC was applied to determine the buck–boost current reference for the supercapacitor as follows

$$i_{cr} = \frac{1}{2}\left(R + \frac{1}{2}B(x_{1,k})^T PB(x_{1,k})\right)^{-1} B(x_{1,k})^T P e_{Vcr,k+1} \tag{40}$$

The control scheme for the regenerative braking system using NIOC and the current reference generator is illustrated in Figure 17.

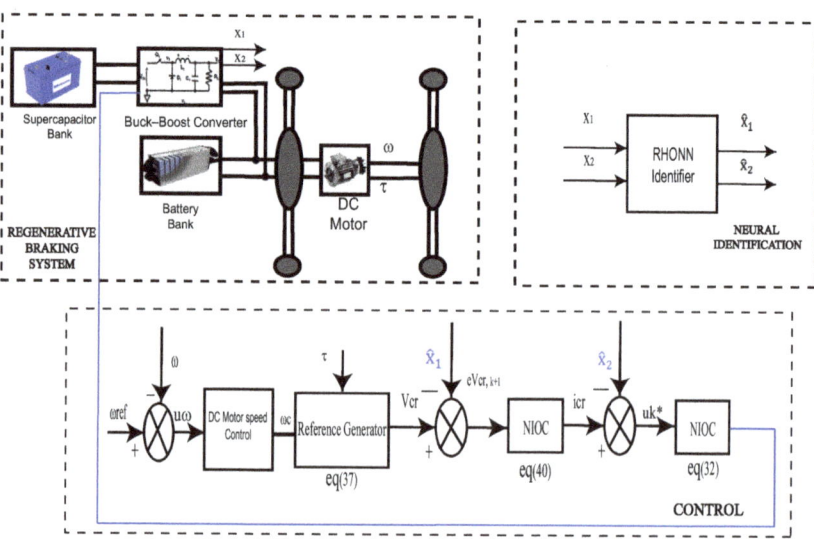

Figure 17. Regenerative braking system control scheme.

The DC motor speed was controlled using a PI controller to track a desired reference

$$u_\omega = k_{\omega_l p}(\omega_{l,ref} - \omega_l) + k_{\omega_l i} \int_0^t (\omega_{l,ref} - \omega_l) \tag{41}$$

where $\omega_{l,ref}$ is the motor reference speed and $k_{\omega_l p}$ and $k_{\omega_l i}$ are proportional and integral controller gains, respectively.

7. Simulation Results

The proposed control scheme as well as the respective MES and AES were implemented and evaluated using the SimPower System toolbox of Matlab (Matlab, Simulink. de 1994–2022, ©The Math Works, Inc.). The parameters of the AES and MES are listed in Table 1.

Table 1. Parameters of the AES and MES.

Description	Unit
Converter resistance R.	$50\ \Omega$
Converter inductance L	13×10^{-3} H
Converter capacitance C_1	2×10^{-3} F
Converter capacitance C_2	1×10^{-6} F
Supercapacitor voltage V_{sc}	350 V
Battery bank voltage V_c	500 V
Initial SOC	80%
Sampling time (t_s)	1×10^{-5} s

7.1. Neural Identification

The realized RHONN identification allowed us to obtain a satisfactory estimation of the system states, which were, in this case, the voltage x_1, k and the current x_2, k during different operation modes. Figure 1 demonstrates the neural identification of the voltage (x_1, k) and its respective neural weights' changes during the operation of the time-varying signals.

Figure 2 presents the neural identification of the current (x_2, k) and its respective neural weights' dynamics that adjust over the operation with the time-varying trajectories.

From the obtained results, it is clear that the proposed RHONN identifiers successfully approximated the voltage and the current dynamics of the AES, even though time-varying trajectories were applied. In addition, all neural weights were bounded. This allowed a correct operation for the recognition of the dynamics of the buck–boost converter because the capability of identifying these variables over time-varying signals let the system operate in different conditions. Furthermore, the implementation of the neural controller did not represent a problem, with the neural identifier working correctly.

7.2. Buck–Boost Trajectories Tracking

In this test, the objective was to demonstrate the trajectories tracking of the AES voltage and current to validate the correct operation of the proposed NIOC scheme. A trajectory was proposed to show the dynamics regulation without connecting the whole regenerative braking system.

The proposed trajectory to be tracked was a time-varying signal, whose amplitude was changed between 340 V and 360 V. Figure 3 presents the obtained results for the voltage (x_1, k) at the output of the buck–boost converter used in the AES system controlled by the proposed NIOC scheme. As is shown, the tracking made by the controller operated correctly and followed the variations of the signal.

Figure 4 demonstrates the behavior of the current (x_2, k), as measured at the inductance of the buck–boost converter. As a result of the good performance of the neural controller for the voltage (x_1, k), the regulation and tracking of the current dynamics worked correctly.

Figure 5a–c display the current, voltage, and state of charge (SOC) of the AES, respectively. Different voltage values were applied to verify the charging and discharging operations of the AES. The voltage had an initial value of 380 V where the buck mode was activated and the supercapacitor was charging. Then, at 25 s, the voltage value was modified to 340 V resulting in the boost mode turning on and the supercapacitor discharging. The conclusion obtained with the voltage values was that the supercapacitor had the capacity to work over charging and discharging operations, which helped extend the life cycle and SOC of the battery when the whole system was connected.

From the obtained results, it is clear that the proposed control scheme (NIOC) ensured the tracking of the proposed trajectories for both the voltage and current of the AES. In addition, the charging and discharging operation modes of the AES were achieved.

7.3. Regenerative Braking System Trajectories Tracking

The objective of this section was to test the complete system functionality including the AES, MES, and the DC motor installed on the EV. The trajectories to be tracked were calculated using the reference generative block (38) where the motor speed was the input and the voltage reference (V_{cr}) was the output. This voltage output was used in the control loop of the AES where a cascade-controller-based NIOC scheme was used to regulate the voltage and the current of the buck–boost converter, respectively, with the objective to ensure the charge and discharge of the AES supercapacitor. The speed of the DC motor was controlled by a PI controller where a time-varying trajectory was tracked as presented in Figure 6.

Figure 7 presents the voltage's desired trajectory tracking, obtained from the reference generator block (38), using NIOC when the EV was fully operated. During acceleration, the voltage value decreased toward 340 V, the boost mode was activated, and the supercapacitor was discharging, allowing the AES's participation in the total EV's needed energy; as a result, the MES's charge duration was enhanced. However, during deceleration the voltage value increased to reach 350 V, the buck mode was turned on, and the supercapacitor was charging, which helped to recuperate the EV energy waste.

Figure 8 illustrates the current trajectory tracking using the proposed NIOC during the regenerative braking, where the reference trajectory was obtained from the NIOC voltage controller.

Figure 10 displays the SOC behavior of the MES battery bank without (blue) and with (red) the AES during the operation of the regenerative braking system. This demonstrated the enhancement of the battery operation using the regenerative braking system. The results obtained showed that the battery SOC decreased slowly with the AES in comparison to when the AES was not implemented.

The SOC of the AES as well the supercapacitor voltage and current are presented in Figure 9, where the supercapacitor is discharging when the EV is in an acceleration state and charging otherwise.

As results of this experiment, the proposed control scheme ensured the trajectory tracking of the AES voltage obtained from the reference regenerative block. The voltage value was automatically changed according to the acceleration or deceleration of the motor EV. In addition, the proposed controller achieved an adequate trajectory tracking of the AES current. On the other hand, the SOC of the MES was largely improved by using the proposed AES control methodology, which helped to recuperate the energy during deceleration and enhance the MES's charge duration. This demonstrated that the regenerative braking system had a good performance and operation when implementing the complete scheme as illustrated in Figure 17. However, it is necessary to add another test to prove even further the good operation of the EV architecture proposed in this article.

7.4. Robustness Test

In this test, the AES parameters were changed to examine the robustness of the proposed NIOC. In addition, a comparison with the classical PI controller was done to illustrate the potential of the proposed neural control scheme. The obtained results of the AES when varying the parameters are presented in Figures 11–13.

The goal of this test was to vary the nominal values of the components that integrate the AES parameters described before in Table 1 and demonstrate the capability of the neural controller to operate over changes in their conditions, which could be considered as parasitic signals that were not part of an ideal electric vehicle's system. Figure 11 illustrates the voltage and current trajectories tracking of the AES when resistor R was changed by 200% of its nominal value.

Figure 12 demonstrates the voltage and current trajectories tracking of the AES when inductance L was changed by 100% of its nominal value.

Figure 13 demonstrates the voltage and current trajectories tracking of the AES when resistor C was changed by 70% of its nominal value.

From the simulation results, we can observe that parameter variations had an important impact on the AES voltage and current controlled by the PI controller, with a high coupling between the control axes and a sluggish response time. However, the proposed controller (NIOC) ensured an adequate performance in the presence of parameter variations, the decoupling was ensured, and the response time was improved compared with that of the PI controller. From this test, we can consider that the proposed controller had better performance and was robust to AES parameter variations and these results were supported by the mean squared error calculated to validate the statement made with these results. Table 2 illustrates the results for the voltage control robustness test while Table 3 describes the mean squared error for the current dynamics. One of the disadvantages of the PID controller was its capacity to reach the trajectory desired, which meant the squared error was farther from zero in comparison with that of the NIOC.

Table 2. Mean squared error in x_1.

Mean Squared Error of Tracking Trajectories in x_1	
Controller	**Mean Value**
PID	16.026×10^{-11}
NIOC	4.8822×10^{-11}

Table 3. Mean squared error in x_2.

Mean Squared Error of Tracking Trajectories in x_2	
Controller	**Mean Value**
PID	140.290×10^{-9}
NIOC	2.2314×10^{-9}

8. Conclusions

This article presented a regenerative braking system for electrical vehicles controlled by a neural inverse optimal controller. The control scheme was used to regulate the dynamics of the AES composed of a buck–boost converter and a supercapacitor, with the objective to enhance the energy recovery during braking and to participate in the delivered MES's energy during acceleration. The proposed controller was developed using a recurrent high-order neural network identifier, and online training by the extended Kalman filter based algorithm, which allowed us to approximate the AES's behavior during the different operation modes. The validation of the correct identification of the dynamics with the RHONN was illustrated correctly with the results obtained in the simulation. This responded to one of the statements mentioned about the implementation and correct operation of recurrent high-order neural networks in nonlinear systems.

The neural controller test with the proposed time-varying trajectories helped to achieve the correct implementation and operation of the dynamics before the complete system was connected. This was considered because working with the complete regenerative braking system before the tuning of the controllers may present some issues that could easily be solved by analyzing the controller separately first. The controller was used to track the desired trajectories of the AES voltage and current, where a reference generator block was utilized to define the voltage's desired value considering the electrical vehicle operation modes. This reference generator block was very important because this helped to achieve the necessary current value for the correct operation of the regenerative braking system. It is important to note the effects that the DC motor had on the EV system and the good performance obtained by the neural controller. Additionally, the proposed controller was compared with the PI controller, regarding reference tracking and robustness against parameter variations. The obtained results illustrated the effectiveness of the proposed control scheme for the AES trajectory tracking even in the presence of time-varying references and disturbances. The mean squared error helped to get a better idea of the improvement that the neural controller presented over a PI controller in this case. The measure of the error showed by far the effectiveness of NIOC even in the presence of disturbances or undesired signals. In addition, the charging and discharging of the AES supercapacitor during acceleration and deceleration was ensured, which helped to recover the wasted energy during braking and to participate in the MES's power budget during acceleration; moreover, it increases the lifetime of the battery bank. As a result, the charge duration of the MES battery bank was largely enhanced, and the electric vehicle's efficiency and operation were improved. Finally, it is necessary to mention that a real-time implementation is very important to consider; thus, the validation of the proposed controller will let us know its real effectiveness in terms of real driving performance. Moreover, new approaches for the inverse optimal and another neural controller such as the neural sliding mode control could be the simulation of a fully electric vehicle model system, where more important variables such as temperature conditions are considered, and the controllers are validated during typical driving conditions.

Author Contributions: Conceptualization, L.D.; methodology, J.A.R.-H. and L.D.; validation, E.N.S. and M.B.; formal analysis, M.A.R.C. and L.D.; investigation, J.A.R.-H., M.A.R.C. and L.D.; software, M.A.R.C. and L.D.; writing—original draft preparation, M.A.R.C. and L.D.; writing—review and editing, M.A.R.C., J.R, L.D., E.N.S. and M.B.; funding acquisition, M.A.R.C. All authors have read and agreed to the published version of the manuscript.

Funding: This research was funded by Consejo Nacional de Ciencia y Tecnología (México) (1085717).

Data Availability Statement: Not applicable.

Conflicts of Interest: The authors declare no conflict of interest.

References

1. Pavlović, T.; Mirjanić, D.; Mitića, I.; Stanković, A. The Impact of Electric Cars Use on the Environment. In *New Technologies, Development and Application II. NT 2019*; Karabegović, I., Ed.; Lecture Notes in Networks and Systems; Springer: Sarajevo, Bosnia and Herzegovina, 2019; Volume 76, pp. 541–548. [CrossRef]
2. Yoong, M.K.; Gan, Y.; Gan, G.; Leong, C.; Phuan, Z.; Cheah, B.; Chew, K. Studies of regenerative braking in electric vehicle. In Proceedings of the 2010 IEEE Conference on Sustainable Utilization and Development in Engineering and Technology, Petaling Jaya, Malaysia, 20–21 November 2010; pp. 40–45.
3. Tran, M.-K.; Bhatti, A.; Vrolik, R.; Wong, D.; Panchal., S.; Flowler, M.; Fraser, R. A Review of Range Extenders in Battery Electric. *World Electr. Veh. J.* **2021**, *12*, 54. [CrossRef]
4. Tie, S.F.; Tan, C.W. A review of energy sources and energy management system in electric vehicles. *Renew. Sustain. Energy Rev.* **2013**, *20*, 82–102. [CrossRef]
5. Villalobos, J.G.; Zamora, I.; Martín, J.S.; Asensio, F.; Aperribay, V. Plug-in electric vehicles in electric distribution networks: A review of smart charging approaches. *Renew. Sustain. Energy Rev.* **2014**, *38*, 717–731. [CrossRef]
6. Hannan, M.A.; Azidin, F.; Mohamed, A. Hybrid electric vehicles and their challenges: A review. *Renew. Sustain. Energy Rev.* **2014**, *29*, 135–150. [CrossRef]

7. Ortuzar, M.; Moreno, J.; Dixon, J. Ultracapacitor-Based Auxiliary Energy System for an Electric Vehicle: Implementation and Evaluation. *IEEE Trans. Ind. Electron.* **2007**, *54*, 2147–2156. [CrossRef]
8. Husain, I.; Ozpineci, B.; Sariful, M.I.; Gurpinar, E.; Su, G.; Yu, W.; Chowdhury, S.; Xue, L.; Rahman, D.; Sahu, R. Electric drive technology trends, challenges, and opportunities for future electric vehicles. *Proc. IEEE* **2021**, *109*, 1039–1059. [CrossRef]
9. Zhang, L.; Cai, X. Control strategy of regenerative braking system in electric vehicles. *Energy Procedia* **2018**, *152*, 496–501. [CrossRef]
10. Xie, J.; Cao, B.; Zhang, H.; Xu, D. Switched robust control of regenerative braking of electric vehicles. In Proceedings of the the 2010 IEEE International Conference on Information and Automation, Harbin, China, 20–23 June 2010; pp. 1609–1612. [CrossRef]
11. Xu, W.; Chen, H.; Zhao, H.; Ren, B. Torque optimization control for electric vehicles with four in-wheel motors equipped with regenerative braking system. *Mechatronics* **2019**, *57*, 95–108. [CrossRef]
12. Kiddee, K.; Keyoonwong, W.; Khan-Ngern, W. An HSC/battery energy storage system-based regenerative braking system control mechanism for battery electric vehicles. *IEEJ Trans. Electr. Electron. Eng.* **2019**, *14*, 457–466. [CrossRef]
13. Indragandhi, V.; Selvamathi, R.; Gunapriya, D.; Balagurunathan, B.; Suresh, G.; Chitra, A. An Efficient Regenerative Braking System Based on Battery-Ultracapacitor for Electric Vehicles. In Proceedings of the 2021 Innovations in Power and Advanced Computing Technologies (i-PACT), Kuala Lumpur, Malaysia, 27–29 November 2021.
14. Manríquez, E.Q.; Sanchez, E.N.; Toledo, M.E.A.; Muñoz, F. Neural control of an induction motor with regenerative braking as electric vehicle architecture. *Eng. Appl. Artif. Intell.* **2021**, *104*, 104275. [CrossRef]
15. Guo, J.; Xiaoping, J.; Guangyu, L. Performance Evaluation of an Anti-Lock Braking System for Electric Vehicles with a Fuzzy Sliding Mode Controller. *Energies* **2014**, *7*, 6459–6476. [CrossRef]
16. Li, X.; Xu, L.; Hua, J.; Li, J.; Ouyang, M. Regenerative braking control strategy for fuel cell hybrid vehicles using fuzzy logic. In Proceedings of the 2008 International Conference on Electrical Machines and Systems, Wuhan, China, 17–20 October 2008; pp. 2712–2716.
17. Rajendran, S.; Spurgeon, S.; Tsampardoukas, G.; Hampson, R. Intelligent Sliding Mode Scheme for Regenerative Braking Control. *IFAC-PapersOnLine* **2018**, *51*, 334–339. [CrossRef]
18. Wu, J.; Lu, Y. Decoupling and optimal control of multilevel buck DC-DC converters with inverse system theory. *IEEE Trans. Ind. Electron.* **2019**, *67*, 7861–7870. [CrossRef]
19. Manriquez, E.Q.; Sanchez, E.N.; Harley, R.G.; Li, S.; Felix, R.A. Neural inverse optimal control implementation for induction motors via rapid control prototyping. *IEEE Trans. Power Electron.* **2018**, *34*, 5981–5992. [CrossRef]
20. Cao, F.; Yang, T.; Li, Y.; Tong, S. Adaptive Neural Inverse Optimal Control for a Class of Strict Feedback Stochastic Nonlinear Systems. In Proceedings of the 2019 IEEE 8th Data Driven Control and Learning Systems Conference (DDCLS), Dali, China, 24–27 May 2019; pp. 432–436. [CrossRef]
21. Sanchez, E.N.; Alanis, A.Y.; Loukianov, A.G. *Discrete-Time High Order Neural Control: Trained with Kalman Filtering*; Springer Science & Business Media: Cham, Switzerland, 2008. [CrossRef]
22. Rios, Y.Y.; Garcia-Rodriguez, J.A.; Sanchez, E.N.; Alanis, A.Y.; Velázquez, E.R. Rapid Prototyping of Neuro-Fuzzy Inverse Optimal Control as Applied to T1DM Patients. In Proceedings of the 2018 IEEE Latin American Conference on Computational Intelligence (LA-CCI), Gudalajara, Mexico, 7–9 November 2018; pp. 1–5. [CrossRef]
23. Djilali, L.; Vega, C.J.; Sanchez, E.N.; Hernandez, J.A.R. Distributed Cooperative Neural Inverse Optimal Control of Microgrids for Island and Grid-Connected Operations. *IEEE Trans. Smart Grid* **2022**, *13*, 928–940. [CrossRef]
24. Franco, M.L.; Sanchez, E.N.; Alanis, A.Y.; Franco, C.L.; Daniel, N.A. Decentralized control for stabilization of nonlinear multi-agent systems using neural inverse optimal control. *Neurocomputing* **2015**, *168*, 81–91. [CrossRef]
25. Cao, J.; Cao, B.; Xu, P.; Bai, Z. Regenerative-Braking Sliding Mode Control of Electric Vehicle Based on Neural Network Identification. In Proceedings of the 2008 IEEE/ASME International Conference on Advanced Intelligent Mechatronics, Xi'an, China, 2–5 July 2008; pp. 1219–1224. [CrossRef]
26. Djilali, L.; Sanchez, E.N.; Ornelas-Tellez, F.; Avalos, A.; Belkheiri, M. Improving Microgrid Low-Voltage Ride-Through Capacity Using Neural Control. *IEEE Syst. J.* **2020**, *14*, 2825–2836. [CrossRef]
27. Alanis, A.Y.; Sanchez, E.N.; Loukianov, A.G. Discrete-time adaptive backstepping nonlinear control via high-order neural networks. *IEEE Trans. Neural Netw.* **2007**, *18*, 1185–1195. [CrossRef] [PubMed]
28. Sanchez, E.; Ornelas, F. *Discrete-Time Inverse Optimal Control for Nonlinear Systems*, 1st ed.; CRC Press: Boca Raton, FL, USA, 2013. [CrossRef]
29. Freeman, R.; Kokotovic, P.V. *Robust Nonlinear Control Design: State-Space and Lyapunov Techniques*; Springer Science & Business Media: Cham, Switzerland, 2008. [CrossRef]
30. Ruiz-Cruz, R.; Sanchez, E.; Loukianov, A.; Ruz-Hernandez, J. Real-time neural inverse optimal control for a wind generator. *IEEE Trans. Sustain. Energy* **2018**, *10*, 1172–1183. [CrossRef]
31. Rovithakis, G.A.; Chistodoulou, M.A. *Adaptive Control with Recurrent High-Order Neural Networks: Theory and Industrial Applications*; Springer Science & Business Media: Cham, Switzerland, 2012. [CrossRef]
32. Utkin, V.; Guldner, J.; Shi, J. *Sliding Mode Control in Electro-Mechanical Systems*; CRC Press: Boca Raton, FL, USA, 2017. [CrossRef]
33. Zohuri, B. *Scalar Wave Driven Energy Applications*; Springer: Cham, Switzerland, 2019. [CrossRef]

Article

A New Bearing Fault Detection Strategy Based on Combined Modes Ensemble Empirical Mode Decomposition, KMAD, and an Enhanced Deconvolution Process

Yasser Damine [1], Noureddine Bessous [2], Remus Pusca [3], Ahmed Chaouki Megherbi [1], Raphaël Romary [3,*] and Salim Sbaa [4]

Citation: Damine, Y.; Bessous, N.; Pusca, R.; Megherbi, A.C.; Romary, R.; Sbaa, S. A New Bearing Fault Detection Strategy Based on Combined Modes Ensemble Empirical Mode Decomposition, KMAD, and an Enhanced Deconvolution Process. *Energies* **2023**, *16*, 2604. https://doi.org/10.3390/en16062604

Academic Editors: Moussa Boukhnifer and Larbi Djilali

Received: 6 February 2023
Revised: 2 March 2023
Accepted: 4 March 2023
Published: 9 March 2023

Copyright: © 2023 by the authors. Licensee MDPI, Basel, Switzerland. This article is an open access article distributed under the terms and conditions of the Creative Commons Attribution (CC BY) license (https://creativecommons.org/licenses/by/4.0/).

[1] Laboratory of Identification, Command, Control and Communication (LI3C), Department of Electrical Engineering, University of Mohamed khider, Biskra 07000, Algeria
[2] Laboratoire de Genie Electrique et des Energies Renouvelables (LGEERE), Department of Electrical Engineering, Faculty of Technology, University of El Oued, El Oued 39000, Algeria
[3] Univ. Artois, UR 4025, Laboratoire Systèmes Electrotechniques et Environnement (LSEE), F-62400 Béthune, France
[4] Department of Electrical Engineering, Faculty of Technology, University of Mohamed Khider, Biskra 07000, Algeria
* Correspondence: raphael.romary@univ-artois.fr

Abstract: In bearing fault diagnosis, ensemble empirical mode decomposition (EEMD) is a reliable technique for treating rolling bearing vibration signals by dividing them into intrinsic mode functions (IMFs). Traditional methods used in EEMD consist of identifying IMFs containing the fault information and reconstructing them. However, an incorrect selection can result in the loss of useful IMFs or the addition of unnecessary ones. To overcome this drawback, this paper presents a novel method called combined modes ensemble empirical mode decomposition (CMEEMD) to directly obtain a combination of useful IMFs containing fault information. This is without needing to pass through the processes of IMF selection and reconstruction, as well as guaranteeing that no defect information is lost. Owing to the small signal-to-noise ratio, this makes it difficult to determine the fault information of a rolling bearing at the early stage. Therefore, improving noise reduction is an essential procedure for detecting defects. The paper introduces a robust process for extracting rolling bearings defect information based on CMEEMD and an enhanced deconvolution technique. Firstly, the proposed CMEEMD extracts all combined modes (CMs) from adjoining IMFs decomposed from the raw fault signal by EEMD. Then, a selection indicator known as kurtosis median absolute deviation (KMAD) is created in this research to identify the combination of the appropriate IMFs. Finally, the enhanced deconvolution process minimizes noise and improves defect identification in the identified CM. Analyzing real and simulated bearing signals demonstrates that the developed method shows excellent performance in extracting defect information. Compared results between selecting the sensitive IMF using kurtosis and selecting the sensitive CM using the proposed KMAD show that the identified CM contains rich fault information in many cases. Furthermore, our comparisons revealed that the enhanced deconvolution approach proposed here outperformed the minimum entropy deconvolution (MED) approach for improving fault pulses and the wavelet de-noising method for noise suppression.

Keywords: combined modes ensemble empirical mode decomposition; KMAD indicator; three-sigma rule; enhanced minimum entropy deconvolution; rolling element bearing faults; fault detection

1. Introduction

The large-scale use of induction machines accounts for 90% of the industry's total energy consumption. Several defects often lead to unexpected failures. These defects can lead to severe damage to the machine if they are overlooked initially. According to previous studies, the high percentage of failures in induction machines is caused by bearing faults. As a result, it is highly recommended to monitor small and medium voltage machines continuously for bearing faults [1,2]. Bearing health condition is commonly monitored by vibration monitoring. The vibration signals provide a wealth of information regarding machine health conditions [3]. Many approaches aim to pick up the characteristic defect information from the rolling bearing's non-stationary and nonlinear vibration signal by employing appropriate signal processing techniques. Huang et al. [4] created a time-frequency analysis approach known as empirical mode decomposition (EMD). EMD differs from short-time Fourier transform and wavelet transform as it is not dependent on the basis function. It is based on adaptive decomposition characteristics and decomposes signals into intrinsic mode functions (IMFs). EMD is suitable for non-stationary and non-linear vibration signals analysis [5], such as bearing faults, and has been widely used for this purpose. However, a significant problem with EMD is the mixing of modes. As a solution to this challenge, an improved version of EMD called ensemble empirical mode decomposition (EEMD) is proposed in [6]. An IMF in the EEMD consists of the average of a set of trials. The results of the EMD decomposition are used for each trial and a finite-amplitude white noise [7]. Compared to the EMD, IMFs produced by the EEMD can better highlight the signal's significant features. The focus of researchers has always been on how to identify EEMD's important IMFs and how to improve the level of noise minimization. These two main issues will be briefly discussed below.

Considering that the decomposed bearing vibration signal contains some IMFs representing defect features, as well as other IMFs containing unused information, researchers have focused on identifying suitable IMFs. Wang et al. [8] suggested the use of the highest value of kurtosis to pick the relevant IMF. Yang et al. [9] selected the effective IMF using mutual information. Li J et al. [10] calculated each IMF's similarity to the input signal based on Spearman's rho to identify the required IMF. A merit index for determining the relevant IMF has been proposed in [11]. However, if only the most suitable IMF is considered, fault information contained in other IMFs may be lost. In contrast, Li Z et al. [12] developed a weighted kurtosis index difference spectrum (WKIDS) to choose the important IMFs. Ma et al. [13] used the correlation coefficient to select the effective IMFs. Luo et al. [14] identified the effective IMFs by using high kurtosis values. However, Damine et al. [15] demonstrated that choosing the most suitable IMF can result in the loss of other important IMFs, and that selecting multiple IMFs can result in the inclusion of unnecessary ones. To address the abovementioned issues, this paper offers a novel approach called combined modes ensemble empirical mode decomposition (CMEEMD). This method is based on the extraction of combined modes (CMs) from the measured vibration signal. After that, a selection indicator is created to identify the combination of suitable IMFs. The purpose of this step is to obtain the most information about the defect directly from the input signal, without having to pass through the IMFs selection and reconstruction processes. It also ensures that no information about the defect is wasted or irrelevant data are included.

Owing to the effect of surrounding noise, extracting bearing fault information at the early stage of damage is challenging. Therefore, it is essential to reveal the defect pulses in the vibration signal. The most commonly used deconvolution process is the minimum entropy deconvolution (MED). The MED is designed to retrieve the bearing defect pulses in the input signal. Pennacchi et al. [16] examined the efficiency of the MED on experimental signals and found that it can detect bearing defects. However, when the original signal contains noise, the efficiency of MED is reduced. In addition, the output of MED will also be affected by noise interference. Therefore, researchers were concentrated on increasing the efficiency of the MED. Chatterton et al. [17] combined EMD with MED to improve bearing defect detection. Ding et al. [18] introduced a deconvolution process using autoregressive MED for extracting bearing features.

In view of the above considerations, this paper presents an enhanced deconvolution approach, which focuses on eliminating the noise interference in the MED output by introducing a de-noising method derived from the three-sigma rule [19]. A new procedure for extracting bearing defect features based on CMEEMD and an enhanced deconvolution process is discussed in this research work. The following describes the originality of these procedures. Firstly, the proposed CMEEMD decomposes the original signal into CMs. An indicator is created to identify the appropriate combination that combines the effective IMFs instead of selecting and reconstructing them. Secondly, an enhanced deconvolution process based on MED and a noise suppression technique using the three-sigma rule is performed on the selected CM. Finally, the envelope spectrum is applied, and the characteristic fault frequency is extracted to diagnose the bearing fault.

The remaining sections of this paper are organised as following: Section 2 is dedicated to the basic theories of EEMD, MED, and the rule of three-sigma de-noising method. Section 3 details the proposed methods of this research. Section 3.1 gives the steps of the CMEEMD. In Section 3.2, the process of selecting an appropriate combination is introduced. In Section 3.3, the enhanced deconvolution strategy is presented. Section 3.4 describes the new bearing fault diagnosis procedure. Section 4 presents the results of applying the proposed method to the simulated signal. In Section 5, the suggested process is performed on the experimental data, and the results are verified. In Section 6, the conclusion of this paper is presented.

2. Theoretical Analysis

2.1. EEMD Method

By comparing EEMD and EMD, it has been concluded that EEMD may be more effective at revealing the characteristic fault information of rolling element bearings [20]. EEMD solves the problem of mode mixing in EMD by adding Gaussian white noise to the original signal. Thus, we can better highlight the signal's intrinsic characteristics. The algorithm of EEMD [7] is given below, and Figure 1 shows the process flow diagram.

(1) Add a random white Gaussian noise $\beta w_i(t)$ to the existing signal:

$$x_i(t) = x(t) + \beta w_i(t) \tag{1}$$

where $\beta w_i(t)$ is the i-th added white noise series, and $x_i(t)$ represents the noise-added signal (i = 1, 2, ... , i).

(2) Divide by EMD the novel signal and obtain N sets of IMFs:

$$x_i(t) = \sum_{j=1}^{N} c_{ij}(t) + r_i \tag{2}$$

where $c_{ij}(t)$ is the IMFs and r_i is the residue.

(3) Using the formula below, determine the ensemble means $c_j(t)$ of the I trials:

$$c_j(t) = \sum_{i=1}^{I} c_{ij}(t) \tag{3}$$

where $c_j(t)$ ($c_1, c_2, ..., c_N$) is the IMFs divided by EEMD.

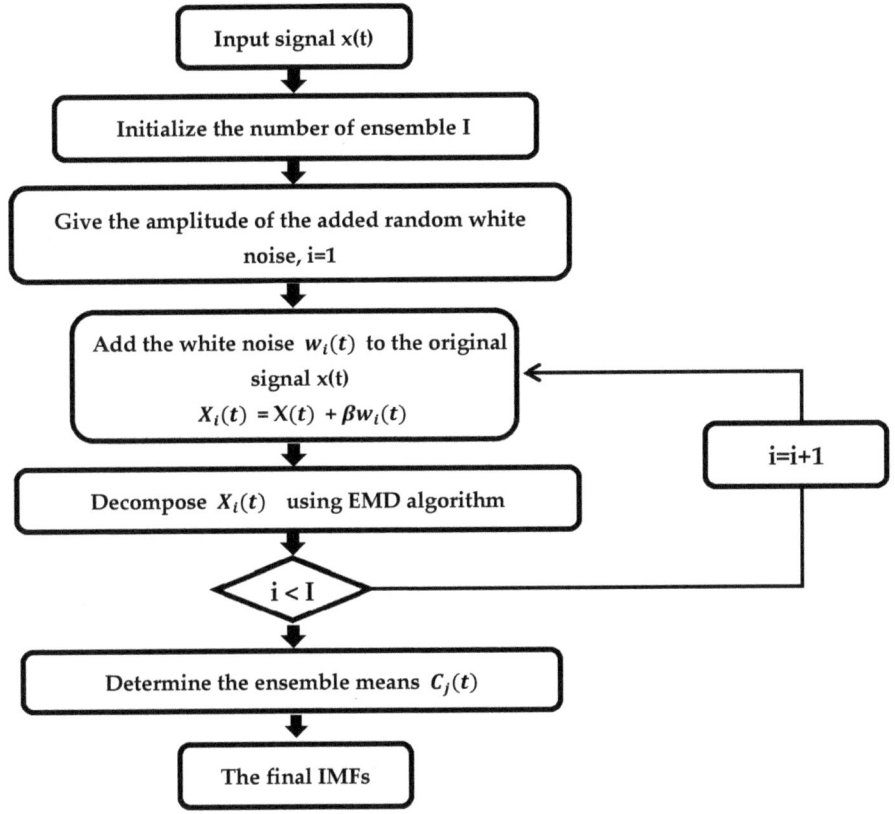

Figure 1. Flow chart of the ensemble empirical mode decomposition (EEMD) algorithm to obtain the intrinsic mode functions (IMFs).

2.2. Minimum Entropy Deconvolution Technique

MED was originally introduced by Ralph [21]. The MED highlights the transient components of the signal with a finite impulse response (FIR) filter. It decreases a signal's randomness by minimizing its entropy. Two terms can represent a general signal x(n):

$$x(n) = z(n) * w(n) + \eta(n) \quad (4)$$

There is a convolution between the defect impulse z and its excitation w, which is the first term in the equation. The second term takes a random noise into account. FIR filter h(n) can be used in minimum entropy deconvolution (MED) to process the original signal. From [22,23], it is possible to obtain:

$$u(n) = x(n) * h(n) = \sum_{i=0}^{M-1} h(i) x(n-i) \quad (5)$$

where $n = 0, 1, \ldots, N$, $N = T + M - 2$. The deconvolution filter length is M, and the input sequence x(n) length is T. In MED, a signal's entropy is minimized by maximizing the Varimax function. The Varimax function for u(n) is:

$$V(u) = \frac{\sum_{n=0}^{N} u^4(n)}{(\sum_{n=0}^{N} u^2(n))^2} \tag{6}$$

The filtering parameters that maximize $V(u)$ are such that:

$$\frac{\partial V(u)}{\partial h(n)} = 0 \tag{7}$$

As a result of substituting Equations (6) in (7) and solving the derivative, we obtain:

$$\sum_{i=0}^{M-1} h(i) \sum_{n=0}^{N} x(n-i)x(n-k) = \sum_{n=0}^{N} \frac{u^3(n)x(n-k)}{V(u) \parallel u \parallel^2} \tag{8}$$

where $k = 0, 1, \ldots, M-1$.

Equation (8) can be written as:

$$R_{XX}h = b \tag{9}$$

where R_{XX} corresponds to a matrix of autocorrelation, h is the filter coefficients vector, and b includes the input of the filter $x(n)$ cross-correlated with the cube of its output $u(n)$. The following steps summarize the optimal inverse filter solution:

- Assume that $h(0)$ is a set of initial filter coefficients;
- Calculate $u(0)$ and $V(u)$;
- Calculate R_{XX}
- Determine $b(1)$ and $h(1)$;
- Repeat the procedure until an optimal filter is obtained.

2.3. The Three-Sigma Rule for Noise Minimization

In probability and statistics, the three-sigma rule states that approximately 99.73% of data following a normal distribution are located inside a range of three standard deviations from the mean [24].

$$P\{\mu - 3\sigma < Y < \mu + 3\sigma\} \approx 99.73\% \tag{10}$$

The mean and standard deviation are represented by μ and σ, respectively. The normal distribution appears with:

$$E(Y) = \mu = 0 \tag{11}$$

$$D(Y) = E(Y^2) - [E(Y)]^2 = E(Y^2) = \sigma^2 \tag{12}$$

The variance and the expectation are represented by $D(Y)$ and $E(Y)$, respectively. Based on Equation (12), the root mean square (RMS) value of Y is:

$$Y_{rms} = \sqrt{\frac{1}{n}\sum_{i=1}^{n}[X_i, -, E(Y)]^2} = \sqrt{\frac{1}{n}\sum_{i=1}^{n} y_i^2} = \sqrt{E(Y)} = \sigma \tag{13}$$

where y_i stands for the sample data of Y and n for the number of samples.

Using Equations (11) and (13), Equation (10) can be written as:

$$P\{-3\sigma < Y < 3\sigma\} = P\{-3Y_{rms} < Y < 3Y_{rms}\} \approx 99.73\% \tag{14}$$

Based on the assumption that a fault-free rolling bearing follows the normal distribution [25], Equation (14) shows that nearly all the noise in the bearing vibration signal Y is distributed within $\pm 3Y_{rms}$. Due to this, it is necessary to remove the components within $\pm 3Y_{rms}$. The steps of the de-noising process are as follows [26]:

1. y(t) is normalized by using zero-mean normalization:

$$Z(t) = \frac{y - \mu}{\sigma} \quad (15)$$

where $Z(t)$ is the normalized signal.
2. Determine Z_{rms} of $Z(t)$;
3. Replace the sampling data z_i of $Z(t)$ falling between $\pm 3Z_{rms}$ with zero while leaving z_i outside of $\pm 3Z_{rms}$ unchanged.

$$w(t) = \begin{cases} 0, & \text{if } |z_i| \leq 3Z_{rms} \\ z_i(t), & \text{otherwise} \end{cases} \quad (16)$$

where $w(t)$ represents $y(t)$ after removing the unnecessary components.

3. Proposed Methods

3.1. Combined Modes Ensemble Empirical Mode Decomposition (CMEEMD)

The proposed CMEEMD aims to extract all the CMs from the adjoining IMFs decomposed from the bearing fault vibration signal using EEMD. This process is described in detail below with a flowchart shown in Figure 2. In this paper, adjoining IMFs are combined using the following expression:

$$CM_{i \to j} = IMF_i + \ldots + IMF_j \quad (17)$$

where $CM_{i \to j}$ is the combined modes of adjoining IMFs from the i-th mode to the j-th mode, IMF_i is the IMF that starts the combination, and IMF_j is the IMF that finishes it. Extraction of CMs is done as follows:

- Divide these CMs into groups. The first group consists of CMs starting with IMF_1. By using Equation (17), we obtain:

$$CM_{1 \to j} = IMF_1 + \ldots + IMF_j \quad 2 \leq j \leq N \quad (18)$$

where $CM_{1 \to j}$ is the combination of adjoining IMFs from IMF_1 to the j-th IMF for $j = 2, \ldots N$, N is the number of IMFs.
- Using Equation (18), extract all CMs starting with IMF_1:

$$\begin{aligned} CM_{1 \to 2} &= IMF_1 + IMF_2 \\ CM_{1 \to 3} &= IMF_1 + IMF_2 + IMF_3 \\ &\vdots \\ CM_{1 \to N} &= IMF_1 + IMF_2 + IMF_3 + \ldots + IMF_N \end{aligned} \quad (19)$$

- The second group is constituted by CMs starting with the second mode. In this case, Equation (17) can be expressed as:

$$CM_{2 \to j} = IMF_2 + \ldots + IMF_j \quad 3 \leq j \leq N \quad (20)$$

where $CM_{2 \to j}$ is the combination of adjoining IMFs from IMF_2 to the j-th IMF for $j = 3, \ldots N$.
- Using Equation (20), extract all CMs starting with IMF_2:

$$\begin{aligned} CM_{2 \to 3} &= IMF_2 + IMF_3 \\ CM_{2 \to 4} &= IMF_2 + IMF_3 + IMF_4 \\ &\vdots \\ CM_{2 \to N} &= IMF_2 + IMF_3 + IMF_4 + \ldots + IMF_N \end{aligned} \quad (21)$$

- The process continues until we reach the $N-1$ group. In this case, the last combination can be represented by the following equation:

$$CM_{N-1 \to N} = IMF_{N-1} + IMF_N \qquad (22)$$

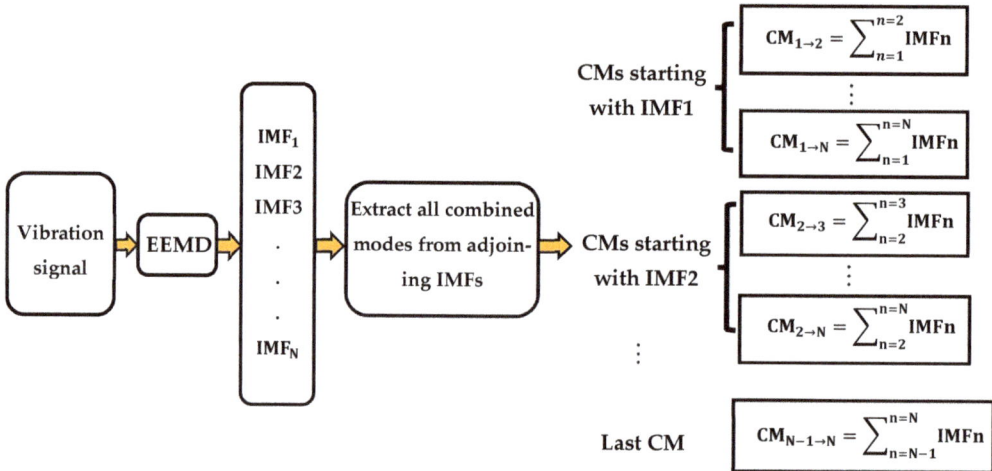

Figure 2. Flow chart of the proposed combined modes ensemble empirical mode decomposition (CMEEMD).

3.2. Sensitive CM Selection Using KMAD Indicator

Once all the CMs have been extracted, we need to identify the appropriate combination of sensitive IMFs. An indicator was required to select this combination among all the other CMs. In many studies, maximum kurtosis was used to identify the most sensitive IMF. However, if we consider only the best IMF, we may lose information about faults contained in other IMFs [27]. Therefore, this paper uses the kurtosis of the combined IMFs. The probability of identifying the appropriate combination is higher when the kurtosis value of the corresponding combination is high. The expression of kurtosis is defined as follows [28]:

$$K = \frac{1}{N} \sum_{i=1}^{N} \frac{(x_i - \mu)^4}{\sigma^4} \qquad (23)$$

where the amplitude of the vibration waveform is indicated by x_i, the mean of the signal by μ, the standard deviation by σ, and the length of the samples by N. According to [29–31], IMFs with high-frequency bands of the vibration signal contain the main fault information about the rolling bearings. It is known that the higher the frequency band, the larger the median absolute deviation (MAD). Therefore, the MAD can be used to identify IMFs with high-frequency bands. The expression of MAD is defined as follows [32]:

$$MAD(y) = median(|y_n - median(y)|) \qquad (24)$$

where y_n represents the n-th sampling of the signal y. To ensure that only sensitive IMFs are combined in the effective combination, the proposed selection indicator aims to prevent unwanted IMFs from being added. Accordingly, as the number of IMFs in the combination decreases, the probability of obtaining the required combination increases. Based on all the above, the paper proposes an indicator (KMAD), which combines kurtosis and MAD to select the appropriate combination of sensitive IMFs.

$$\text{KMAD}_{i \to j} = \frac{K_{i \to j} \cdot \text{MAD}_i}{\sum_i^j \text{MAD}_n} \quad (25)$$

In this equation, $K_{i \to j}$ is the kurtosis value of $CM_{i \to j}$, where $CM_{i \to j}$ is the combined modes of adjoining IMFs from the i-th IMF to the j-th IMF, MAD_i is the mean absolute deviation of the i-th IMF that starts the combination, and $\sum_i^j \text{MAD}_n$ means the sum of MADs of IMFs from the i-th IMF to the j-th IMF. A combination with fewer IMFs has a lower value of $\sum_i^j \text{MAD}_n$, which increases the probability of obtaining the combination of useful IMFs. For each $CM_{i \to j}$, $\text{KMAD}_{i \to j}$ is calculated, where the highest value corresponds to the required combination.

3.3. The Enhanced Deconvolution Process

One of the most commonly used methods for this is MED. However, when the input signal contains noise, the effectiveness of the MED will be reduced. For this reason, noise will affect the MED output. Therefore, an enhanced deconvolution approach is presented in this paper, which aims to minimize noise interference in the MED output by integrating the three-sigma rule (see Section 2.3). Figure 3 is a flow chart illustrating the enhanced MED strategy, and the steps are as follows:

1. Apply the MED technique to the input signal;
2. Perform the de-noising method derived from the three-sigma rule on the MED output. It consists of the following steps:

- Normalize the MED output using zero-mean normalization;
- Calculate the root mean square value Y_{rms} of the normalized signal $Y(t)$;
- Replace the sampling data y_i of $Y(t)$ falling between $\pm 3Y_{rms}$ with zero while keeping y_i outside of $\pm 3Y_{rms}$ unmodified.

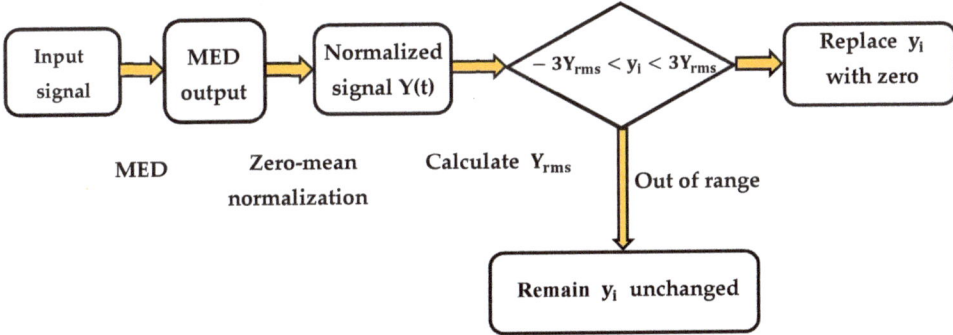

Figure 3. Proposed enhanced deconvolution process flowchart.

3.4. The Proposed Strategy for Bearing Fault Detection

This paper describes a novel feature extraction method based on CMEEMD and proposes a deconvolution process to diagnose the bearing fault from the vibration signals. Figure 4 illustrates the flowchart of the proposed method for detecting bearing defects. The detailed process of the feature extraction method proposed is as follows:

1. Perform CMEEMD on the fault vibration signal as follows:

- Decompose the fault vibration signal with the defect into IMFs by EEMD;
- Extract all combined modes (CMs) from adjoining IMFs (see Section 3.1).

2. Select the appropriate combination using the KMAD indicator (see Section 3.2):
   - Calculate the KMAD value of each CM;
   - Select the required combination based on the highest value of KMAD.
3. Perform the enhanced deconvolution process on the selected CM (see Section 3.3).

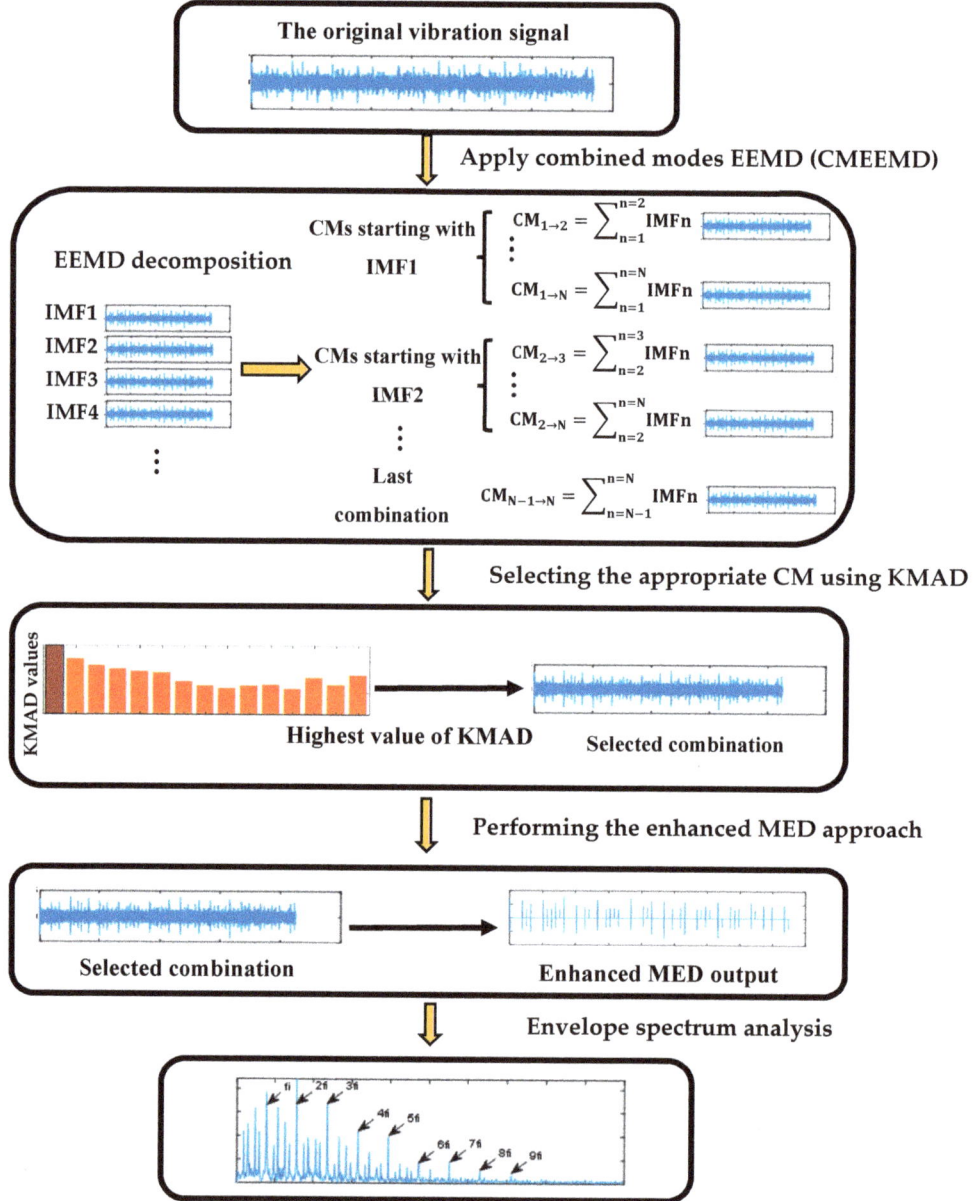

Figure 4. Proposed strategy using CMEEMD, kurtosis median absolute deviation (KMAD) and an Enhanced Deconvolution Process for diagnosing bearing faults.

4. The Simulation Validation

A simulation of an inner ring defect bearing is presented in this section to illustrate the effectiveness and usefulness of the suggested method for extracting fault characteristics. The periodic impulses represent the vibration waveform caused by a local failure in the bearing. However, these impulses are usually buried in white noise. As a result, we can obtain the simulated signal of the rolling bearing from [33]. In this paper, the sampling frequency is 12,000 Hz, the resonant frequency is 3000 Hz, the inner-race fault frequency is 79 Hz, the time lag is zero, the rotational frequency is 28 Hz, and the damping ratio B = 500. The random noise has a zero mean and variance of $\sigma^2 = 0.7^2$. The data length of the signal is 10,240. The simulated signal y(t) is plotted in Figure 5a. It can be seen that the noise effect prevents the extraction of periodic impulses. From the envelope spectrum in Figure 5b, although the fault characteristic fi and the first harmonic 2fi can be extracted, the remaining harmonics are covered by noise interference. To improve fault detection, this signal needs to be pre-processed.

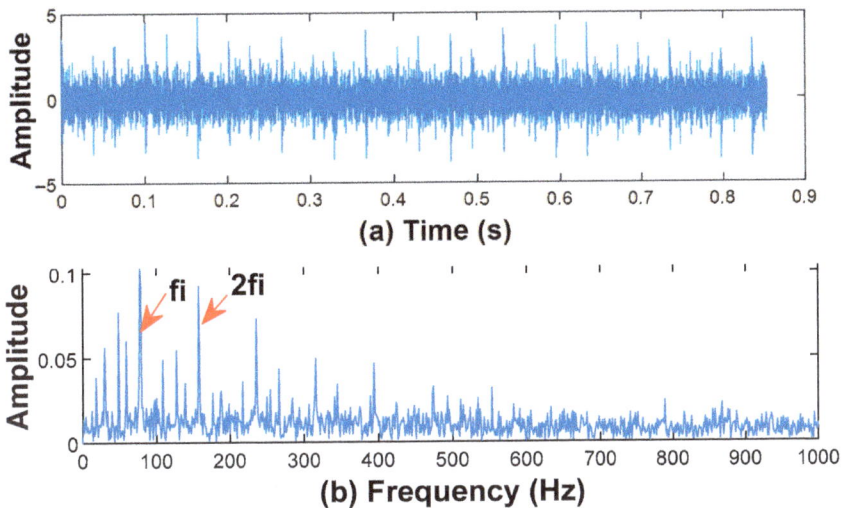

Figure 5. Inner ring fault simulated signal: (**a**) waveform; (**b**) envelope spectrum.

4.1. Analysis of the Proposed Method

Based on the detailed flowchart of the proposed feature extraction method described in Figure 4, the following processes are followed.

4.1.1. CMEEMD Analysis

According to [29–31], the significant defect information about rolling bearings is included in IMFs with high-frequency bands. Therefore, the proposed CMEEMD uses the EEMD to decompose this simulated signal into six IMFs. Then, one extracts all the CMs from the adjoining IMFs. Based on the recommended method for extracting combined modes CMs detailed in Section 3.1, fifteen CMs are generated from the six IMFs. The obtained IMFs are plotted in Figure 6, and the extracted CMs are illustrated in Figure 7. The next step identifies the most sensitive combination.

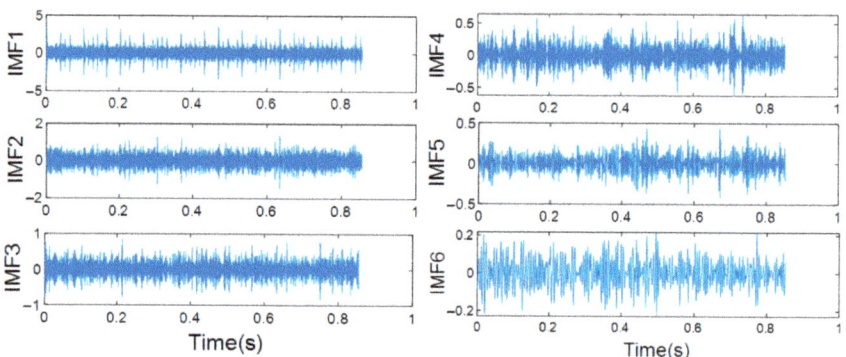

Figure 6. Decomposed result of the simulated signal by EEMD.

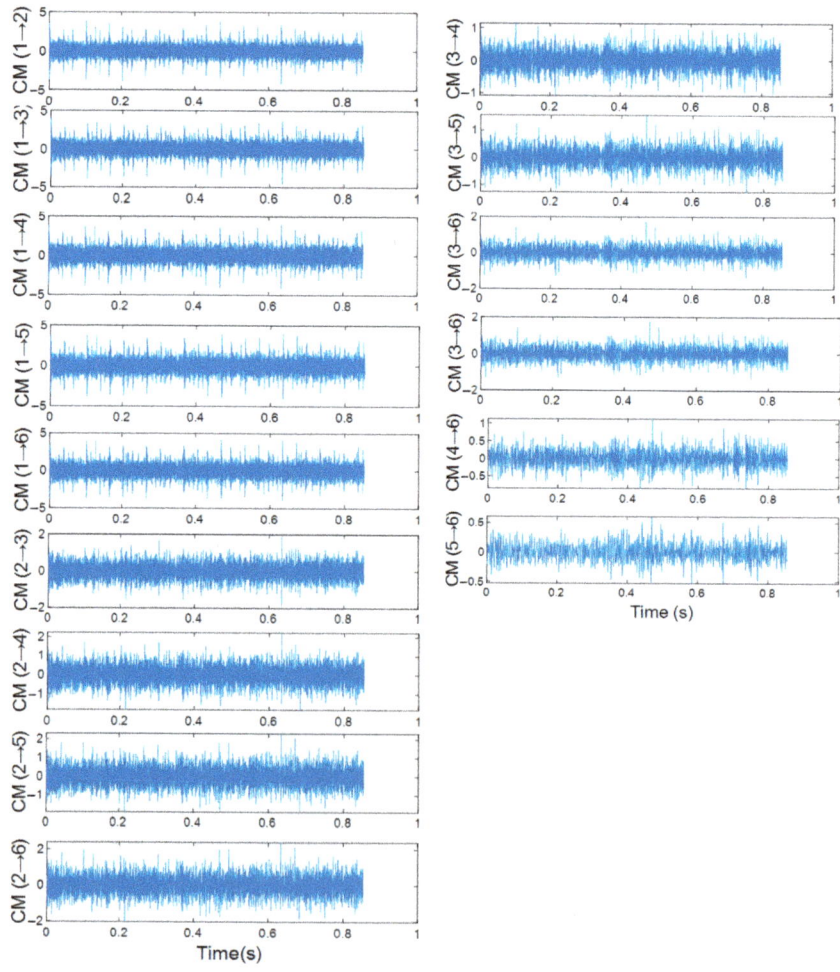

Figure 7. Extracted Combined modes (CMs) from the adjoining IMFs.

4.1.2. Selecting the Appropriate CM

Based on the time-domain waveforms given in Figure 7, the differences between the CMs are insignificant. Therefore, the most effective combination is selected using the proposed KMAD indicator. Based on Equation (25), the KMAD values of each combination are illustrated in Figure 8. It is observed that the combination $CM_{1\rightarrow2}$ has the highest value among all the other combinations. This indicates that it is the appropriate combination of the sensitive IMFs, i.e., IMF1 and IMF2.

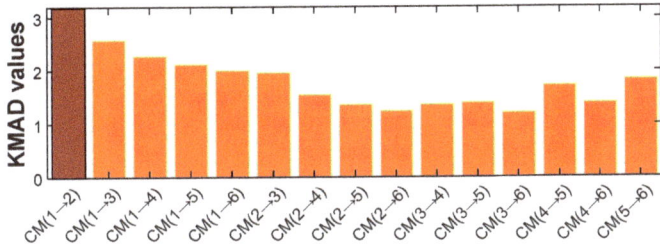

Figure 8. Kurtosis median absolute deviation (KMAD) values of each combination.

4.1.3. Performing the Proposed Deconvolution Process

This method focuses on minimizing noise interference in the MED output. The first step is to highlight the fault impulses in selected combination $CM_{1\rightarrow2}$ using MED. Following that, we minimize the noise using the rule of three-sigma. As illustrated in Figure 9a, the noise is minimized, and the fault impulses are emphasized. From the envelope spectrum in Figure 9b, we can efficiently and accurately extract the inner race fault characteristic frequency f_i and nine harmonics ($2f_i$, $3f_i$, $4f_i$, $5f_i$, $6f_i$, $7f_i$, $8f_i$, $9f_i$, and $10f_i$). This indicates that the rolling bearing fault feature extraction method proposed in this paper can extract fault information excellently.

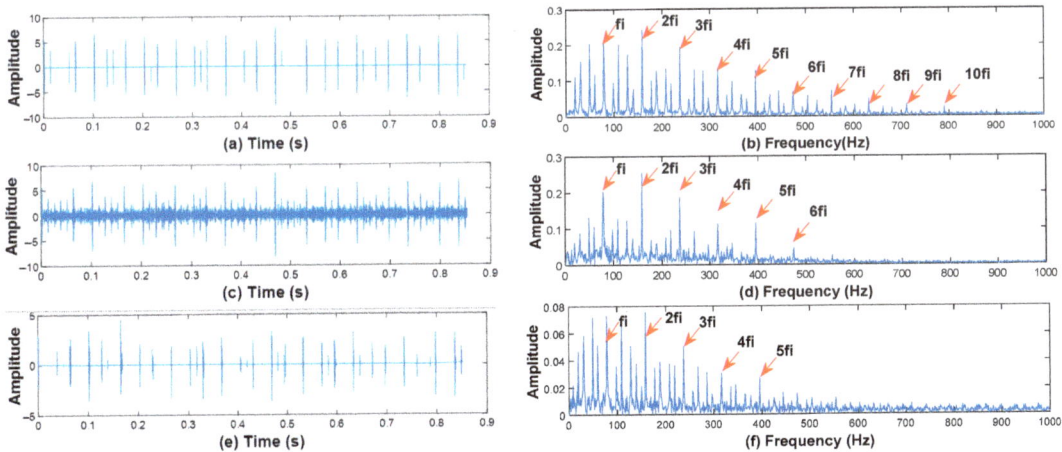

Figure 9. Simulated bearing fault diagnosis results for processing the selected CM using: proposed method (**a**,**b**); conventional MED (**c**,**d**); and wavelet denoising (**e**,**f**).

4.2. Advantages of the Proposed Methods for the Diagnosis of the Simulated Signal

To demonstrate the superiority of the proposed enhanced deconvolution process, the conventional MED is performed on the selected combination $CM_{1\rightarrow2}$. Figure 9c,d shows the results of processing the selected combination $CM_{1\rightarrow2}$. by the MED. As shown in

Figure 9c, the fault impulses are highlighted, and the noise level is decreased. However, some noise interference can still be seen. By comparing it with Figure 9a, it is clear that noise interference has been reduced significantly. From the envelope spectrum in Figure 9d, we can extract only the inner race fault characteristic frequency f_i and five harmonics ($2f_i$, $3f_i$, $4f_i$, $5f_i$, and $6f_i$). In comparison with Figure 9b, it is apparent that we can get more fault information. The comparison results demonstrate that the proposed enhanced MED outperformed the MED for improving fault detection. To demonstrate the superiority of the enhanced MED approach in minimizing noise, the wavelet de-noised method is performed on the selected combination $CM_{1\rightarrow 2}$. Figure 9e shows that the noise interference is reduced to some extent; however, the extracted fault frequency and its harmonics in Figure 9f are not as good as in Figure 9b. In this case, the wavelet de-noising method is less efficient in suppressing noise, making it difficult to extract fault information from the combination $CM_{1\rightarrow 2}$. The results demonstrate that the proposed enhanced MED outperformed the wavelet de-noising method in suppressing noise. The inter-harmonics (inter-characteristic frequencies of the faults) present the harmonics of the rotational frequency which are considered as extracted information. In Figure 9b, one can see that the harmonics multiple of the rotational frequency are obvious, while they are hidden in Figure 9d. This is due to the fact that the noise has been minimised in Figure 9a. In addition, a comparison of the conventional IMF selection method using maximum kurtosis with the proposed KMAD selection indicator is presented to illustrate its advantages. Table 1 shows the kurtosis values of the first six IMFs. It can be seen that IMF1 has the highest value of all the decomposition results, so it is selected as a sensitive IMF. IMF1 was treated using the enhanced deconvolution approach. As shown in the envelope spectrum of Figure 10a, the extracted fault information is weaker than the extracted fault information in Figure 10b. This indicates that the combination $CM_{1\rightarrow 2}$ contains rich fault feature information. The KMAD indicator identified $CM_{1\rightarrow 2}$ as a combination of suitable IMFs, i.e., IMF1 and IMF2. Consequently, if we choose only IMF1, the information contained in IMF2 will be lost. This proves that selecting the appropriate combination using the KMAD selection indicator overcomes the drawback of the IMF selection method using kurtosis to ensure that no information about the defect is lost.

Table 1. Kurtosis values of each intrinsic mode function (IMF).

IMF	Kurtosis
IMF1	4.6127
IMF2	3.2595
IMF3	3.0748
IMF4	3.0268
IMF5	2.9883
IMF6	2.8621

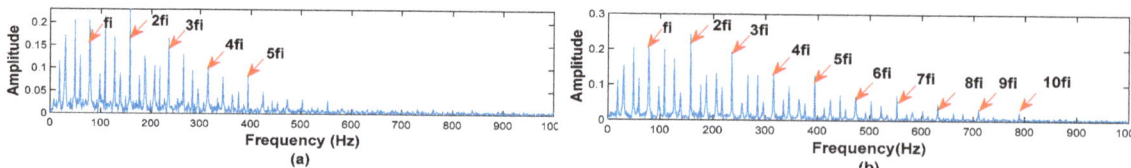

Figure 10. Diagnosis results of the simulated bearing fault using: (**a**) sensitive IMF-based Kurtosis; (**b**) sensitive CM-based KMAD.

5. Experimental Validation

Experimental data from the Case Western Reserve University [34] was used to validate the proposed method's effectiveness for detecting rolling bearing faults. As shown in Figure 11, the experimental set is composed of a 2 hp motor, a torque sensor/encoder, a dynamometer, and control electronics. Single point faults were introduced using electro-discharge machining, providing defects in the outer ring, the ball, and the inner ring. The rotating speed of the shaft varied from 1730 to 1797 RPM. We used the time signal of the drive end bearing in this study, recorded for the inner race, outer race, and ball fault. The data were gathered with 12,000 Hz. The deep groove ball bearing 6205-2RS JEM SKF was used in this experimental test. The bearing parameters are detailed in [34]. The bearing defect is localized in the early stages: a crack or spall. Rolling elements generate shock impulses every time they hit a local fault in the inner or outer ring. These repeated shock pulses produce a vibration at the frequency associated with the faulty element. This frequency is usually called the fault characteristics frequency, for example, BPFI (ball passing frequency inner race), BPFO (ball passing frequency outer race), and BFF (ball fault frequency), which are related to the inner race, the outer race, and the ball, respectively. The following are their mathematical equations [35]:

Figure 11. Experimental test rig from the Case Western Reserve University (CWRW) [34].

$$BPFI = \frac{f_r}{2} N_b \left(1 + \frac{D_b \cos\beta}{D_c}\right) \quad (26)$$

$$BPFO = \frac{f_r}{2} N_b \left(1 - \frac{D_b \cos\beta}{D_c}\right) \quad (27)$$

$$BFF = \frac{f_r}{2} \frac{D_c}{D_b} \left[1 - \left(\frac{D_b \cos\beta}{D_c}\right)^2\right] \quad (28)$$

F_r, N_b, D_c, D_b, and β correspond to the frequency of rotation, rolling element number, pitch diameter, ball diameter, and angle of contact, respectively.

5.1. Case 1: Diagnosis of the Inner Race Fault

In this case, the vibration signal emanates from the inner race fault. The shaft speed is 1772 rpm, the load is 1hp, and the fault size is 0.007 inches. According to Equation (26), the calculated fault characteristic frequency for the inner race is 159.9 Hz. Taking 24,000 data points for analysis, I measured original bearing signal with an inner race fault signal is plotted in Figure 12a. The periodic impulses cannot be extracted due to the noise effect. From the envelope spectrum in Figure 12b, the fault characteristic *fi* and the first harmonic can be extracted. However, the other harmonics are surrounded by noise interference. Therefore, this signal requires pre-processing to improve fault detection.

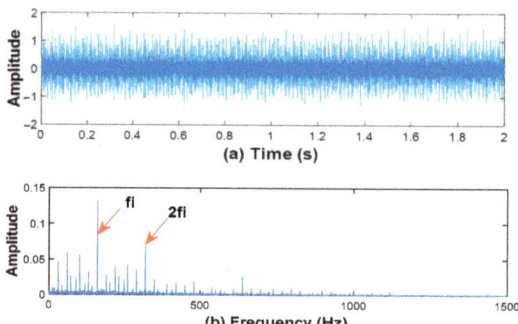

Figure 12. Experimental inner race defect: (**a**) waveform; (**b**) envelope spectrum.

5.1.1. Analysis of the Proposed Method

First, the proposed CMEEMD is used to extract the CMs from the experimental inner race fault signal. From the first six IMFs, fifteen combined modes (CMs) are generated using the method detailed in Section 3.1. The obtained IMFs are plotted in Figure 13, and the extracted CMs are illustrated in Figure 14. The next step is determining the most appropriate combination. By looking at the time domain waveform of each combination in Figure 14, it can be seen that the difference between the CMs is not significant. It is impossible to recognize directly which combination contains the most information about the fault. Therefore, the appropriate combination is selected using the proposed KMAD indicator. Based on Equation (25), Figure 15 illustrates the KMAD values of each combination. The combination $CM_{1 \to 2}$ has the highest value among all the other combinations, indicating that it is the best combination of the sensitive IMFs, including IMF1 and IMF2. Following this, the enhanced MED approach is executed on the selected combination. First, MED is used to minimize the entropy of $CM_{1 \to 2}$. After that, the output MED noise is minimized using the three-sigma rule. As illustrated in Figure 16a, the noise is restricted, and the fault impulses are highlighted. From the envelope spectrum in Figure 16b, we can extract the inner race fault characteristic frequency f_i and ten harmonics ($2f_i$, $3f_i$, $4f_i$, $5f_i$, $6f_i$, $7f_i$, $8f_i$, $9f_i$, $10f_i$, and $11f_i$). This suggests that the rolling bearing fault feature extraction method proposed in this paper is able to extract rich fault information.

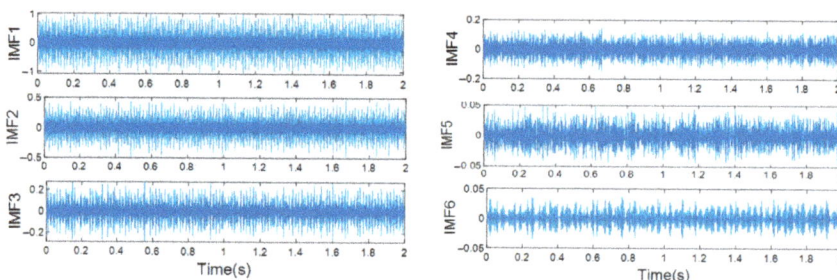

Figure 13. Decomposed result by EEMD.

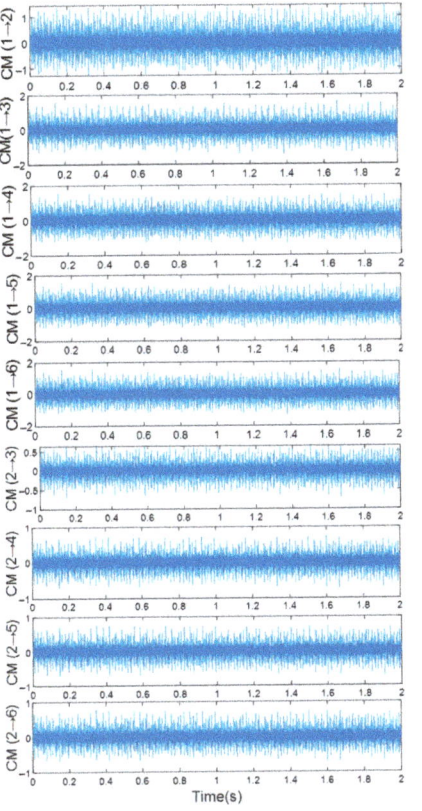
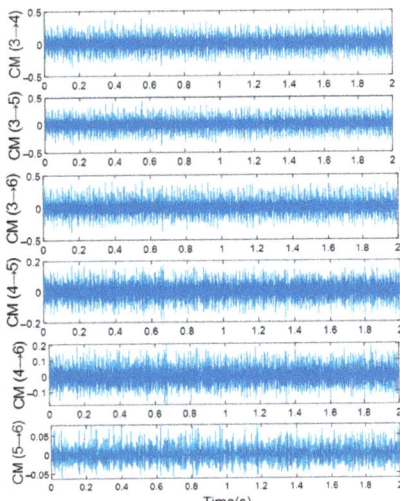

Figure 14. Extracted CMs result.

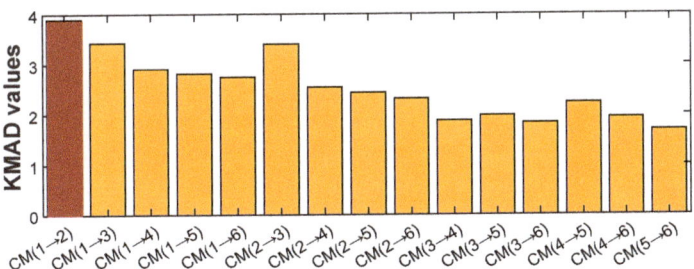

Figure 15. Sensitive CM selection using KMAD.

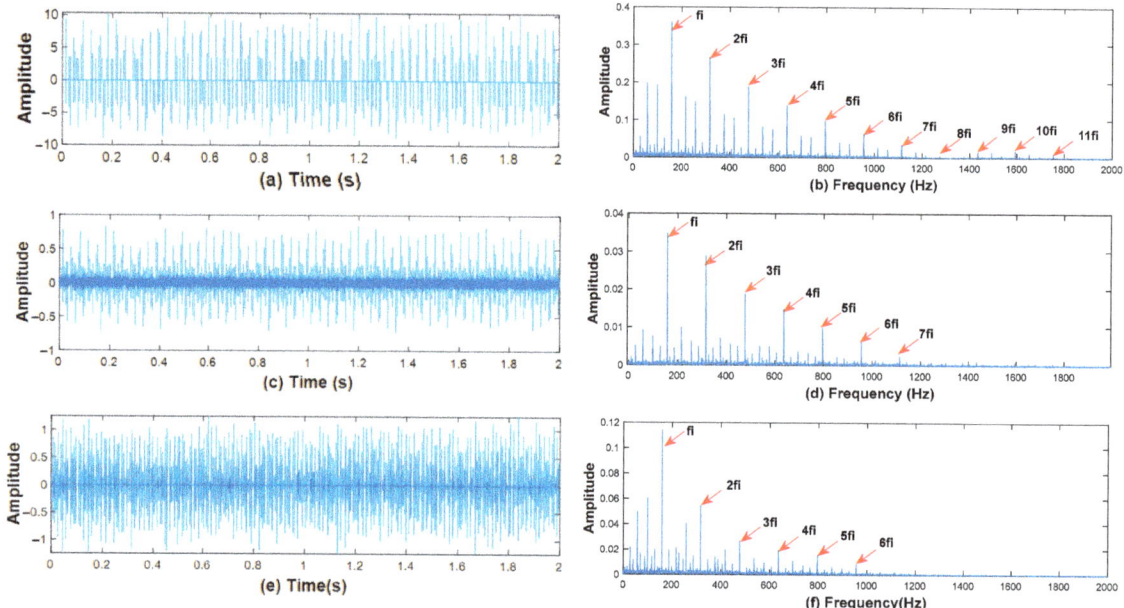

Figure 16. Inner race fault diagnosis results for processing the selected CM using: proposed method (**a**,**b**); conventional MED (**c**,**d**); and wavelet denoising (**e**,**f**).

5.1.2. Advantages of the Proposed Techniques for Inner Race Fault Diagnosis

Figure 16c,d shows the results of processing $CM_{1\rightarrow2}$ by the conventional MED. As shown in Figure 16c, the fault impulses are emphasized, and the noise level is reduced. However, it can be seen that some noise interference still exists. According to Figure 16a, noise interference has been reduced effectively. From the envelope spectrum in Figure 16d, we can distinguish only the inner race fault characteristic frequency f_i and six harmonics ($2f_i$, $3f_i$, $4f_i$, $5f_i$, $6f_i$, and $7f_i$). By comparing it with Figure 16b, it is clear that we can get more fault information. The comparison results show that the enhanced MED performs better than the MED in improving defect detection. To show the enhanced MED approach's superiority in eliminating noise, the wavelet de-noised method is performed on the selected combination $CM_{1\rightarrow2}$. As shown in Figure 16e, although the noise is reduced, the fault impulses are not highlighted as in Figure 16a. In addition, the extracted fault frequency and its harmonics in Figure 16f are not as excellent as those in Figure 16b. In this case, it can be said that the inability of the wavelet de-noising method to reduce noise effectively makes it difficult to extract rich fault information from the combination $CM_{1\rightarrow2}$. The comparison results demonstrate that the enhanced MED performs better than the wavelet de-noising method in eliminating noise. The amplitudes of the inter-harmonics shown in Figure 16b,d,f are much smaller than those shown in Figure 9b,d,f, respectively. This is due to the fact that a signal with high noise ($\sigma^2 = 0.7^2$) is created in the simulation. This makes it more difficult to eliminate noise interference in the simulated signal than in the experimental signal. As a result, the amplitude of the noise interference will mix with the inter-harmonics. To illustrate the advantages of the KMAD selection indicator, this paper conducted a comparison with the IMF selection method using kurtosis. Table 2 shows the kurtosis values of the first six IMFs. It is evident that IMF2 has the highest value among all the decomposition results, so it is selected as the sensitive IMF. IMF2 was processed using the enhanced MED approach. From the envelope spectrum of Figure 17a, it is clear that the extracted fault information is less than the extracted fault information in Figure 17b. This shows that the combination $CM_{1\rightarrow2}$ holds rich fault feature information.

The KMAD indicator selected $CM_{1\to 2}$ as an appropriate combination of suitable IMFs, namely IMF1 and IMF2. As a result, if we take only IMF2, the information in IMF1 will be lost. This demonstrates that utilizing the KMAD selection indicator to select the appropriate combination overcomes the disadvantage of using kurtosis to choose the sensitive IMF and guarantees no information about the fault is lost.

Table 2. Kurtosis values of each IMF for Inner Race Fault Diagnosis.

IMF	Kurtosis
IMF1	4.6903
IMF2	4.7682
IMF3	4.3248
IMF4	3.0268
IMF5	2.5395
IMF6	2.6651

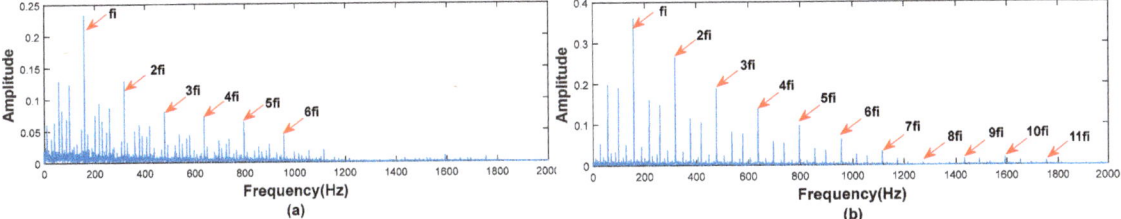

Figure 17. Diagnosis results using: (a) sensitive IMF-based Kurtosis; (b) sensitive CM-based KMAD.

5.2. Case 2: Diagnosis of the Outer Race Fault

The vibration signal in this case is caused by an outer race fault, with the shaft rotating at 1797 rpm and no load applied. The size of the fault is 0.021 inches, and the calculated fault characteristic frequency is 107.01 Hz. Taking 24,000 data points for analysis, Figure 18a shows the measured bearing signal with an outer race fault. It can be seen that the noise prevents the periodic impulses from being extracted. From the envelope spectrum in Figure 18b, although the fault characteristic frequency f_o and the first harmonic $2f_o$ can be extracted, the remaining harmonics are shrouded in noise interference. Therefore, this fault signal necessitates pre-processing to improve fault detection.

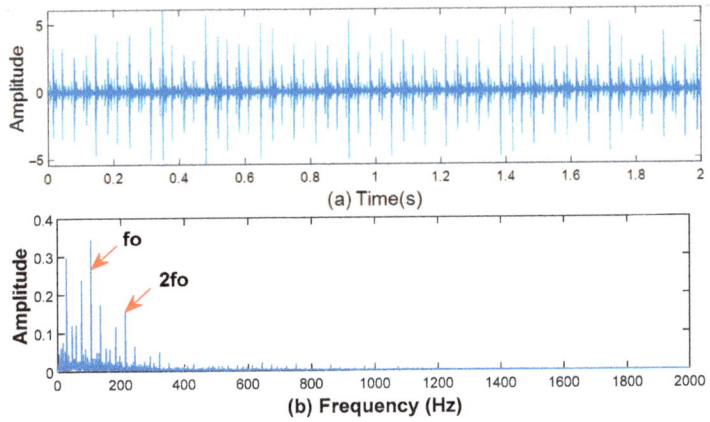

Figure 18. Experimental outer race defect: (a) waveform; (b) envelope spectrum.

5.2.1. Analysis of the Proposed Method

In the first step, CMEEMD extracts the CMs from the experimental outer race fault signal. Using the CMs extraction technique described in Section 3.1, fifteen CMs are created from the first six IMFs. Figure 19 shows the resulting IMFs, and Figure 20 shows the extracted CMs. The next step is to determine which combination is the most sensitive. The time-domain waveforms of each combination in Figure 20 show that there is no noticeable difference between the CMs. It is impossible to directly recognize the combination that combines only the useful IMFs. As a result, the suggested KMAD indicator is used to identify the appropriate combination. Figure 21 shows the KMAD values for each combination. The combination $CM_{1 \to 2}$ has the highest value. This indicates that it is a combination of sensitive IMFs, i.e., IMF1 and IMF2. Following that, the combination $CM_{1 \to 2}$ was processed using the enhanced MED approach. First, MED highlights the fault impulses of $CM_{1 \to 2}$. Then, the MED output is treated to the de-noised method derived from the three-sigma rule. As shown in Figure 22a, the noise is minimized, and the fault impulses are prominent. From the envelope spectrum in Figure 22b, we can accurately extract the outer race fault characteristic frequency f_o and nine harmonics ($2f_o$, $3f_o$, $4f_o$, $5f_o$, $6f_o$, $7f_o$, $8f_o$, $9f_o$, and $10f_o$). This implies that the proposed method for bearing fault feature extraction can effectively extract rich fault information.

5.2.2. Advantages of the Proposed Techniques for Outer Race Fault Diagnosis

The results of processing $CM_{1 \to 2}$ by MED are shown in Figure 22c,d. As seen in Figure 22c, the noise level is decreased, and the fault impulses are accentuated. However, there still exists noise interference. Compared to Figure 22a, noise interference has been significantly reduced. Analyzing the envelope spectrum in Figure 22d, it can be seen that we can extract less fault information than we can in Figure 22b. It is evident from the comparison results that the enhanced MED is more effective in improving fault detection compared to the MED. The wavelet de-noising method is performed on the selected combination, and the results are shown in Figure 22e,f. Although the noise has been reduced to a certain extent in Figure 22e, the extracted fault frequency and its harmonics in Figure 22f are less accurate than those extracted in Figure 22b. In this case, the inability of the wavelet de-noising to successfully decrease noise prevents the extraction of rich fault information from the combination $CM_{1 \to 2}$. The results of the comparison confirm that the proposed enhanced MED eliminates noise better than the wavelet de-noising method. To show the advantages of the CM selection method using KMAD, this paper performs a comparison with the IMF selection method using kurtosis. The kurtosis values for the first six IMFs are presented in Table 3. It appears that IMF2 has the highest value, so it is selected as a sensitive IMF. Next, IMF2 was treated using the enhanced MED approach. Based on the envelope spectrum of Figure 23a, we can extract only the outer race fault characteristic frequency f_o and three harmonics ($2f_o$, $3f_o$, $4f_o$). By comparing it with Figure 23b, it is clear that we can extract more fault information (f_o, $2f_o$, $3f_o$, $4f_o$, $5f_o$, $6f_o$, $7f_o$, $8f_o$, $9f_o$, and $10f_o$). This indicates that the selected combination contains rich defect information. The KMAD indicator identified $CM_{1 \to 2}$ as an appropriate combination of suitable IMFs, i.e., IMF1 and IMF2. Therefore, if we only select IMF2, the fault information in IMF1 will be wasted. This demonstrates that selecting the appropriate combination using the proposed indicator overcomes the disadvantage of the IMF selection using kurtosis to assure that no defect information is wasted.

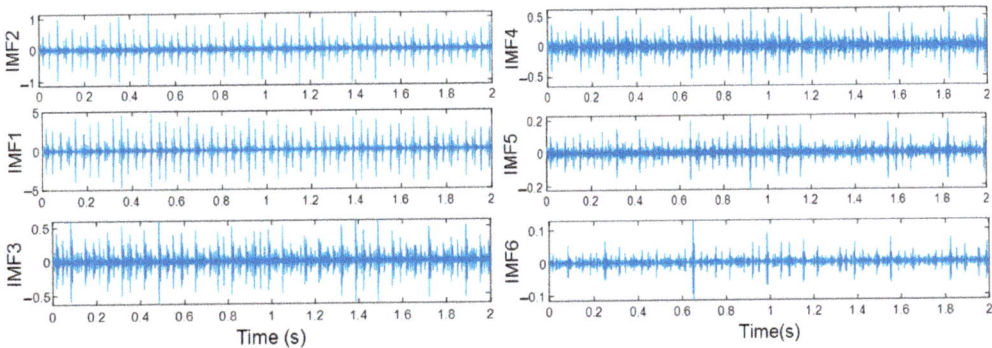

Figure 19. Decomposed result by EEMD.

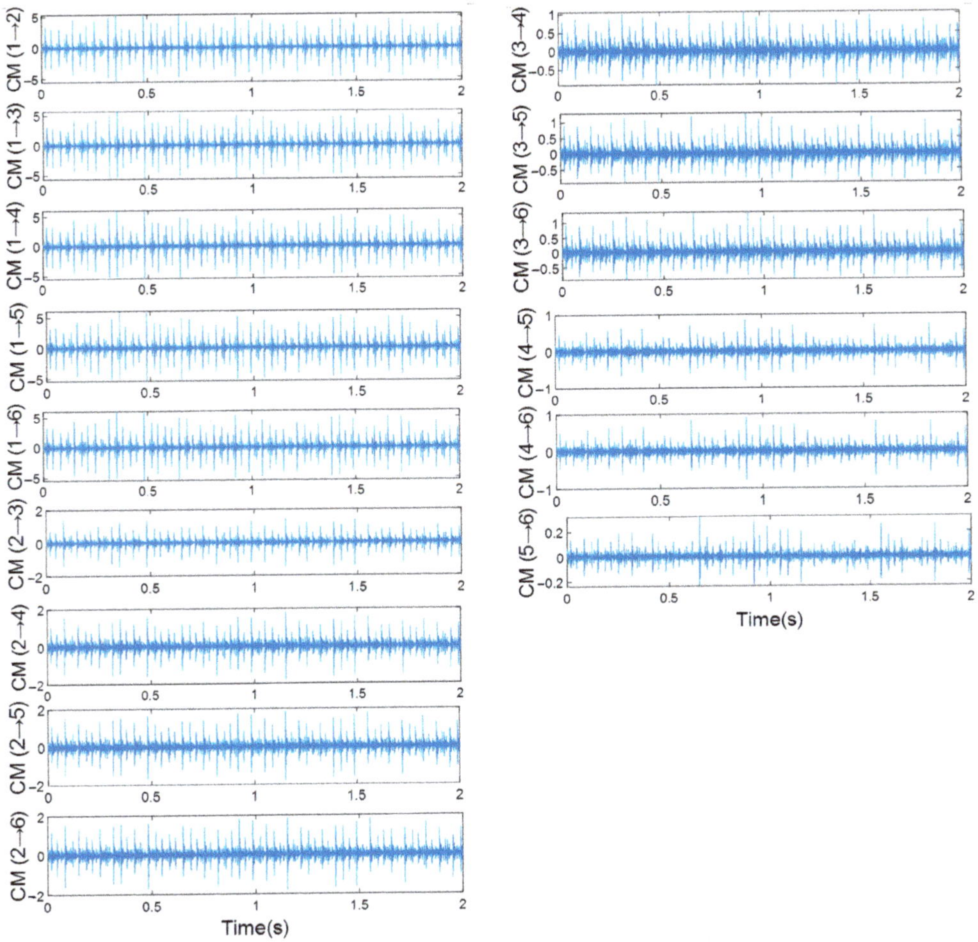

Figure 20. Extracted CMs result.

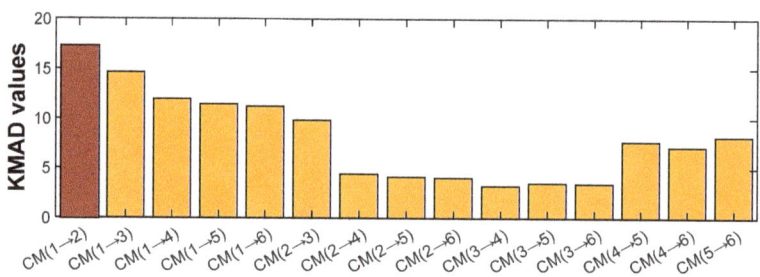

Figure 21. Sensitive CM selection using KMAD.

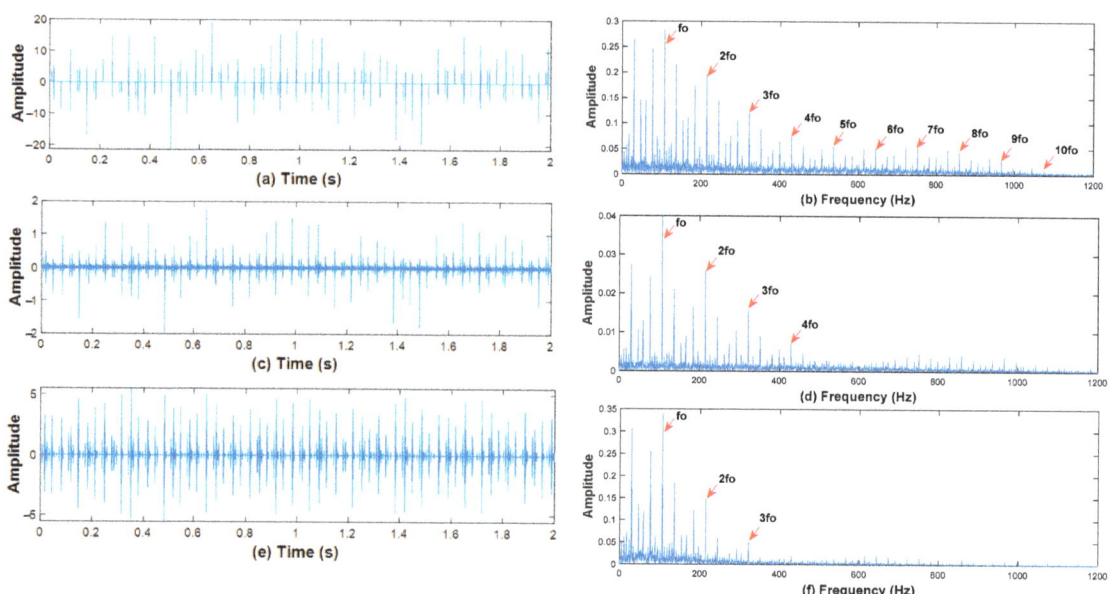

Figure 22. Outer race fault diagnosis results for processing the selected CM using: proposed method (**a**,**b**); conventional MED (**c**,**d**); and wavelet denoising (**e**,**f**).

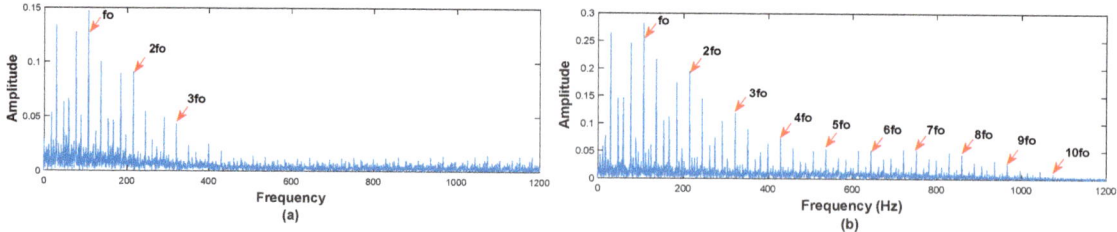

Figure 23. Diagnosis results using: (**a**) sensitive IMF-based Kurtosis; (**b**) sensitive CM-based KMAD.

Table 3. Kurtosis values of each IMF for Outer Race Fault Diagnosis.

IMF	Kurtosis
IMF1	17.7045
IMF2	25.1902
IMF3	10.6024
IMF4	8.8478
IMF5	10.0602
IMF6	11.0606

5.3. Case 3: Diagnosis of the Ball Bearing Fault

The ball race fault in this case generates the vibration signal. The shaft speed is 1772 rpm, the load is 1 hp, and the fault size is 0.028 inches. The calculated fault characteristic frequency for the ball race is 139.18 Hz based on Equation (28). Analyzing 24,000 data points, the bearing signal with a ball race fault is shown in Figure 24a. Due to the noise, it is difficult to distinguish the impact characteristics. From the envelope spectrum in Figure 24b, although the fault characteristic frequency f_b can be distinguished, its harmonics are masked by noise interference. To improve fault detection, this fault signal requires a pre-processing step.

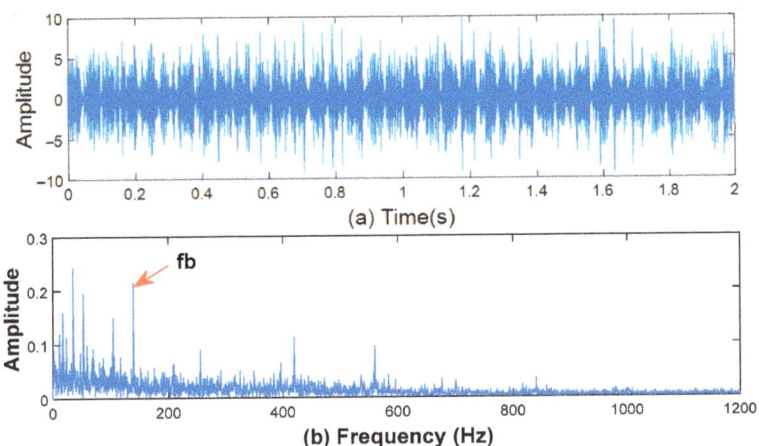

Figure 24. Experimental ball bearing defect: (**a**) waveform; (**b**) envelope spectrum.

5.3.1. Analysis of the Proposed Method

First, CMEEMD extracts the CMs of adjoining modes resulting from the decomposition of the ball defect vibration signal. The first six IMFs produce fifteen CMs using the CMs extraction technique described in Section 3.1. The obtained IMFs are shown in Figure 25, and the extracted CMs are shown in Figure 26. Identifying the most sensitive combination is the next step. According to Figure 26, there is no noticeable difference between the CMs based on their time-domain waveforms. Directly identifying the combination of useful IMFs is impossible. Therefore, the suggested KMAD indicator is used to identify the appropriate combination. According to Figure 27, the combination $CM_{1\to 2}$ has the highest KMAD value. Accordingly, it indicates that it combines sensitive IMFs, i.e., IMF1 and IMF2. The combination $CM_{1\to 2}$ was then performed using the enhanced deconvolution approach presented here. The noise is reduced considerably as shown in Figure 28a, and rich fault information (f_b, $2f_b$, $3f_b$, $4f_b$, $5f_b$, $6f_b$, $7f_b$, $8f_b$, and $9f_b$) can be extracted from the envelope spectrum presented in Figure 28b. This suggests that the proposed strategy can greatly enhance fault identification. Additionally, this demonstrates the validity of the proposed strategy for bearing fault feature extraction.

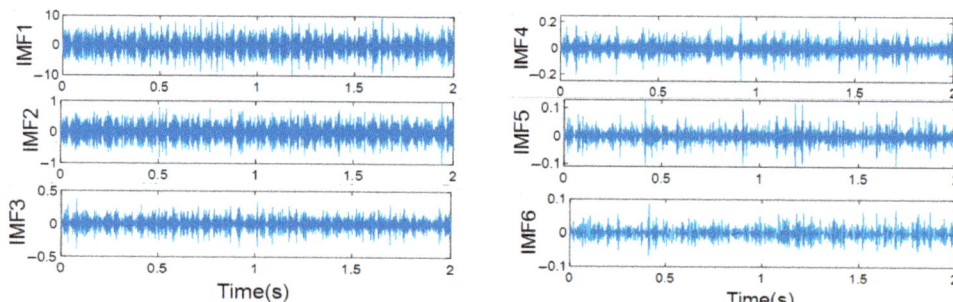

Figure 25. Decomposed result by EEMD.

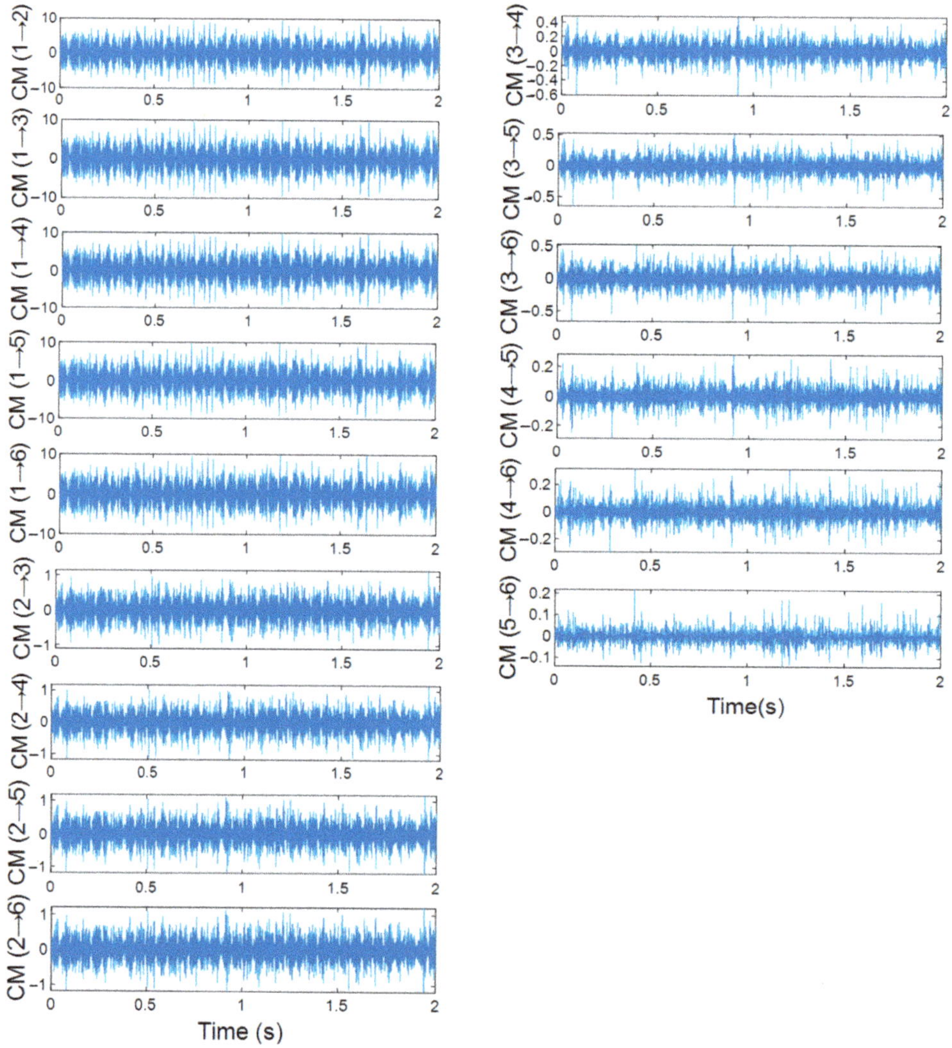

Figure 26. Extracted CMs result.

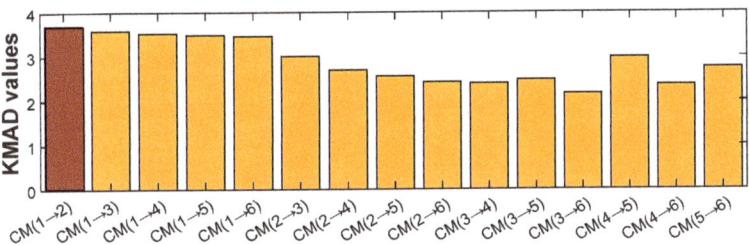

Figure 27. Effective CM selection based on KMAD.

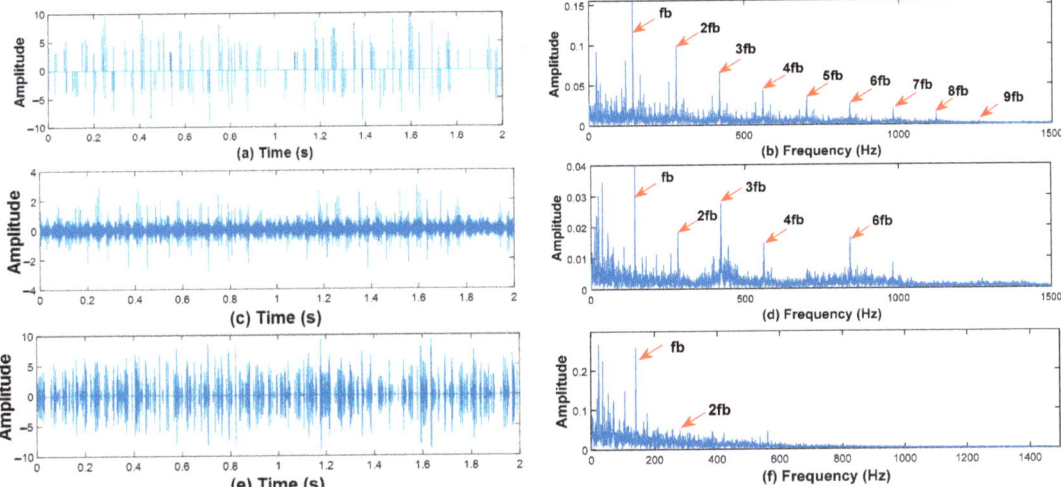

Figure 28. Ball fault diagnosis results for processing the selected CM using: proposed method (**a**,**b**); conventional MED (**c**,**d**); and wavelet denoising (**e**,**f**).

5.3.2. Advantages of the Proposed Techniques for Ball Bearing Fault Diagnosis

Figure 28c,d shows the results of processing the combination $CM_{1\rightarrow 2}$ by the MED. As shown in Figure 28c, despite the noise level reduction, noise interference is still present. Compared to Figure 28a, the noise interference has been successfully minimized. Based on the envelope spectrum in Figure 28d, we can distinguish only the characteristic frequency of the ball race fault f_b and five harmonics ($2f_b$, $3f_b$, $4f_b$, $6f_b$, $7f_b$). Comparing it with Figure 28b, it is clear that the fault frequency with its multiplication components are extracted perfectly. It is evident from the results of the comparison that the enhanced MED is better than the MED for improving fault detection. The wavelet de-noised method is performed on the selected combination, and the results are shown in Figure 28e,f. Although the noise has been reduced to a certain extent in Figure 28e, the envelope spectrum presented in Figure 28f shows that we can distinguish only the characteristic frequency f_b and the first harmonic, whereas Figure 28b shows that we can perfectly extract fault information (f_b, $2f_b$, $3f_b$, $4f_b$, $5f_b$, $6f_b$, $7f_b$, $8f_b$, and $9f_b$). In this case, the inability of the wavelet de-noising approach to successfully decrease noise prevents the extraction of rich fault information from the combination $CM_{1\rightarrow 2}$. It is evident from the comparison results that the enhanced MED suppresses noise more effectively than the wavelet de-noising technique. As an illustration of the advantages of the proposed CM selection method, we have compared it to the IMF selection method using maximum kurtosis. From Table 4, it can be seen that IMF5 has the highest value, so it is selected as a sensitive IMF. This IMF was processed using the proposed enhanced MED, and the envelope spectrum is shown in Figure 29a.

It is clear that no information about the defect can be extracted. This is due to the fact that the use of maximum kurtosis to select the sensitive IMF failed in this case, while the envelope spectrum in Figure 29b illustrated rich fault information. This is because the KMAD indicator proposed here succeeds in selecting the combination of valuable IMFs and proves its superiority for choosing the appropriate combination of useful IMFs.

Table 4. Kurtosis values of each IMF.

IMF	Kurtosis
IMF1	3.9782
IMF2	3.6592
IMF3	4.4182
IMF4	4.4670
IMF5	5.0951
IMF6	3.5202

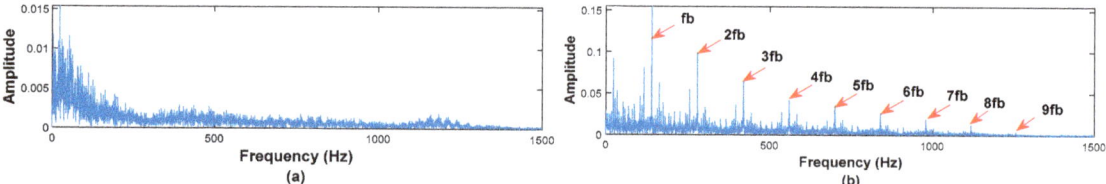

Figure 29. Diagnosis results using: (**a**) sensitive IMF-based kurtosis; (**b**) sensitive CM-based KMAD.

6. Conclusions

A novel rolling bearing fault feature extraction method is presented here, composed of the following proposed ideas: CMEEMD, the KMAD selection indicator, and an enhanced deconvolution approach. Firstly, the proposed CMEEMD extracts all the CMs from the original bearing vibration signal. A selection indicator named KMAD is proposed to identify the appropriate combination of suitable IMFs. This step aims to directly obtain a signal containing the most characteristic information about the fault, without going through the IMFs selection and reconstruction processes, and guaranteeing that no defect information is lost. Secondly, due to the effect of background noise, it is difficult to obtain rich fault information. Therefore, the proposed enhanced MED is performed on the selected combination. The principle of the enhanced MED is to minimize the noise of the MED output to obtain better analysis results. The selection method used in this paper has been applied to several other bearing vibration signals. From these experimental data, we found that the selected combination is most often $CM_{1\to2}$; however, in rare cases it can also be $CM_{1\to3}$ and $CM_{2\to3}$. On the other hand, several researchers confirm that the bearing defect information is included in the first IMFs. This supports and confirms the validity of the presented method.

The analysis of the simulated signal (presented in Section 4) and experimental rolling bearing cases (inner race, outer race, and ball race presented in Section 5) leads to the following results being concluded:

1. Compared to the MED technique, the enhanced MED presented in this paper is more robust in revealing defect pulses (taking Figure 9 as an example).
2. Comparison with the wavelet de-noising method demonstrated that the enhanced MED performs well with noise suppression and is more effective in revealing fault information (taking Figure 28 as an example).
3. Compared results between the sensitive IMF using maximum kurtosis and the sensitive CM using the proposed KMAD indicate that the CM selected contains rich fault feature information (taking Figure 17 as an example).

4. CMEEMD and KMAD proposed herein solves the drawback of the IMF selection method by using the maximum kurtosis value to ensure that no information about the defect is wasted (taking Figure 23 as an example).
5. In contrast to the conventional IMF selection method that failed to identify the appropriate IMF for the ball defect, the KMAD indicator was successful in selecting the appropriate combination of useful IMFs (see Figure 29).
6. The analysis of simulated and experimental rolling bearing signals confirms that the proposed strategy for bearing fault diagnosis can greatly enhance fault detection and effectively extract rich fault information (taking Figures 9 and 28 as examples).

Author Contributions: Conceptualization, Y.D.; methodology, Y.D., N.B., R.P., A.C.M., R.R. and S.S.; validation, Y.D., N.B. and A.C.M.; resources, Y.D., N.B. and A.C.M.; writing—original draft preparation, Y.D., N.B., A.C.M., R.R., R.P. and S.S.; writing—review and editing, Y.D., N.B., R.P., A.C.M., R.R. and S.S; visualization, Y.D., A.C.M., R.R., N.B., R.P. and N.B.; supervision, Y.D., R.R., R.P., N.B., A.C.M. and S.S.; paper submitting, R.P., R.R., N.B. and Y.D. All authors have read and agreed to the published version of the manuscript.

Funding: This research received no external funding.

Data Availability Statement: Not applicable.

Conflicts of Interest: The authors declare no conflict of interest.

References

1. Deekshit Kompella, K.C.; Venu Gopala Rao, M.; Srinivasa Rao, R. Bearing fault detection in a 3 phase induction motor using stator current frequency spectral subtraction with various wavelet decomposition techniques. *Ain Shams Eng. J.* **2018**, *9*, 2427–2439. [CrossRef]
2. Kumar, P.S.; Kumaraswamidhas, L.A.; Laha, S.K. Selecting effective intrinsic mode functions of empirical mode decomposition and variational mode decomposition using dynamic time warping algorithm for Rolling Element Bearing Fault diagnosis. *Trans. Inst. Meas. Control* **2018**, *41*, 1923–1932. [CrossRef]
3. Chen, L.; Xu, G.; Zhang, S.; Yan, W.; Wu, Q. Health indicator construction of machinery based on end-to-end trainable convolution recurrent neural networks. *J. Manuf. Syst.* **2020**, *54*, 1–11. [CrossRef]
4. Huang, N.E.; Shen, Z.; Long, S.R.; Wu, M.C.; Shih, H.H.; Zheng, Q.; Yen, N.-C.; Tung, C.C.; Liu, H.H. The empirical mode decomposition and the Hilbert spectrum for nonlinear and non-stationary time series analysis. *Proc. R. Soc. Lond. Ser. A Math. Phys. Eng. Sci.* **1998**, *454*, 903–995. [CrossRef]
5. Zheng, J.; Huang, S.; Pan, H.; Jiang, K. An improved empirical wavelet transform and refined composite multiscale dispersion entropy-based fault diagnosis method for rolling bearing. *IEEE Access* **2020**, *8*, 168732–168742. [CrossRef]
6. Wu, Z.; Huang, N. Ensemble empirical mode decomposition: A noise-assisted data analysis method. *Adv. Adapt. Data Anal.* **2009**, *1*, 1–41. [CrossRef]
7. Lei, Y.; Lin, J.; He, Z.; Zuo, M.J. A review on empirical mode decomposition in fault diagnosis of rotating machinery. *Mech. Syst. Signal Process.* **2013**, *35*, 108–126. [CrossRef]
8. Wang, H.; Chen, J.; Dong, G. Feature extraction of rolling bearing's early weak fault based on EEMD and tunable Q-factor wavelet transform. *Mech. Syst. Signal Process.* **2014**, *48*, 103–119. [CrossRef]
9. Yang, F.; Kou, Z.; Wu, J.; Li, T. Application of mutual information-sample entropy based Med-ICEEMDAN de-noising scheme for weak fault diagnosis of hoist bearing. *Entropy* **2018**, *20*, 667. [CrossRef]
10. Li, J.; Tong, Y.; Guan, L.; Wu, S.; Li, D. A UV-visible absorption spectrum denoising method based on EEMD and an improved universal threshold filter. *RSC Adv.* **2018**, *8*, 8558–8568. [CrossRef]
11. Ricci, R.; Pennacchi, P. Diagnostics of gear faults based on EMD and automatic selection of intrinsic mode functions. *Mech. Syst. Signal Process.* **2011**, *25*, 821–838. [CrossRef]
12. Li, Z.; Shi, B. Research of fault diagnosis based on sensitive intrinsic mode function selection of EEMD and Adaptive Stochastic Resonance. *Shock Vib.* **2016**, *2016*, 2841249. [CrossRef]
13. Ma, J.; Wu, J.; Wang, X. Incipient fault feature extraction of rolling bearings based on the MVMD and teager energy operator. *ISA Trans.* **2018**, *80*, 297–311. [CrossRef]
14. Luo, C.; Jia, M.P.; Wen, Y. The Diagnosis Approach for Rolling Bearing Fault Based on Kurtosis Criterion EMD and Hilbert Envelope Spectrum. In Proceedings of the 2017 IEEE 3rd Information Technology and Mechatronics Engineering Conference (ITOEC), Chongqing, China, 3–5 October 2017.
15. Damine, Y.; Megherbi, A.C.; Sbaa, S.; Bessous, N. Study of the IMF Selection Methods Using Kurtosis Parameter for Bearing Fault Diagnosis. In Proceedings of the 2022 IEEE 19th International Multi-Conference on Systems, Signals & Devices (SSD), Setif, Algeria, 6–10 May 2022.

16. Pennacchi, P.; Ricci, R.; Chatterton, S.; Borghesani, P. Effectiveness of med for fault diagnosis in roller bearings. *Springer Proc. Phys.* **2011**, *139*, 637–642. [CrossRef]
17. Chatterton, S.; Ricci, R.; Pennacchi, P.; Borghesani, P. Signal Processing Diagnostic Tool for rolling element bearings using EMD and Med. *Lect. Notes Mech. Eng.* **2013**, 379–388. [CrossRef]
18. Ding, J.; Huang, L.; Xiao, D.; Jiang, L. A fault feature extraction method for rolling bearing based on intrinsic time-scale decomposition and AR minimum entropy deconvolution. *Shock Vib.* **2021**, *2021*, 1–19. [CrossRef]
19. Zhao, H.; Min, F.; Zhu, W. Test-cost-sensitive attribute reduction of data with normal distribution measurement errors. *Math. Probl. Eng.* **2013**, *2013*, 6673965. [CrossRef]
20. Fang, K.; Zhang, H.; Qi, H.; Dai, Y. Comparison of EMD and EEMD in Rolling Bearing Fault Signal Analysis. In Proceedings of the 2018 IEEE International Instrumentation and Measurement Technology Conference (I2MTC), Houston, TX, USA, 14–17 May 2018.
21. Wiggins, R.A. Minimum entropy deconvolution. *Geoexploration* **1978**, *16*, 21–35. [CrossRef]
22. González, G.; Badra, R.E.; Medina, R.; Regidor, J. Period estimation using minimum entropy deconvolution (MED). *Signal Process.* **1995**, *41*, 91–100. [CrossRef]
23. Sawalhi, N.; Randall, R.B.; Endo, H. The enhancement of fault detection and diagnosis in rolling element bearings using minimum entropy deconvolution combined with spectral kurtosis. *Mech. Syst. Signal Process.* **2007**, *21*, 2616–2633. [CrossRef]
24. Shojae Chaeikar, S.; Manaf, A.A.; Alarood, A.A.; Zamani, M. PFW: Polygonal fuzzy weighted—An SVM kernel for the classification of overlapping data groups. *Electronics* **2020**, *9*, 615. [CrossRef]
25. Qin, B.; Luo, Q.; Zhang, J.; Li, Z.; Qin, Y. Fault frequency identification of rolling bearing using reinforced ensemble local mean decomposition. *J. Control Sci. Eng.* **2021**, *2021*, 2744193. [CrossRef]
26. Wang, H.-D.; Deng, S.-E.; Yang, J.-X.; Liao, H. A fault diagnosis method for rolling element bearing (REB) based on reducing Reb Foundation vibration and noise-assisted vibration signal analysis. *Proc. Inst. Mech. Eng. Part C J. Mech. Eng. Sci.* **2018**, *233*, 2574–2587. [CrossRef]
27. Zhen, D.; Guo, J.; Xu, Y.; Zhang, H.; Gu, F. A novel fault detection method for rolling bearings based on non-stationary vibration signature analysis. *Sensors* **2019**, *19*, 3994. [CrossRef] [PubMed]
28. Dibaj, A.; Hassannejad, R.; Ettefagh, M.M.; Ehghaghi, M.B. Incipient fault diagnosis of bearings based on parameter-optimized VMD and Envelope Spectrum Weighted Kurtosis index with a new sensitivity assessment threshold. *ISA Trans.* **2021**, *114*, 413–433. [CrossRef]
29. Chen, J.; Yu, D.; Yang, Y. The application of energy operator demodulation approach based on EMD in machinery fault diagnosis. *Mech. Syst. Signal Process.* **2007**, *21*, 668–677. [CrossRef]
30. Yang, Y.; Yu, D.; Cheng, J. A fault diagnosis approach for roller bearing based on IMF envelope spectrum and SVM. *Measurement* **2007**, *40*, 943–950. [CrossRef]
31. Cheng, Y.; Wang, Z.; Chen, B.; Zhang, W.; Huang, G. An improved complementary ensemble empirical mode decomposition with adaptive noise and its application to rolling element bearing fault diagnosis. *ISA Trans.* **2019**, *91*, 218–234. [CrossRef]
32. Sun, Y.; Yu, J. Fault detection of rolling bearing using sparse representation-based adjacent signal difference. *IEEE Trans. Instrum. Meas.* **2021**, *70*, 1–16. [CrossRef]
33. Liu, T.; Chen, J.; Dong, G.; Xiao, W.; Zhou, X. The fault detection and diagnosis in rolling element bearings using frequency band entropy. *Proc. Inst. Mech. Eng. Part C J. Mech. Eng. Sci.* **2012**, *227*, 87–99. [CrossRef]
34. Download a Data File: Case School of Engineering: Case Western Reserve University. Available online: https://engineering.case.edu/bearingdatacenter/download-data-file (accessed on 24 October 2020).
35. Saruhan, H.; Sandemir, S.; Çiçek, A.; Uygur, I. Vibration analysis of rolling element bearings defects. *J. Appl. Res. Technol.* **2014**, *12*, 384–395. [CrossRef]

Disclaimer/Publisher's Note: The statements, opinions and data contained in all publications are solely those of the individual author(s) and contributor(s) and not of MDPI and/or the editor(s). MDPI and/or the editor(s) disclaim responsibility for any injury to people or property resulting from any ideas, methods, instructions or products referred to in the content.

Article

Analytical Modeling, Analysis and Diagnosis of External Rotor PMSM with Stator Winding Unbalance Fault

Ahmed Belkhadir [1,2], Remus Pusca [1], Driss Belkhayat [2], Raphaël Romary [1,*] and Youssef Zidani [2]

1. Univ. Artois, UR 4025, Laboratoire Systèmes Electrotechniques et Environnement (LSEE), F-62400 Béthune, France
2. Univ. Cadi Ayyad, P.O. Box 549, Laboratoire des Systèmes Electriques, Efficacité Energétique et Télécommunications (LSEEET), Faculty of Sciences and Technologies, Marrakech 40000, Morocco
* Correspondence: raphael.romary@univ-artois.fr

Abstract: Multiple factors and consequences may lead to a stator winding fault in an external rotor permanent magnet synchronous motor that can unleash a complete system shutdown and impair performance and motor reliability. This type of fault causes disturbances in operation if it is not recognized and detected in time, since it might lead to catastrophic consequences. In particular, an external rotor permanent magnet synchronous motor has disadvantages in terms of fault tolerance. Consequently, the distribution of the air-gap flux density will no longer be uniform, producing fault harmonics. However, a crucial step of diagnosis and controlling the system condition is to develop an accurate model of the machine with a lack of turns in the stator winding. This paper presents an analytical model of the stator winding unbalance fault represented by lack of turns. Here, mathematical approaches are used by introducing a stator winding parameter for the analytical modeling of the faulty machine. This model can be employed to determine the various quantities of the machine under different fault levels, including the magnetomotive force, the flux density in the air-gap, the flux generated by the stator winding, the stator inductances, and the electromagnetic torque. On this basis, a corresponding link between the fault level and its signature is established. The feasibility and efficiency of the analytical approach are validated by finite element analysis and experimental implementation.

Keywords: stator winding unbalance fault; external rotor permanent magnet synchronous motor; fault harmonics; diagnosis; lack of turns; analytical approach; finite element analysis

Citation: Belkhadir, A.; Pusca, R.; Belkhayat, D.; Romary, R.; Zidani, Y. Analytical Modeling, Analysis and Diagnosis of External Rotor PMSM with Stator Winding Unbalance Fault. *Energies* **2023**, *16*, 3198. https://doi.org/10.3390/en16073198

Academic Editors: Moussa Boukhnifer and Larbi Djilali

Received: 3 March 2023
Revised: 25 March 2023
Accepted: 28 March 2023
Published: 1 April 2023

Copyright: © 2023 by the authors. Licensee MDPI, Basel, Switzerland. This article is an open access article distributed under the terms and conditions of the Creative Commons Attribution (CC BY) license (https:// creativecommons.org/licenses/by/ 4.0/).

1. Introduction

In recent years, external rotor permanent magnet synchronous motors (ER-PMSMs) mounted directly in the wheels of vehicles has been one of the trends in drive systems employed in hybrid and electrical vehicle (HEV) powertrains. The design of this type of motor presents a challenge, as it must be characterized by high durability and energy efficiency [1–3].

Interest in continually developing techniques for the diagnosis of faults in electrical machines is related to several factors. Firstly, the overall number of embedded motors that are employed in different applications, such as industrial systems, renewable energy generating systems, and HEVs [4]. However, owing to the ageing of materials, manufacturing faults, or sever conditions, various types of electrical, mechanical, and magnetic faults can occur in the machine [5], for example: open phase faults, interturn short circuit faults, lack of turns (LTs) faults in the stator winding, eccentricity, demagnetization, and magnetic circuit faults [6]. Hence, the integration of detection strategies, diagnosis, and fault-tolerant control becomes unavoidable. Moreover, during the real operation of the ER-PMSM, LTs faults may emerge due to manufacturing tolerances or ageing issues. Therefore, once the LTs fault appears, the stator current increases to generate enough torque, which leads to torque and speed ripples, and further exacerbates thermal problems. As a result, early diagnosis of the LTs fault is critical for preventing deterioration of ER-PMSM performance and reducing eventual losses by implementing the most effective corrective measures. These measures may consist of repairing the faulty machine or, in certain situations, appropriately reconfiguring the control strategy [6].

Currently, several studies have been conducted on the diagnosis and localization of stator winding faults in ER-PMSMs. Modeling and experiments have been used as the first step in these studies, and research aims to extract electrical signals and quantities such as voltage and currents, mechanical quantities such as speed, torque and vibration, and magnetic field signals such as magnetic flux and density [7–9]. Secondly, appropriate signal processing methods were employed to extract fault characteristics from various signals, identify the mode, assess the severity, and classify the fault [10,11].

The lack of turns (LTs) faults can be detected via different approaches based on signals, data, and models. The first technique aims to identify characteristic fault frequencies in measured ER-PMSM signals [12,13], which are processed using time-frequency signal analysis tools, such as the Fourier transform [14,15], wavelet transforms [16], and the Hilbert–Huang transform [17]. Unlike the Fourier transform, the drawback of wavelets and Hilbert–Huang transforms is their incompatibility with real-time analysis. Additionally, a detection technique using an analysis of the external field, provided by sensors positioned around the machine, using information fusion methods is proposed in [18–21], but these methods require an accurate knowledge of the external stray flux.

Advanced machine learning method are also employed to detect ER-PMSM stator faults. These methods are attractive due to their advanced data processing capabilities combined with external machine signals, such as vibration, acoustic noise, and torque [22]. A one-dimensional convolutional neural network model, which analyzes torque and current signals to diagnose the motor across a wide range of speeds, variable loads, and fault levels, is proposed in [23]. These kinds of advanced algorithms are very efficient, but they require a large amount of computation and historical data to form models and classify localized defects, as well as extremely high hardware requirements.

This research focuses on the diagnosis of LTs faults in ER-PMSMs by examining and analyzing the current and speed spectrum, allowing a simple and powerful implementation of an online fault diagnosis approach. To better understand the influence and consequences of the fault, an analytical approach and finite element model validation were employed. The finite element analysis (FEA) offers the benefits of a well-established application and high computational precision and accuracy. The analytical approach involves the development of a mathematical model that replicates the behavior of the machine in the presence of the fault. This model can be used to predict the impact of the fault on various machine quantities, and can contribute to the development of effective fault diagnosis methods. Overall, the combination of the analytical approach and FEA validation provides comprehensive knowledge of the impact of the LTs fault on ER-PMSMs employed in electric mobility.

The main contributions of this research are as follows: a novel technique and approach for modeling LTs faults in ER-PMSMs. The suggested analytical model requires less computational time and can subsequently provide an accurate reference for real-time fault diagnosis, accuracy, and maintenance. Then, the experimental validation for an ER-PMSM operating in the case of a motor is presented. Experimental measurements that must be taken to ensure a reliable diagnosis are also presented.

The main structure of this paper is as follows: in Section 2, the healthy and faulty analytical models are established to examine how the LTs fault impacts the various electrical and mechanical quantities of the machine, and the FEA is used to verify the effectiveness and accuracy of the proposed analytical model. The experimental setup is provided in Section 3, and the experimental results are in Section 4. Finally, Section 5 summarizes the conclusions and prospects of the presented work.

2. Exhaustive Analysis of LTs Fault

After evoking the secondary consequences at the origin of the stator winding unbalance fault illustrated by LTs, this section aims to provide a global comprehension of the machine in healthy and faulty conditions using an analytical approach and numerical validation.

The development of a mathematical model enabling simulation of the behavior of the ER-PMSM in the two healthy and faulty operating modes is indicated below. This model is based on a 2D extension of the winding function approach to determine the different inductances of the machine, taking into consideration all the space harmonics, the real geometry of the ER-PMSM, as well as the distribution of the windings in the stator slots. The model of the 24 slots and 22 poles of the ER-PMSM with concentric winding is shown in Figure 1. The major specifications of the machine are listed in Table 1.

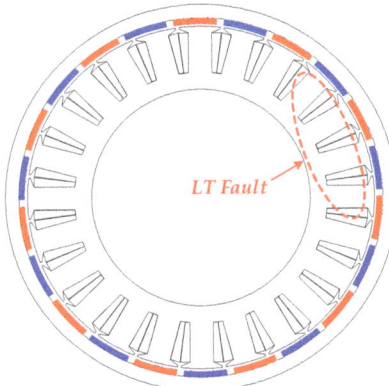

Figure 1. The model of the double-layer fractional-slot concentrated-wound ER-PMSM.

Table 1. ER-PMSM parameters.

Parameter	Symbol	Value	Parameter	Symbol	Value
Rated power (kW)	P	1.5	Inner rotor diameter (mm)	–	183
Rated speed (rpm)	w	600	Length (mm)	L_{axe}	35
DC bus voltage (V)	V_{DC}	150	Air-gap length (mm)	g	1.365
Rated current (A_{RMS})	I_N	11	Stator slot width (mm)	l_e^s	10.37
Rated torque (Nm)	Γ_e	24	Stator tooth width (mm)	l_d^s	21.1431
Number of poles	p	22	Stator fictive slot depth (mm)	p^s	2.0740
Number of phases	m	3	Slot opening width (mm)		2
Number of slots	N^s	24	Magnet thickness (mm)	h_m	3
Number of turns per coil	N_T	22	Residual flux density of PM (T)	B_r	1.26
Outer stator diameter (mm)	–	112	Magnet-arc to pole-pitch ratio (%)	α_p	85.55
Inner stator radius (mm)	R_s	88.6350	Permanent magnets	–	NdFeB N38SH
Outer rotor diameter (mm)	–	186	Magnetic steel	–	M530-50A

The analytical model is developed based on the assumptions given throughout this section, as follow [24]:

(1) Magnetic saturation is negligible;
(2) Ideal ferromagnetic steel and the magnetic energy is concentrated in the air-gap;
(3) Small air-gap relative to the internal diameter of the stator and radial magnetic field (tangential magnetic fields are negligible);
(4) Neglected conductivity and eddy current effects.

2.1. Analytical Approach

2.1.1. Distribution Function of the ER-PMSM

The distribution function of a coil placed in the magnetic circuit delivers information about its position. Therefore, the objective is to apply the winding function approach to compute the inductances of the 24 slots and 22 poles of a double-layer fractional-slot concentrated-wound (FSCW) ER-PMSM (24/22) [24]. Their waveforms are shown in Figure 2.

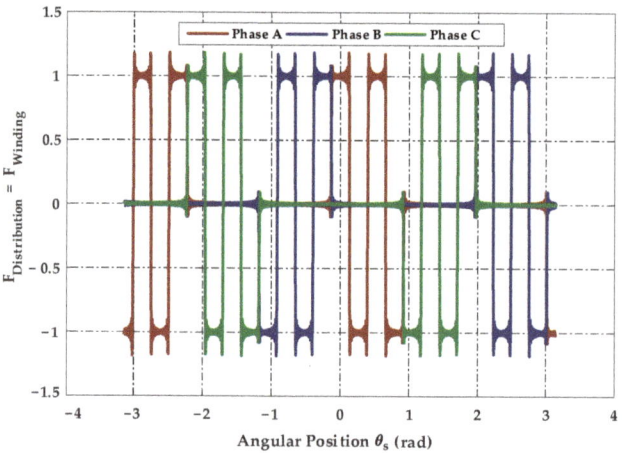

Figure 2. Distribution function of ER-PMSM.

The Fourier decomposition of the distribution function can be given by:

$$F_{distribution} = \frac{P_e}{2\pi} O_c + \frac{2}{\pi} \sum_{h=1}^{+\infty} \frac{1}{h} \sin\left(h P_e \frac{O_c}{2}\right) \cos(h P_e \theta_s) \quad (1)$$

The total distribution function of a winding is the sum of the distribution functions of a winding, and the sum of the distribution functions of the coils in a series of the same phase per pair of poles of this winding.

$$F_{distribution,T}(\theta_s) = \sum_{i=1}^{q} F_{distribution}(\theta_s) \quad (2)$$

where P_e is the winding periodicity, O_c is the coil opening angle, q is the coil number, and θ_s is the angular position in relation to the stator reference axis.

2.1.2. Winding Function of the ER-PMSM

The winding function presents the magnetomotive force (MMF) of a single-turn winding carrying a unit current. We may represent these functions as a series of harmonics due to the winding's periodicity and each phase coil's function, which contains periodic square pulses. In this analysis instance, the mean value of the distribution function is null. However, Figure 2 illustrates the winding function, which is proportional to the distribution function.

$$F_w(\theta_s) = F_{distribution,T}(\theta_s) - \langle F_{distribution,T}(\theta_s) \rangle \quad (3)$$

The spatial harmonic amplitudes of the winding function, obtained from Equation (3), for a stator with concentrated winding around a double-layer tooth is illustrated in Figure 3.

Evidently, the harmonic of rank $h = 11$ had the highest amplitude of 0.4396 for a machine (24/22).

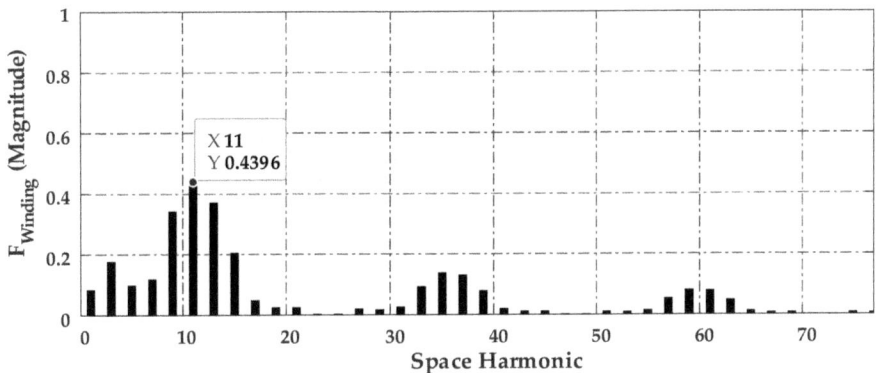

Figure 3. Harmonic spectrum of the winding function.

2.1.3. Winding Factor of the ER-PMSM

The winding factor presents the efficiency of the coil arrangement to create a MMF and determines the capacity of the electromagnetic torque production. We can write a general presentation of the total winding factor K_w of harmonic h, using the voltage phasor diagram [25]:

$$K_w = K_d K_p \quad (4)$$

where K_d is the distribution factor, K_p is the pitch factor, $\alpha_u = p\frac{2\pi}{Q}$ is the slot angle, m is the number of phases, $Q = 2qpm$ is the number of slots, $\tau_p = \frac{\pi D}{2p}$ is the pole pitch, and D is the diameter of the air-gap.

$$K_d = \frac{\sin\left(h\frac{q\alpha_u}{2}\right)}{q\sin\left(h\frac{\alpha_u}{2}\right)} = \frac{\sin\left(h\frac{Q}{2pm}\frac{2\pi p}{2Q}\right)}{q\sin\left(h\frac{2\pi p}{2q2pm}\right)} = \frac{\sin\left(h\frac{\pi}{6}\right)}{q\sin\left(h\frac{\pi}{6q}\right)} \quad (5)$$

$$K_p = \sin\left(h\frac{O_c}{\tau_p}\frac{\pi}{2}\right) \quad (6)$$

According to Figure 4, the spectrum represents the spatial distribution of the winding factor. The maximum torque of the machine generated in the air-gap will be created by the $h11$ component, with a winding factor of 0.949. This component corresponds to the fundamental, with a rank $h = p$.

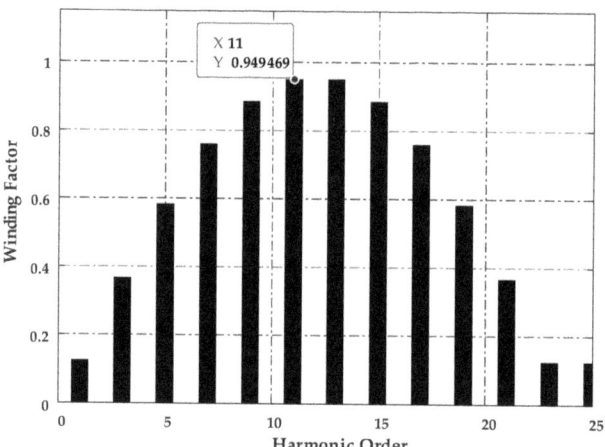

Figure 4. Harmonic spectrum of the winding factor.

2.1.4. Magnetomotive Force (MMF)

A. Healthy MMF of the ER-PMSM

The MMF of a double-layer FSCW stator is rich in harmonics. However, this harmonic content leads to torque ripples, unbalanced saturation, and iron losses [26,27]. For this reason, the combination of slots and number of poles has specific features that must be examined and studied before that the machine is designed. The harmonics of the winding factor (Figure 4) are frequently employed as an indication of the properties of this combination [28].

The spatial MMF distribution of a phase winding is obtained by superimposing the MMF of all its coils. Due to the periodicity of the winding, the healthy MMF of each phase j includes periodic square pulses, and can be represented by a Fourier series decomposition, as follows:

$$\varepsilon_j^s(t,\theta_s) = K_w N_T i_j^s(t) F_w(\theta_s) \tag{7}$$

where N_T is the turns number and $i_j^s(t) = \sqrt{2} I_j^s \sin(\omega t + \varphi_j)$ is the temporal expression of sinusoidal current.

The total healthy MMF generated by the 22 pole, 24 slot FSCW stator is determined using the following formula:

$$\varepsilon^s(t,\theta_s) = \sum_{j=1}^{m} \left(\varepsilon_j^s(t,\theta_s) \right) \tag{8}$$

$$\varepsilon^s(t,\theta_s) = \frac{1}{\pi} m I_{\max} N_T \sum_{h=1}^{+\infty} \frac{1}{h} K_w \sin\left(h P_e \frac{O_c}{2}\right) \cos(h P_e \theta_s - (\omega t - \varphi)) \tag{9}$$

The distribution of the total healthy MMF of the three-phase stator winding is shown in Figure 5 when it is supplied with a balanced three-phase current of $f = 110$ Hz. The harmonic spectrum of the total MMF distribution in the healthy case is determined from Equations (8) and (9), and illustrated in Figure 6. We notice that the harmonic of rank $h11$ is the most dominating of ER-PMSM (24/22).

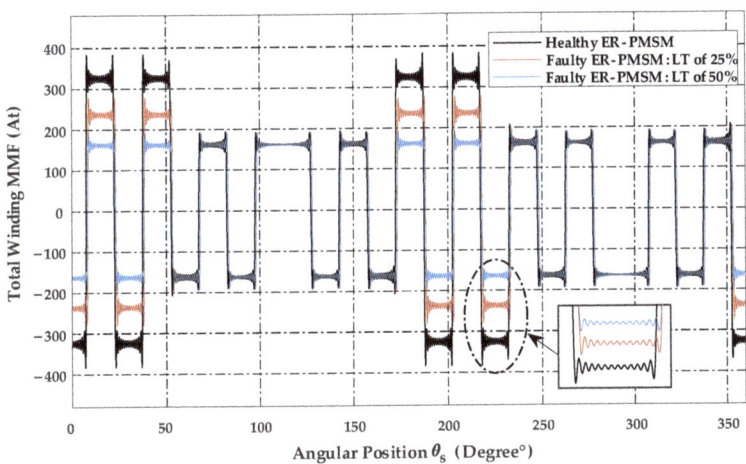

Figure 5. Healthy and faulty stator MMF distribution.

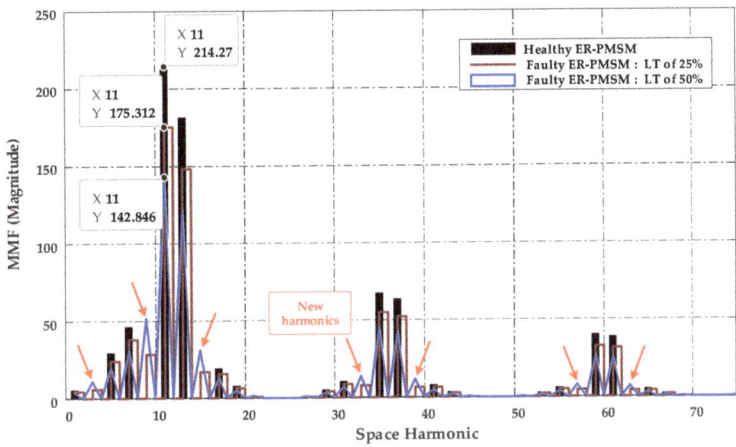

Figure 6. Spatial harmonic spectrum of the total healthy and faulty MMF.

B. Faulty MMF of the ER-PMSM

When a stator winding unbalanced fault of LTs occurs in the machine, the stator MMF will be impacted by the fault. The general analytic model of the faulty MMF is obtained using the previously stated methodologies and is shown in Figure 5, which can be represented by the following expressions:

$$\varepsilon^s_{j,faulty}(t,\theta_s) = K_w F_{w,faulty}(\theta_s) N_T i^s_j(t) \quad (10)$$

$$F_{w,faulty}(\theta_s) = \frac{N_f}{N_T} F_{w,a}(\theta_s) + F_{w,b}(\theta_s)\cos\left(\omega t - \frac{2\pi}{3}\right) + F_{w,c}(\theta_s)\cos\left(\omega t + \frac{2\pi}{3}\right) \quad (11)$$

$$\varepsilon^s_{faulty}(t,\theta_s) = \sum_{j=1}^{+\infty} \varepsilon^s_{j,faulty}(t,\theta_s) \quad (12)$$

The analytical modeling in the faulty state is performed for two fault levels: 25% and 50% LTs in phase A.

143

Where $F_{w,faulty}$ is the winding function in the faulty state, and N_f is the number of faulty stator turns.

A comparison between the harmonic content of the MMF is presented in Figure 6. Evidently, when the machine is subjected to the LTs fault, all of the odd harmonics are present in the spectrum, as well as the appearance of new harmonics that are due to the LTs fault levels (Red arrows).

2.1.5. Air-Gap Flux Density

The permeance function is proportional to the inverse of the air-gap thickness, and depends only on the average permeance of the air-gap and the form of the stator slots.

For determination of the air-gap permeance, a simplified geometry of the motor shown in Figure 7 is considered, with a fictitious model for the slots resulting from the assumptions for a nonsalient FSCW ER-PMSM, and with radial field lines.

$$\wp(\theta_s) = P_0 + \sum_{K_s=1}^{+\infty} P_{K_s} \cos(K_s N^s \theta_s) \quad (13)$$

$$P_0 = \frac{\mu_0}{g + p^s}\left(1 + \frac{p^s r_d^s}{g}\right) \quad (14)$$

$$P_{K_s} = 2\mu_0 \frac{p^s}{g(g + p^s)} \frac{\sin(K_s r_d^s \pi)}{K_s \pi} \quad (15)$$

where $\mu_0 = 4\pi 10^{-7}$ is the permeability of a vacuum approximately equal to that of the air, $p^s = l^e{}_s/5$ is the stator fictive slot depth [18], r_d^s is the stator toothing ratio, and K_s is the permeance rank.

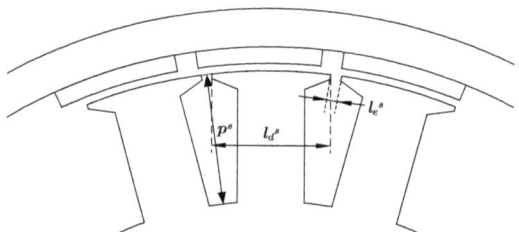

Figure 7. Geometry adopted for the stator slots of the ER-PMSM.

The healthy and faulty air-gap flux density for a sinusoidal current system can be written as the following expression:

$$b^s(t, \theta_s) = \wp(\theta_s)\varepsilon^s(t, \theta_s) \quad (16)$$

$$b^s_{faulty}(t, \theta_s) = \wp(\theta_s)\varepsilon^s_{faulty}(t, \theta_s) \quad (17)$$

Here, it is assumed that the magnetic field distribution would be affected in the fault condition. Depending on the two levels of the LTs fault, there is a reduction in the flux inside the stator core. Based on Figure 9, the magnitude of the flux density is considerably impacted by the fault. The spectrum of the flux density harmonic in the air-gap is depicted in Figure 10. The amplitude of the fundamental component in the healthy condition and no-load case is 0.0652 T.

2.1.6. Inductances in Healthy and Faulty FSCW ER-PMSMs

To be able to determine the different parameters of the machine in healthy and damaged modes for examining the structure of the ER-PMSM, it is crucial to calculate the values of the self and mutual inductances. In order to examine several configurations, these calculations are performed using an analytical approach, as well as a numerical and experimental validation model.

Inductances have a crucial role in this modeling technique since they account for the many effects that might occur in the machine. Accurate modeling will lead to additional information of the signals, and a good compromise in terms of model accuracy.

The inductances of the machine will be determined analytically using the previously stated winding function.

A. Healthy Magnetic Flux

The magnetic flux produced by a phase i flowing through a phase j winding is provided by the following equations:

$$\Phi_{ji,h}(t) = N_T \iint b_i^s(t, \theta_s) ds_j \tag{18}$$

$$\Phi_{ji,h}(t) = (N_T)^2 I_i(t) R_s L_{axe} K_w \int_0^{2\pi} F_{distribution,j}(\theta_s) \wp(\theta_s) F_{w,i}(\theta_s) d\theta_s \tag{19}$$

The self-magnetic flux of phase A is given by:

$$\Phi_{aa,h}(t) = \frac{2}{P_e} R_s L_{axe} (N_T)^2 \frac{2I_{max}}{\pi} \wp(\theta_s) \sum_{h=1}^{+\infty} \left[\frac{1}{h^2} K_w \sin^2\left(hP_e \frac{O_c}{2}\right) \cos(\omega t - \varphi) \right] \tag{20}$$

The mutual magnetic flux between phase A and phase B can be expressed by:

$$\Phi_{ab,h}(t) = \frac{2}{P_e} R_s L_{axe} (N_T)^2 \frac{2I_{max}}{\pi} \wp(\theta_s)$$
$$\sum_{h=1}^{+\infty} \left[\frac{1}{h^2} K_w \sin\left(hP_e \frac{O_c}{2}\right) \left(\sin\left(hP_e \frac{3O_c}{2}\right) - \sin\left(hP_e \frac{O_c}{2}\right)\right) \cos(\omega t - \varphi) \right] \tag{21}$$

where R_s is the inner stator radius.

B. Faulty Magnetic Flux

The self-magnetic flux of phase A in the faulty state is given by:

$$\Phi_{a11,faulty}(t) = \frac{2}{P_e} R_s L_{axe} (N_f)^2 \frac{2I_{max}}{\pi} \wp(\theta_s) \sum_{h=1}^{+\infty} \left[\frac{1}{h^2} K_w \sin^2\left(hP_e \frac{O_c}{2}\right) \cos(\omega t - \varphi) \right] \tag{22}$$

$$\Phi_{a12,faulty}(t) = -\frac{1}{P_e} R_s L_{axe} (N_f)^2 \frac{2I_{max}}{\pi} \wp(\theta_s)$$
$$\sum_{h=1}^{+\infty} \left[\frac{1}{h^2} K_w \sin\left(hP_e \frac{O_c}{2}\right) \left(\sin\left(hP_e \frac{3O_c}{2}\right) - \sin\left(hP_e \frac{O_c}{2}\right)\right) \cos(\omega t - \varphi) \right] \tag{23}$$

The total flux linkage of a phase winding is the sum of the aforementioned, as follows:

$$\Phi_{a,faulty}(t) = \sum_{n=1}^{8} \Phi_{an,faulty}(t) \tag{24}$$

A faulty mutual magnetic flux between phase A and phase b is given by:

$$\Phi_{a1b8,faulty}(t) = \frac{1}{P_e} R_s L_{axe} N_T N_f \frac{2I_{max}}{\pi} \wp(\theta_s)$$
$$\sum_{h=1}^{+\infty} \left[\frac{1}{h^2} K_w \sin\left(hP_e \frac{O_c}{2}\right) \left(\sin\left(hP_e \frac{3O_c}{2}\right) - \sin\left(hP_e \frac{O_c}{2}\right)\right) \cos(\omega t - \varphi) \right] \tag{25}$$

$$\Phi_{a5b4,faulty}(t) = \frac{1}{P_e} R_s L_{axe} N_T N_f \frac{2I_{max}}{\pi} \wp(\theta_s)$$
$$\sum_{h=1}^{+\infty} \left[\frac{1}{h^2} K_w \sin\left(hP_e \frac{O_c}{2}\right) \left(\sin\left(hP_e \frac{3O_c}{2}\right) - \sin\left(hP_e \frac{O_c}{2}\right)\right) \cos(\omega t - \varphi) \right] \tag{26}$$

$$\Phi_{ab,faulty}(t) = \Phi_{a1b8,faulty}(t) + \Phi_{a5b4,faulty}(t) \tag{27}$$

Figure 11 depicts the magnetic flux produced by the three phases with a fault of 25% and 50% of LTs of phase A.

C. Inductances in Healthy and Faulty States

The ER-PMSM inductances are calculated using the approach provided in [25,29], which uses the winding function corresponding to the MMF produced by the stator winding. Accordingly, the self or mutual inductances are determined from the following relation:

$$L_{ji,h} = (N_T)^2 R_s L_{axe} K_w \int_0^{2\pi} F_{distribution,j}(\theta_s) \wp(\theta_s) F_{w,i}(\theta_s) d\theta_s \tag{28}$$

According to the analytical modeling of the faulty magnetic flux proposed and calculated from Equations (25) and (28), the general form of the faulty self and mutual inductance can be expressed as follows:

$$L_{a-A25\%/50\%} = \frac{\Phi_{a,faulty}(t)}{I_a(t)}; \quad M_{a-A25\%/50\%-b} = \frac{\Phi_{ab,faulty}(t)}{I_a(t)} \tag{29}$$

where the faulty inductances depend on the severity of the LTs fault, and the number of faulty stator turns N_f. The analytical calculation of self and mutual inductances is conducted as an example to explain the complete procedure. The other inductances can be derived similarly. Figure 12a–c exhibit, respectively, the healthy and faulty inductances according to the two LTs fault levels. All of the results obtained for healthy and degraded operations by the analytical and numerical approaches are close.

An AC immobilization test is employed for the experimental measurement (Figure 12). In this method, an AC current flows in phase A, while the other two phases are in an open circuit. The self-inductance is then obtained using the RMS voltage and current measured at different rotor positions [30].

2.2. Numerical Validation: Finite Element Analysis

In order to evaluate and validate the analytical approach, a two-dimensional FEA-2D model of the motor is tested and simulated (Figure 1). The analysis of the behavior for the LTs fault, with consideration of the electrical and mechanical quantities of the machine in operational mode, is performed in an open-loop system. All of the testing and analysis will be focused on the intrinsic features of the LTs fault. The main structural parameters and specifications are listed in Table 1. Figure 8 illustrates the study ER-PMSM machine with faulted turns and the mesh of the FEA model.

The impact of the LTs fault is at first examined without the influence of the controllers under two fault severities. The ER-PMSM with 24 slots and 22 poles is used for the fault analysis. The following analyses are performed at a level of 25% and 50% of LTs in phase A of the machine.

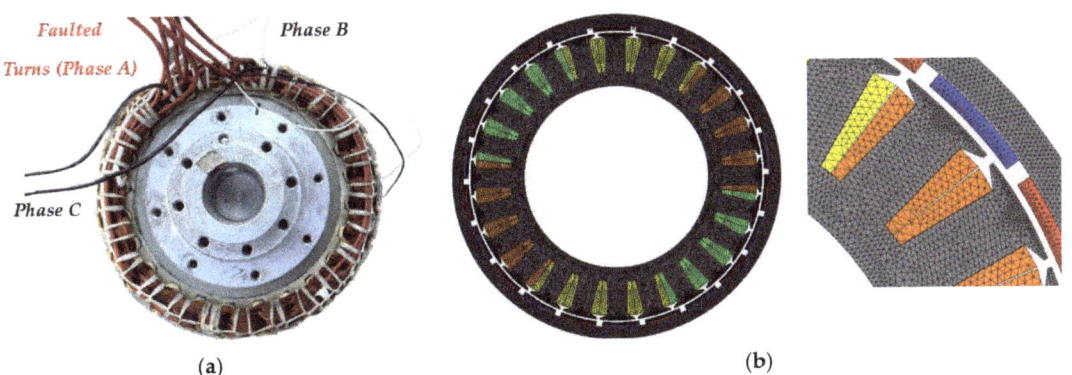

Figure 8. ER-PMSM with LTs in phase A. (**a**) ER-PMSM stator winding with LTs fault in phase A; (**b**) the mesh of the ER-PMSM FEA analysis.

Performance Analysis of ER-PMSM with LTs Fault

Figure 9 illustrates the radial air-gap flux density at no load obtained by the analytical model and the FEA model in the healthy and faulty states according to the two LT fault levels. In the case of an unbalanced stator winding fault, the amplitudes of the no-load flux density waveforms drop. However, all analyses are consistent with the analytical analyses.

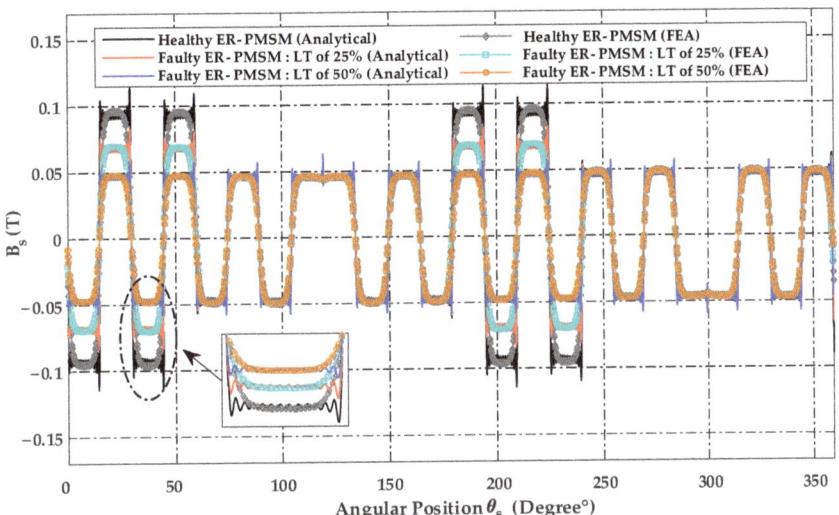

Figure 9. Air-gap flux density distribution in the healthy and faulty cases.

The harmonic spectrum is depicted in Figure 10. The flux density in healthy and faulty circumstances is mainly composed of the fundamental and other harmonics, as well as the appearance of new harmonics attributable to the faults (red arrows). In addition, the amplitude of the fundamental suffers a diminution in the faulty cases compared to the healthy condition.

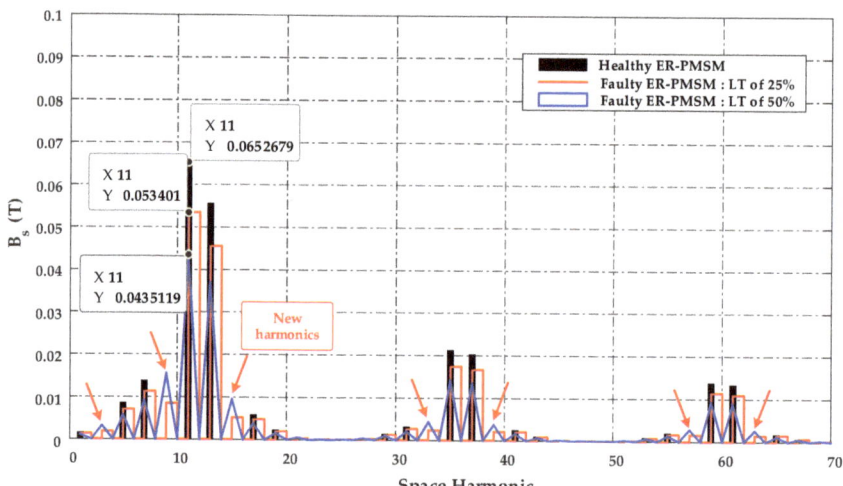

Figure 10. Spatial harmonic spectrum of the healthy and faulty air-gap flux density.

Figure 11 shows the total fluxes generated by the stator power supply in the two situations of the LTs fault of 25% and 50% of phase A, developed theoretically using the analytical model described previously, and the finite element method at a speed of 600 rpm. Due to the change in the stator winding induced by the fault in phase A, an unbalance arises depending on the level of the fault, which leads to torque ripples (Figure 13). This analysis will be used to determine the inductance of the machine in the presence of the fault. The inductances take a critical role in the modeling process, since they consider the many phenomena that might appear within the machine. Accurate modeling will lead to additional information of the signals and a good compromise in terms of the accuracy of the model.

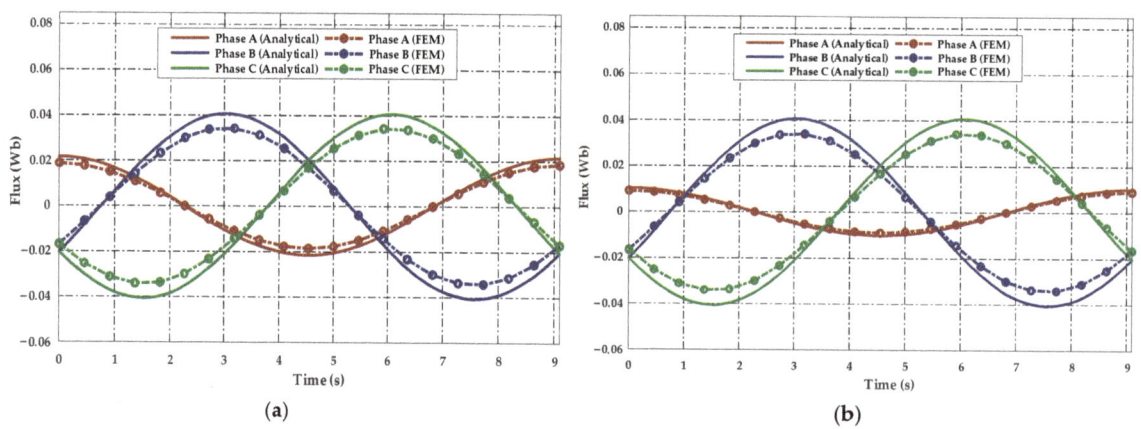

Figure 11. Total flux created by the stator winding, (**a**) Total flux with 25% of LTs fault; (**b**) total flux with 50% of LTs fault.

The inductances of the ER-PMSM are reported in Figure 12. The values given for the inductances, calculated using the winding function approach from the real distribution of windings in the stator slots, are closer to those of the finite element analysis and the experimental ones. On the other hand, a considerable discrepancy is noted between the FEA and analytical findings for the analytically calculated inductance. This discrepancy can be associated with the inadequacy of the winding function to take into account the real flux paths, geometry, and nonuniformly saturated iron in an FSCW ER-PMSM machine.

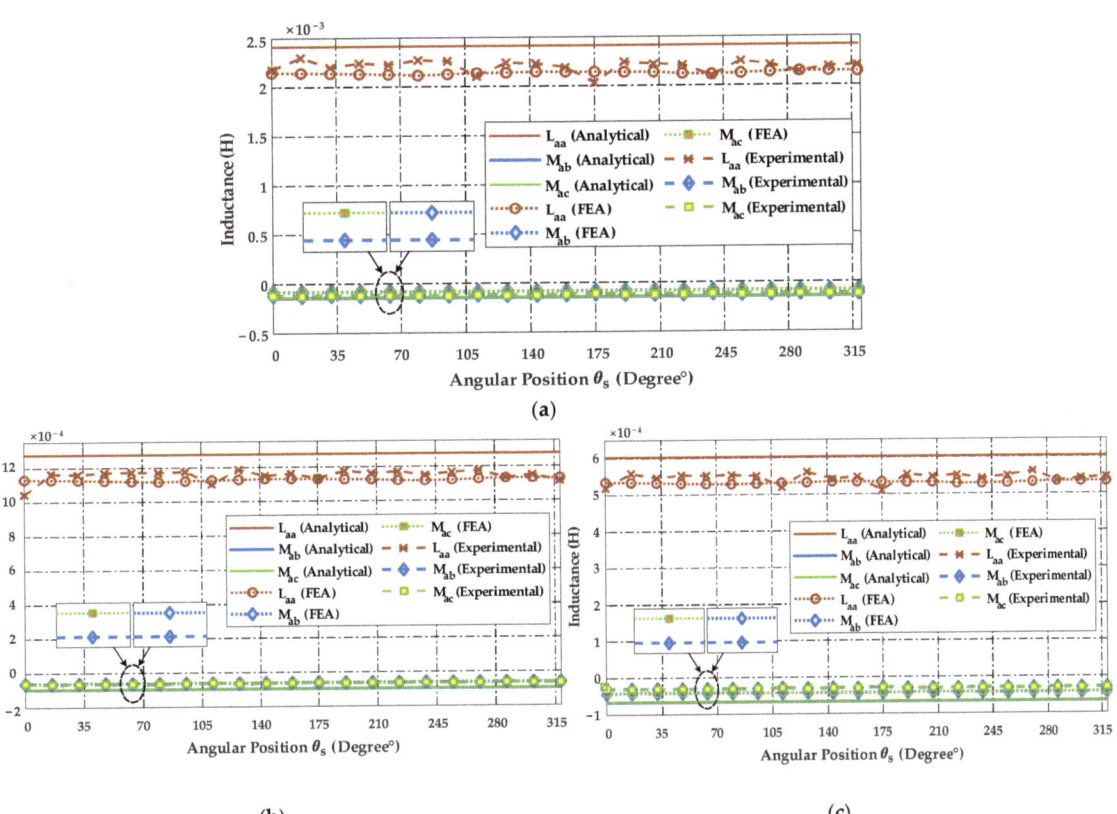

Figure 12. Comparison between inductances derived by the analytical approach, FEA, and experimental measurements. (**a**) Healthy inductances; (**b**) faulty inductances with 25% of LTs fault; (**c**) faulty inductances with 50% of LTs fault.

In Figure 13, the ER-PMSM is operated at nominal load. The temporal representation of the output torque of the machine by FEA for the three cases is provided. In the LTs fault circumstances of 25% and 50% of phase A, the torque decreases accordingly concerning the healthy state, according to the fault's severity level. Moreover, due to the change of its symmetry, torque ripples appear, which are due to the growth of harmonics. The evolution of the faulty torque reveals torque ripples corresponding to the double of the frequency, owing to the decrease of the resistance and the inductance of a phase, which makes the bandwidth wider. The results concerning the torque are provided in Table 2. It can be observed that the level of the torque ripples is related to the severity of the fault.

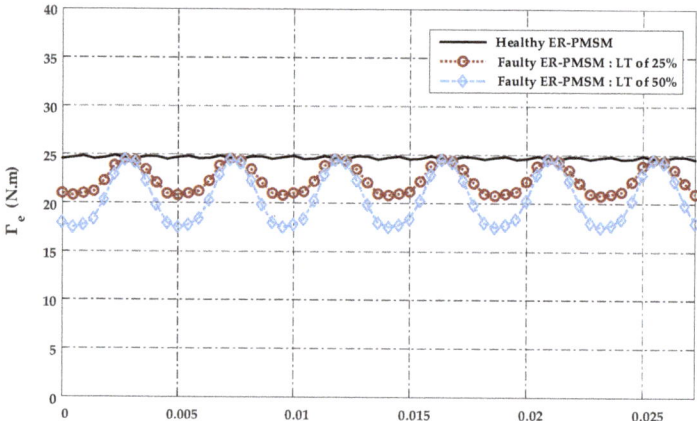

Figure 13. Electromagnetic torque comparison.

Table 2. Summary of simulation results of electromagnetic torque.

State	Torque Ripples (N·m)
Healthy ER-PMSM	$\begin{cases} \Gamma_{e-\min} = 24.5023 \\ \Gamma_{e-\max} = 24.8956 \end{cases}$
Faulty state: LTs of 25%	$\begin{cases} \Gamma_{e-\min} = 20.7869 \\ \Gamma_{e-\max} = 24.5755 \end{cases}$
Faulty state: LTs of 50%	$\begin{cases} \Gamma_{e-\min} = 17.5335 \\ \Gamma_{e-\max} = 24.4895 \end{cases}$

3. Experimental Setup

This section describes the configuration and experimental campaign of the test bench devoted to the control and diagnosis of the LTs fault of the ER-PMSM, to validate and verify the theoretical and numerical methodologies. The parameters of the faulty machine are reported in Table 1. The test bench is illustrated in Figure 14a. The structure of the stator winding of the machine has been designed in such a way as to offer the possibility of achieving various levels of LTs and interturn short circuit ITSC defects (Figure 14b). The architecture of the experimental platform is given in Figure 14c. The experimental bench for control and diagnosis consists mainly of a three-phase ER-PMSM of 22 poles and a power of 1.5 kW, coupled to a permanent magnet synchronous generator (PMSG). This PMSG is equipped with an incremental encoder connected to the interface with a specific cable to measure the rotation speed, as well as to capture the exact position of the rotor, whose stator is connected to a three-phase resistive load. The torque is measured via a T22/50 Nm torque meter connected to an MX440B universal amplifier module from HBM's QuantumX. The machine is driven in a closed-loop and the motor is fielded by a Semikron three-phase IGBT inverter at 5 kHz. The control and signal acquisition are performed via a dSpace-MicroLabBox 1202 platform. The DC bus voltage V_{DC} is maintained at 150 V by the grid through the three-phase autotransformer and a diode bridge rectifier.

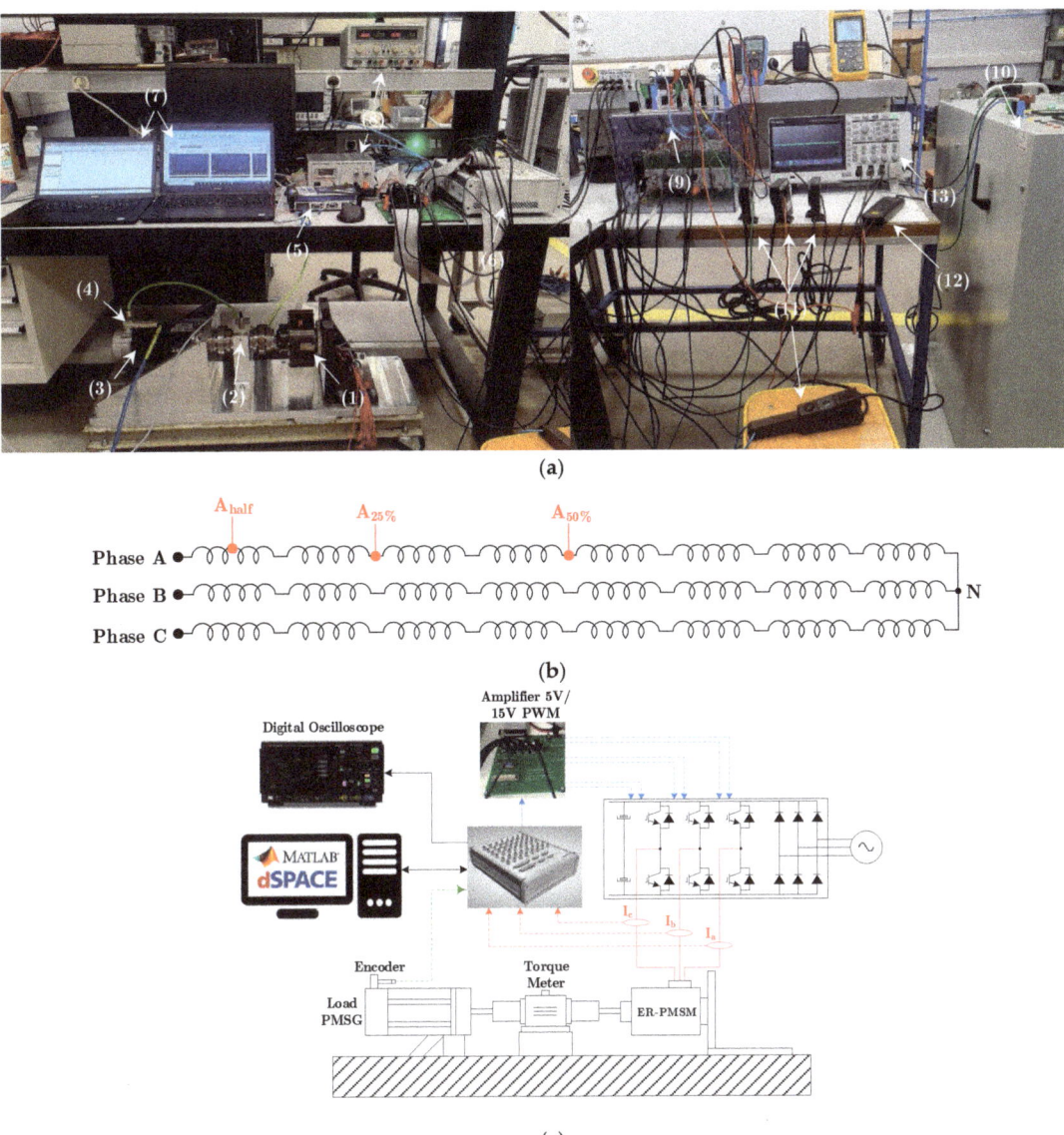

Figure 14. Experimental setup. (**a**) Global view of the test bench setup for control and diagnosis of the ER-PMSM; (1) ER-PMSM 1.5 kW; (2) torque meter T22/50 Nm; (3) load PMSG; (4) encoder; (5) MX440B module; (6) dSpace 1202 MicroLabBox; (7) MATLAB/Simulink ControlDesk platform; (8) DC power supply; (9) Semikron 3Φ inverter; (10) power supply 3Φ/50 Hz; (11) current sensors; (12) voltage sensor; (13) digital oscilloscope. (**b**) Structure of the stator winding of the ER-PMSM; (**c**) block diagram of the setup.

The faulty harmonics are aimed at kf_s, with the values measured in decibels (dB). As a result, for fault sensitivity, we consider the case of the LTs fault in half of an elementary coil A_{half}, of 25% and 50%, corresponding, respectively, to 11 turns, 44 turns, and 88 turns. In reality, the 25% and 50% cases are rarely possible, and the fault is often apparent on some turns. Nonetheless, these values, although large, provide a vision of the tendency to monitor the machine with the fault and the evolution of these harmonics.

4. Experimental Results

Figure 14c depicts a synoptic diagram of the experimental implementation that indicates the experimental validation of the ER-PMSM control results in the healthy condition with an LTs fault of 25% and 50%. The experimental results are acquired by a digital oscilloscope linked to the real-time interface. The choice of the sampling frequency significantly impacts the quality of the signals, particularly the phase currents, speeds, and electromagnetic torque, and whatever the control algorithm used. For each healthy/faulty scenario, a load torque condition is tested: 8 Nm for a speed of 600 rpm. In this section, we present the experimental results illustrating the behavior of the ER-PMSM impacted by 25% and 50% LTs fault in its stator winding. Figure 15 depicts the flow chart of the control loop used in this paper with the LTs fault diagnosis.

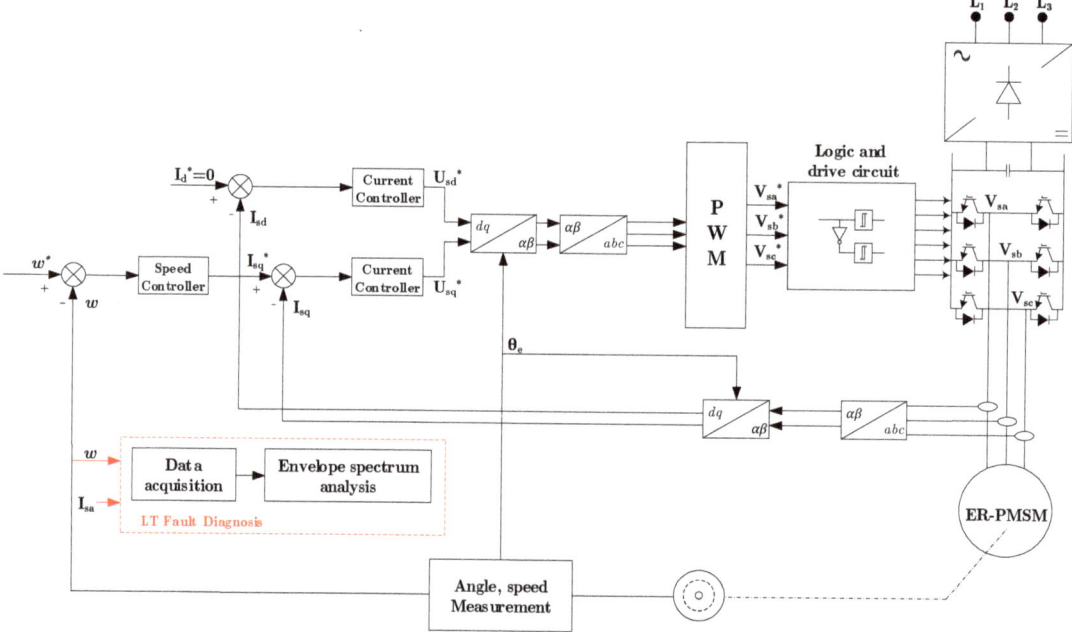

Figure 15. The flow chart of the proposed strategy.

4.1. Healthy State of the ER-PMSM

The following section examines the behavior of the machine in the healthy state. Figure 16 depicts the electrical and mechanical quantities recorded in real-time. The presentation of the experimental results will be restricted to the intersective PWM. We are interested in the rotational speed and the waveform of the stator currents. Figure 16a shows that the rotor speed follows its reference. The three-phase stator currents are shown in Figure 16b and in the d-q rotating frame in Figure 16c. The spectral analysis of the signals offers a way to diagnose this type of stator winding fault. We present the spectral analysis of the rotational speed and the stator current, using the fast Fourier transform (FFT) in steady state at rated load. We will show how this method makes it possible to determine the frequency content of the rotational speed and the stator current, and thus to find the lines associated with the LTs faults of 25% and 50%.

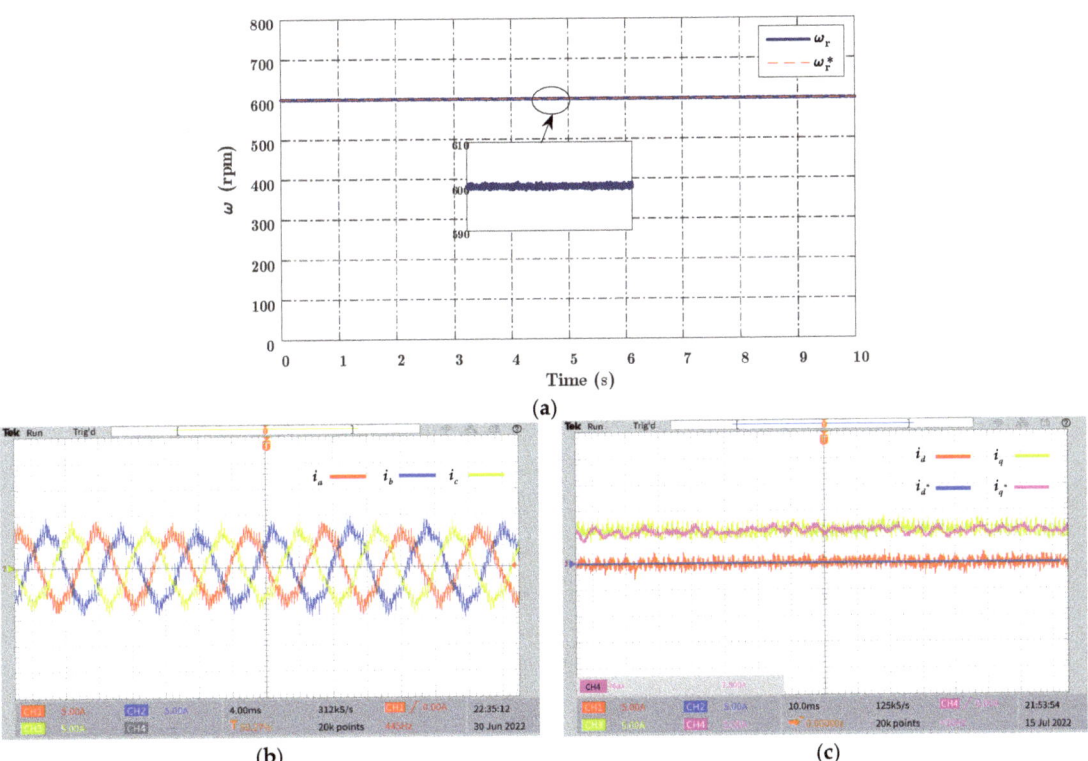

Figure 16. Electrical and mechanical characteristics in the healthy state. (**a**) Rotor speed; (**b**) stator phase currents; (**c**) direct and quadratic current component.

4.2. Faulty State of the ER-PMSM

The following section provides the experimental results of the machine with the LTs fault. Figures 17 and 18 exhibit the results obtained from the control of the different electrical and mechanical parameters of the machine with a fault in the stator winding. The presence of the LTs fault in stator phase A, according to the severity level, shows:

- The rotor speed is not substantially influenced by the LTs fault because of the control loop that hides and compensates for the effect of the fault;
- High ripples arise in the stator current of phase A, the direct current i_d, and the quadratic current i_q. The influence of the fault generates an unbalance and a noticeable variation in the current envelope.

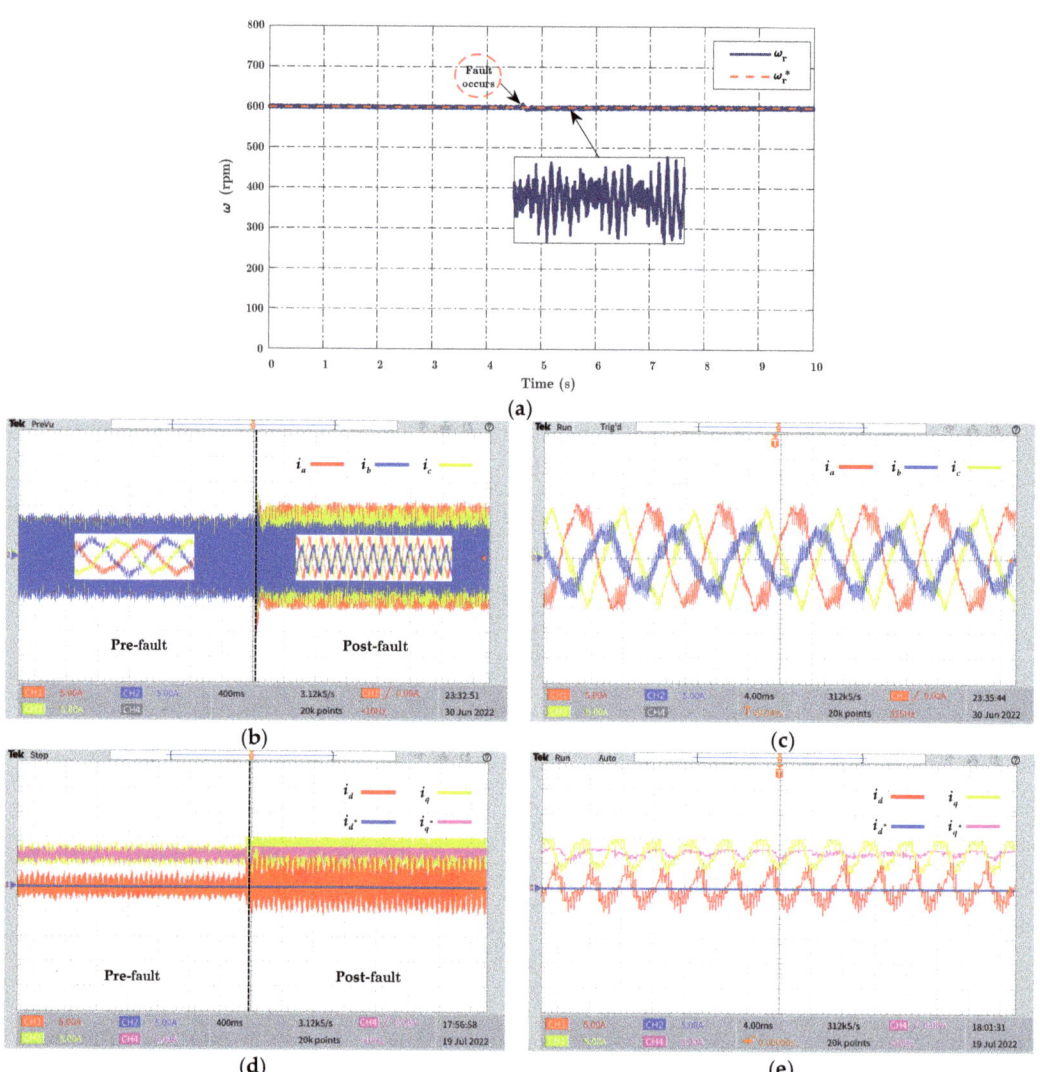

Figure 17. Electrical and mechanical characteristics in the faulty state with an LTs fault of 25%. (**a**) Rotor speed; (**b**) stator phase currents; (**c**) zoom of the stator phase currents; (**d**) direct and quadratic current component; (**e**) zoom of the direct and quadratic current component.

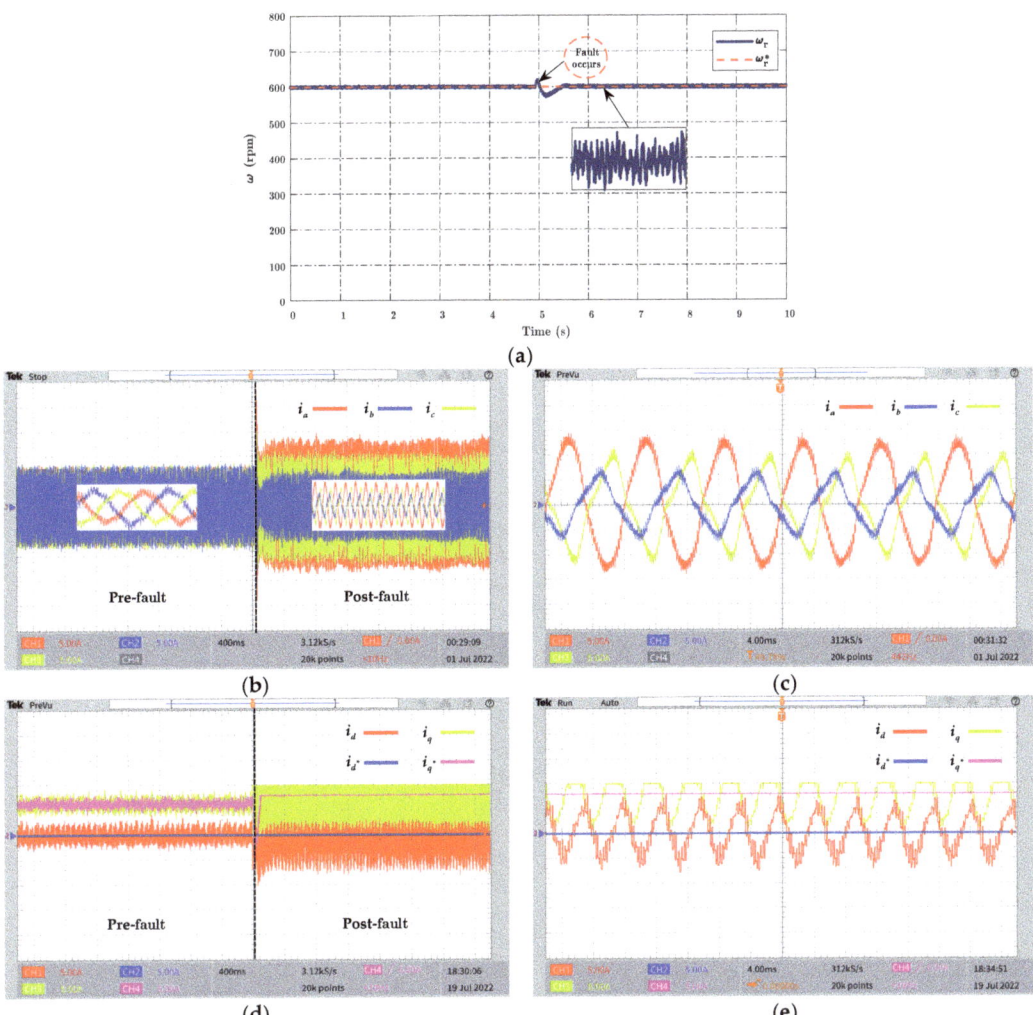

Figure 18. Electrical and mechanical characteristics in the faulty state with an LTs fault of 50%. (**a**) Rotor speed; (**b**) stator phase currents; (**c**) zoom of the stator phase currents; (**d**) direct and quadratic current component; (**e**) zoom of the direct and quadratic current component.

4.2.1. Motor Speed Signature Analysis

The examination of the motor speed signature analysis (MSSA) may provide a noninvasive method applied for the detection of stator winding faults. It is a nonparametric approach devoted to the analysis of stationary phenomena [31,32]. Figure 19 illustrates the spectrum analysis of the speed in the presence of an LTs fault of 25% to 50%. According to this study, we notice the appearance or the presence of several components having a direct relation with the defect according to the specified degree of severity. The appearance of the lines is an indicator of the existence of the LTs fault.

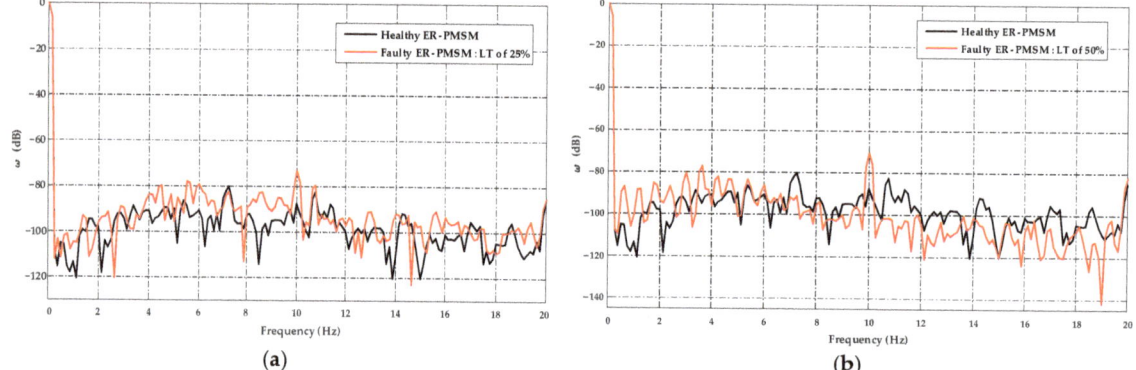

Figure 19. Experimental speed spectrum of the ER-PMSM at 8 Nm load and 600 rpm, in healthy and faulty conditions. (**a**) LTs fault of 25%; (**b**) LTs fault of 50%.

4.2.2. Motor Current Signature Analysis

The next stage is to evaluate the application of the spectral analysis of the stator phase current envelope signal to LTs detection. The steady-state stator phase current envelope spectra for the motor operating with f_s = 110 Hz at 8 Nm load for 25% LTs fault level is illustrated in Figure 20. The examination of the motor current signature analysis (MCSA) reveals the influence of the LTs fault in the appearance of harmonics around the fundamental, which increases with the fault intensity. The influence of the fault is manifested by the presence of new visible frequency components in the current spectrum around the fundamental at $2f_s$ and $3f_s$. Based on this analysis, it can be inferred that the rise in amplitudes induced by the fault is significant. We can also highlight more typical criteria, such as the occurrence of kf_s frequency lines near the fundamental (k = 1, 2, 3, 4 ...) on the stator current spectrum. Table 3 illustrates the magnitudes and frequency of the ER-PMSM stator current analysis fault.

Table 3. Experimental magnitude of current components generated by the ER-PMSM at 8 Nm load and f_s = 110 Hz.

ER-PMSM Magnitude (dB)	Current (dB)				
	f_s = 110 Hz	$2f_s$ = 220 Hz	$3f_s$ = 330 Hz	$4f_s$ = 440 Hz	$5f_s$ = 550 Hz
Healthy state	0	−43.361	−41.096	−55.616	−23.937
Faulty state: LTs of A_{half}	0	−43.177	−35.001	−45.819	−22.828
Faulty state: LTs of 25%	0	−31.636	−20.759	−38.989	−21.159
Faulty state: LTs of 50%	0	−37.708	−20.689	−42.237	−31.881

According to Figure 20, we notice that after the presence of the fault at a level of 25%, an appearance and increase of harmonics is reflected in the frequency domain. The presence of the LTs fault causes torque and speed ripples, which leads to significant mechanical vibrations in the machine, as well as an unbalance that manifests itself in the form of an important increase in the current of the faulty phase and a less significant increase for the other two phases. The spectral analysis of the MCSA indicates a visible rise in amplitude at $2f_s$, $3f_s$, and $4f_s$ for both fault levels. However, the fifth harmonic of the spectrum (550 Hz) will not be influenced by the 50% LTs fault. According to these results, we observe the existence of proportionality between the severity level of the LTs fault and the amplitude of the characteristic harmonic of the fault. Therefore, we can subsequently detect an incipient fault, which is the main objective of the LTs fault diagnosis. All information acquired from the spectrum analysis can be employed in an automated fault detection process, while analyzing the presence of new harmonics and setting detection thresholds using adaptive observers for reconfiguration and fault isolation.

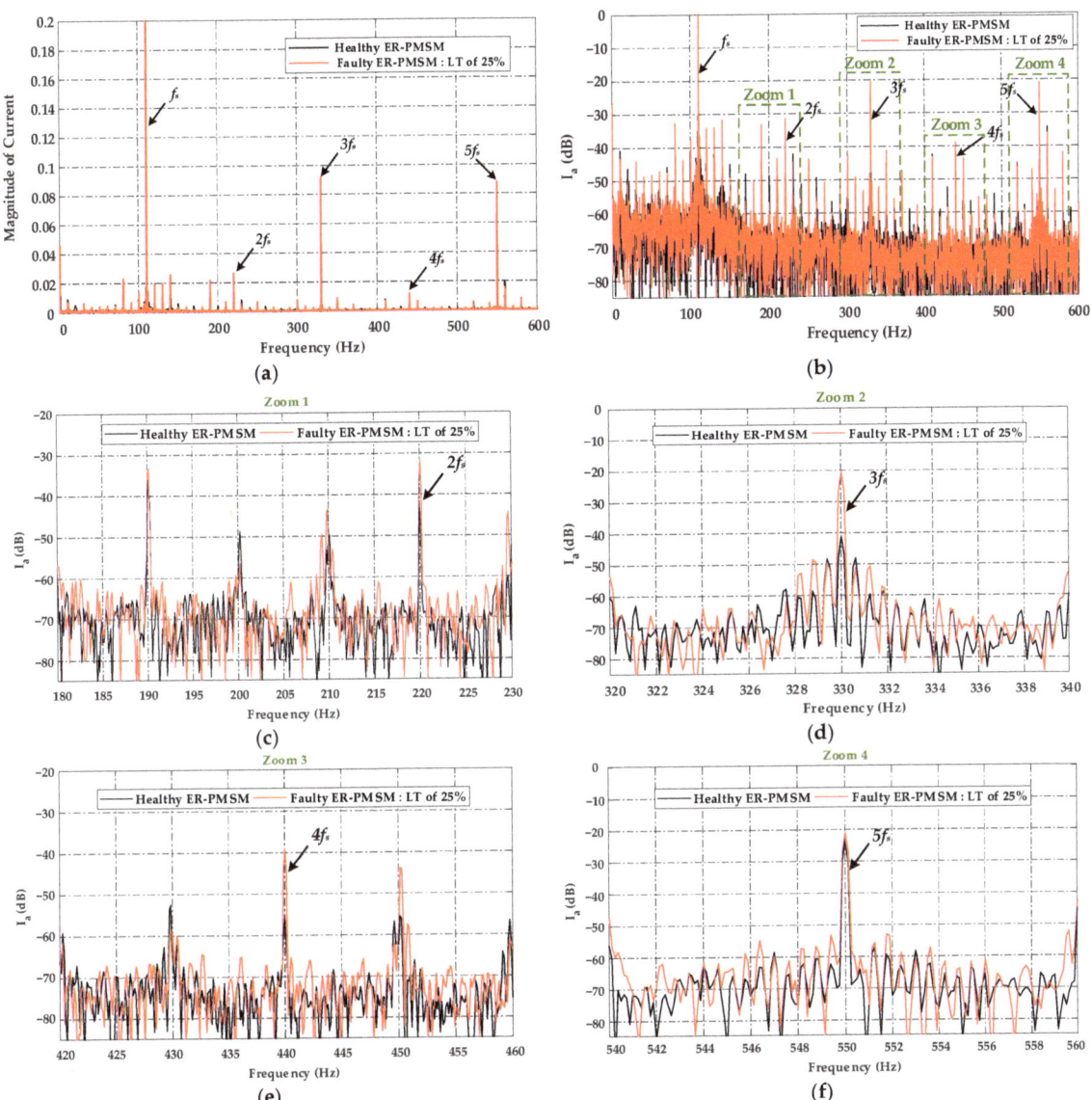

Figure 20. (**a**–**f**) Experimental current spectrum of the ER-PMSM at 8 Nm load and 600 rpm, in healthy and faulty conditions with an LTs fault of 25%.

The validity of the analytical model was proven by two approaches: FEA and experimental measurement of inductances and electromotive forces of the healthy and faulty machines. The occurrence of new harmonics can be attributed to several factors connected to the LTs fault, including torque and speed ripples, system nonlinearity, and unbalance of the ER-PMSM electromotive force. These factors can induce variations in the electrical and mechanical behavior of the machine, resulting in spectral alterations that can be detected and analyzed to diagnose the fault. While validation of the analytical model by FEA and experimental measurements is crucial, understanding the underlying reasons for the spectrum changes is equally important for effective fault diagnosis in real-world applications.

5. Conclusions

This paper presents a new approach for modeling lack of turns (LTs) faults in the stator winding of external rotor permanent magnet synchronous motors (ER-PMSMs), which are of major importance in permanent magnet (PM) electrical machines since, in rotation, the induction effect generated by PMs aggravates the effects of these faults. Different operating conditions of ER-PMSMs with LTs fault provide a reference for real-time diagnosis, prediction, and maintenance planning. The suggested method has substantial advantages, such as fast calculation, good precision, and explicit physical correlations between different factors. For this purpose, theoretical operational performances of various operating situations have been evaluated.

The paper also examines the problem of diagnosing LTs faults in ER-PMSMs at their early stage by analyzing the spectral content of the speed and current signatures under various operating situations. The provided results confirm the efficiency of the diagnosis and the application of spectral analysis for the extraction of LTs fault indicators.

Future research will focus on the development and combination of adaptive fault-tolerant control of LTs faults, advanced fault diagnosis methodologies for ER-PMSMs, and the implementation of automated fault detection processes in automotive applications.

Author Contributions: Conceptualization, A.B., R.R., R.P., and D.B.; methodology, A.B., R.R., and R.P.; software, A.B.; validation, R.R., R.P., D.B., and Y.Z.; formal analysis, A.B.; investigation, A.B.; resources, A.B.; data curation, A.B.; experimental measurements, A.B.; writing—original draft preparation, A.B.; writing—review and editing, A.B., R.R., R.P., and D.B.; visualization, A.B., R.P., Y.Z., D.B., and R.R.; supervision, R.R., D.B., R.P., and Y.Z.; project administration, R.R., D.B., R.P., and Y.Z.; funding acquisition, R.R., and R.P. All authors have read and agreed to the published version of the manuscript.

Funding: This research work received no external funding.

Data Availability Statement: Data sharing not applicable.

Conflicts of Interest: The authors declare no conflict of interest.

References

1. Łebkowski, A. Design, Analysis of the Location and Materials of Neodymium Magnets on the Torque and Power of in-Wheel External Rotor PMSM for Electric Vehicles. *Energies* **2018**, *11*, 2293. [CrossRef]
2. Pop, C.V.; Fodorean, D.; Husar, C.; Irimia, C. Structural Behavior Evaluation of an In-Wheel Motor Based on Numerical and Experimental Approach. *Electr. Eng.* **2020**, *102*, 65–74. [CrossRef]
3. Ma, C.; Gao, Y.; Degano, M.; Wang, Y.; Fang, J.; Gerada, C.; Zhou, S.; Mu, Y. Eccentric Position Diagnosis of Static Eccentricity Fault of External Rotor Permanent Magnet Synchronous Motor as an In-wheel Motor. *IET Electr. Power Appl.* **2020**, *14*, 2263–2272. [CrossRef]
4. Ebrahimi, S.H.; Choux, M.; Huynh, V.K. Real-Time Detection of Incipient Inter-Turn Short Circuit and Sensor Faults in Permanent Magnet Synchronous Motor Drives Based on Generalized Likelihood Ratio Test and Structural Analysis. *Sensors* **2022**, *22*, 3407. [CrossRef] [PubMed]
5. Ebrahimi, B.M.; Javan Roshtkhari, M.; Faiz, J.; Khatami, S.V. Advanced Eccentricity Fault Recognition in Permanent Magnet Synchronous Motors Using Stator Current Signature Analysis. *IEEE Trans. Ind. Electron.* **2014**, *61*, 2041–2052. [CrossRef]
6. Belkhadir, A.; Belkhayat, D.; Zidani, Y.; Pusca, R.; Romary, R. Torque Ripple Minimization Control of Permanent Magnet Synchronous Motor Using Adaptive Ant Colony Optimization. In Proceedings of the 2022 8th International Conference on Control, Decision and Information Technologies (CoDIT), Istanbul, Turkey, 17–20 May 2022; pp. 629–635.
7. Orlowska-Kowalska, T.; Wolkiewicz, M.; Pietrzak, P.; Skowron, M.; Ewert, P.; Tarchala, G.; Krzysztofiak, M.; Kowalski, C.T. Fault Diagnosis and Fault-Tolerant Control of PMSM Drives–State of the Art and Future Challenges. *IEEE Access* **2022**, *10*, 59979–60024. [CrossRef]

8. Kudelina, K.; Asad, B.; Vaimann, T.; Rassõlkin, A.; Kallaste, A.; Khang, H. Van Methods of Condition Monitoring and Fault Detection for Electrical Machines. *Energies* **2021**, *14*, 7459. [CrossRef]
9. Ullah, Z.; Hur, J. A Comprehensive Review of Winding Short Circuit Fault and Irreversible Demagnetization Fault Detection in PM Type Machines. *Energies* **2018**, *11*, 3309. [CrossRef]
10. Chen, Y.; Liang, S.; Li, W.; Liang, H.; Wang, C. Faults and Diagnosis Methods of Permanent Magnet Synchronous Motors: A Review. *Appl. Sci.* **2019**, *9*, 2116. [CrossRef]
11. Solís, R.; Torres, L.; Pérez, P. Review of Methods for Diagnosing Faults in the Stators of BLDC Motors. *Processes* **2022**, *11*, 82. [CrossRef]
12. Liang, H.; Chen, Y.; Liang, S.; Wang, C. Fault Detection of Stator Inter-Turn Short-Circuit in PMSM on Stator Current and Vibration Signal. *Appl. Sci.* **2018**, *8*, 1677. [CrossRef]
13. Hang, J.; Zhang, J.; Cheng, M.; Huang, J. Online Interturn Fault Diagnosis of Permanent Magnet Synchronous Machine Using Zero-Sequence Components. *IEEE Trans. Power Electron.* **2015**, *30*, 6731–6741. [CrossRef]
14. Yepes, A.G.; Fonseca, D.S.B.; Antunes, H.R.P.; Lopez, O.; Marques Cardoso, A.J.; Doval-Gandoy, J. Discrimination Between Eccentricity and Interturn Faults Using Current or Voltage-Reference Signature Analysis in Symmetrical Six-Phase Induction Machines. *IEEE Trans. Power Electron.* **2023**, *38*, 2421–2434. [CrossRef]
15. Yang, M.; Chai, N.; Liu, Z.; Ren, B.; Xu, D. Motor Speed Signature Analysis for Local Bearing Fault Detection with Noise Cancellation Based on Improved Drive Algorithm. *IEEE Trans. Ind. Electron.* **2020**, *67*, 4172–4182. [CrossRef]
16. Haje Obeid, N.; Battiston, A.; Boileau, T.; Nahid-Mobarakeh, B. Early Intermittent Interturn Fault Detection and Localization for a Permanent Magnet Synchronous Motor of Electrical Vehicles Using Wavelet Transform. *IEEE Trans. Transp. Electrif.* **2017**, *3*, 694–702. [CrossRef]
17. Espinosa, A.G.; Rosero, J.A.; Cusido, J.; Romeral, L.; Ortega, J.A. Fault Detection by Means of Hilbert–Huang Transform of the Stator Current in a PMSM With Demagnetization. *IEEE Trans. Energy Convers.* **2010**, *25*, 312–318. [CrossRef]
18. Pusca, R.; Romary, R.; Touti, E.; Livinti, P.; Nuca, I.; Ceban, A. Procedure for Detection of Stator Inter-Turn Short Circuit in Ac Machines Measuring the External Magnetic Field. *Energies* **2021**, *14*, 1132. [CrossRef]
19. Irhoumah, M.; Pusca, R.; Lefèvre, E.; Mercier, D.; Romary, R. Stray Flux Multi-Sensor for Stator Fault Detection in Synchronous Machines. *Electronics* **2021**, *10*, 2313. [CrossRef]
20. Irhoumah, M.; Pusca, R.; Lefevre, E.; Mercier, D.; Romary, R.; Demian, C. Information Fusion with Belief Functions for Detection of Interturn Short-Circuit Faults in Electrical Machines Using External Flux Sensors. *IEEE Trans. Ind. Electron.* **2018**, *65*, 2642–2652. [CrossRef]
21. Irhoumah, M.; Pusca, R.; Lefevre, E.; Mercier, D.; Romary, R. Detection of the Stator Winding Inter-Turn Faults in Asynchronous and Synchronous Machines Through the Correlation Between Harmonics of the Voltage of Two Magnetic Flux Sensors. *IEEE Trans. Ind. Appl.* **2019**, *55*, 2682–2689. [CrossRef]
22. Shih, K.-J.; Hsieh, M.-F.; Chen, B.-J.; Huang, S.-F. Machine Learning for Inter-Turn Short-Circuit Fault Diagnosis in Permanent Magnet Synchronous Motors. *IEEE Trans. Magn.* **2022**, *58*, 1–7. [CrossRef]
23. Pietrzak, P.; Wolkiewicz, M.; Orlowska-Kowalska, T. PMSM Stator Winding Fault Detection and Classification Based on Bispectrum Analysis and Convolutional Neural Network. *IEEE Trans. Ind. Electron.* **2023**, *70*, 5192–5202. [CrossRef]
24. Farshadnia, M. *Advanced Theory of Fractional-Slot Concentrated- Wound Permanent Magnet Synchronous Machines*; Springer: Berlin/Heidelberg, Germany, 2018; ISBN 978-981-10-8708-0.
25. Pyrhonen, J. Juha Pyrhönen, Tapani Jokinen, Valéria Hrabovcová. In *Design of Rotating Electrical Machines*; Wiley: Hoboken, NJ, USA, 2014; Volume 614, ISBN 9781118581575.
26. Bianchi, N.; Fornasiero, E. Impact of MMF Space Harmonic on Rotor Losses in Fractional-Slot Permanent-Magnet Machines. *IEEE Trans. Energy Convers.* **2009**, *24*, 323–328. [CrossRef]
27. EL-Refaie, A.M. Fractional-Slot Concentrated-Windings Synchronous Permanent Magnet Machines: Opportunities and Challenges. *IEEE Trans. Ind. Electron.* **2010**, *57*, 107–121. [CrossRef]
28. Di Tommaso, A.O.; Genduso, F.; Miceli, R. A Software for the Evaluation of Winding Factor Harmonic Distribution in High Efficiency Electrical Motors and Generators. In Proceedings of the 2013 8th Eighth International Conference and Exhibition on Ecological Vehicles and Renewable Energies (EVER), Monte Carlo, Monaco, 27–30 March 2013. [CrossRef]
29. Hamiti, T.; Lubin, T.; Baghli, L.; Rezzoug, A. Modeling of a Synchronous Reluctance Machine Accounting for Space Harmonics in View of Torque Ripple Minimization. *Math. Comput. Simul.* **2010**, *81*, 354–366. [CrossRef]
30. *IEEE Std 115a; IEEE Standard Procedures for Obtaining Synchronous Machine Parameters by Standstill Frequency Response Testing*. IEEE: Piscataway, NJ, USA, 1987.
31. Shi, P.; Chen, Z.; Vagapov, Y.; Zouaoui, Z. A New Diagnosis of Broken Rotor Bar Fault Extent in Three Phase Squirrel Cage Induction Motor. *Mech. Syst. Signal Process.* **2014**, *42*, 388–403. [CrossRef]
32. Moumene, I.; Ouelaa, N. Application of the Wavelets Multiresolution Analysis and the High-Frequency Resonance Technique for Gears and Bearings Faults Diagnosis. *Int. J. Adv. Manuf. Technol.* **2016**, *83*, 1315–1339. [CrossRef]

Disclaimer/Publisher's Note: The statements, opinions and data contained in all publications are solely those of the individual author(s) and contributor(s) and not of MDPI and/or the editor(s). MDPI and/or the editor(s) disclaim responsibility for any injury to people or property resulting from any ideas, methods, instructions or products referred to in the content.

Article

Impact of Inter-Turn Short Circuit in Excitation Windings on Magnetic Field and Stator Current of Synchronous Condenser under Unbalanced Voltage

Junqing Li [1], Chengzhi Zhang [1], Yuling He [2,*], Xiaodong Hu [1], Jiya Geng [1] and Yapeng Ma [1]

[1] School of Electrical and Electronic Engineering, North China Electric Power University, Baoding 071003, China
[2] Department of Mechanical Engineering, North China Electric Power University, Baoding 071003, China
* Correspondence: heyuling1@163.com

Abstract: Inter-turn short circuit in the excitation windings of synchronous condensers is a common fault that directly impacts their normal operation. However, current fault analysis and diagnosis of synchronous condensers primarily rely on voltage-balanced conditions, while research on short-circuit faults under unbalanced voltage conditions is limited. Therefore, this paper aims to analyze the fault characteristics of inter-turn short circuits in the excitation windings of synchronous condensers under unbalanced grid voltage. Mathematical models were developed to represent the air gap flux density and stator parallel currents for four operating conditions: normal operation and inter-turn short circuit fault under balanced voltage, as well as a process without fault and with inter-turn short circuit fault under unbalanced voltage. By comparing the harmonic content and amplitudes, various aspects of the fault mechanism of synchronous condensers were revealed, and the operating characteristics under different conditions were analyzed. Considering the four aforementioned operating conditions, finite element simulation models were created for the TTS-300-2 synchronous condenser in a specific substation as a case study. The results demonstrate that the inter-turn short circuit fault in the excitation windings under unbalanced voltage leads to an increase in even harmonic currents in the stator parallel currents, particularly the second and fourth harmonics. This validates the accuracy of the theoretical analysis findings.

Keywords: synchronous condenser; unbalanced voltage; inter-turn short circuit in excitation windings; finite element; fault analysis; stator parallel currents

Citation: Li, J.; Zhang, C.; He, Y.; Hu, X.; Geng, J.; Ma, Y. Impact of Inter-Turn Short Circuit in Excitation Windings on Magnetic Field and Stator Current of Synchronous Condenser under Unbalanced Voltage. *Energies* 2023, 16, 5695. https://doi.org/10.3390/en16155695

Academic Editor: Gianluca Brando

Received: 6 June 2023
Revised: 23 July 2023
Accepted: 26 July 2023
Published: 29 July 2023

Copyright: © 2023 by the authors. Licensee MDPI, Basel, Switzerland. This article is an open access article distributed under the terms and conditions of the Creative Commons Attribution (CC BY) license (https://creativecommons.org/licenses/by/4.0/).

1. Introduction

Currently, the ultra-high voltage direct current (UHVDC) system is rapidly developing, imposing more significant requirements on reactive power within the power grid. Large-scale synchronous condensers (LSSC), with high capacity, exhibit exceptional transient reactive power support and short-term overload capabilities. By positioning LSSC at the inverter end of a weak AC grid, commutation failures can be effectively prevented, and fault clearing speed can be accelerated [1]. Consequently, ensuring the reliable operation of LSSC is crucial in enhancing the dynamic voltage stability of power systems and ensuring the stable operation of UHVDC transmission [2].

Inter-turn short circuit in the excitation windings of synchronous condensers is a common fault not only limits the reactive power capability but also increases the excitation current, power losses, and local temperature of the synchronous condenser. In severe cases, it can exacerbate rotor vibrations, generate significant axial magnetization, and even completely shut down the synchronous condenser [3]. Analyzing characteristic patterns and accurately diagnosing inter-turn short circuit faults in the excitation windings are complex tasks within the field of system engineering. Fault diagnosis is typically carried out using methods such as the DC resistance method [4], AC impedance method [5], and repetitive pulse method [6]. Since synchronous condensers share structural similarities with synchronous generators, research findings related to synchronous generators can be utilized for synchronous condensers [7–11]. M. Xu et al. [12] comprehensively analyzed the stator circulating current within parallel branches, considering various degrees and positions of rotor inter-turn short circuits. The analysis was performed using the finite element method, yielding valuable insights into the behavior of the circulating current. Currently, several studies have been conducted on inter-turn short circuit faults in the excitation windings of synchronous condensers. G. Xu et al. [13] present the single-phase short circuit faults' electromagnetic and temperature field calculation and the experimental validation. M. Ma et al. [14] investigated inter-turn short circuit faults in the rotor windings of synchronous condensers by analyzing commutation failures in high-voltage direct current transmission. Finite element analysis revealed that commutation failures can cause abnormal vibrations in the rotor side of the synchronous condenser, which positively correlates with the severity of the fault. Y. Zhang et al. [15] applied wavelet transform to preprocessed excitation current signals extracted from normal and faulty states of the synchronous condenser. The extracted features were input into a radial basis function neural network for fault diagnosis. Z. Chen et al. [16] analyzed inter-turn short circuit faults in the rotor windings of synchronous condensers from the perspective of temperature distribution and validated the analysis through finite element simulations. C. Wei et al. [17] investigated the relationship between the number of short-circuited turns in the excitation windings of synchronous condensers and the magnetic field current in the rotor windings, proposing an online monitoring fault diagnosis strategy. A. N. Novozhilov et al. [18] established a mathematical model for inter-turn short circuit faults in the excitation windings of synchronous condensers, achieving an accuracy range of 5% to 10%. Most existing research on the mentioned faults primarily focuses on the short circuit between turns in the excitation winding. However, the voltage waveform of the grid deviates from a standard sinusoidal shape due to the non-standard sinusoidal waveform generated by most generators. As a result, when a short circuit occurs in the excitation winding, it can be seen as a combination of unbalanced voltage and a short circuit between turns in the excitation winding. The prevalence of unbalanced voltage further complicates the fault analysis, as the fault environments studied in literature may not accurately reflect real-world scenarios. Consequently, matching the observed fault characteristics of the synchronous condenser with known fault patterns during on-site diagnostics becomes challenging, often leading to incorrect or even misdiagnosis. Therefore, it is crucial to research the compound fault of unbalanced voltage and short circuits in the excitation winding. This research aims to identify distinctive fault characteristics specific to this type of fault and differentiate them from fault characteristics caused by individual faults. Such investigations are essential for achieving accurate diagnosis and establishing reliable diagnostic criteria in scenarios involving these compound faults. J. LI et al. [19] conducted a simulation study on the inter-turn short circuit fault in the stator winding of a doubly-fed induction generator. They used finite element modeling to analyze the changes in stator line voltage, rotor line current, and electromagnetic torque when the excitation winding experiences an inter-turn short circuit fault, considering the presence of inherent grid voltage imbalance and static eccentricity. However, their study [19] focused solely on simulation modeling analysis and did not investigate the theoretical research on mathematical expressions of the relevant fault characteristic quantities after the occurrence of the fault. As a result, their study has certain limitations that need to be addressed.

Among the electrical characteristic-oriented methods, the current-based method is most widely employed since it does not require extra equipment and can make full use of the stator winding as search coils for further processing. In addition, the parallel branch circulating current signal in the stator winding carries valuable fault information and, in certain cases, offers more effective fault diagnosis than rotor vibration signals. To address the issue of voltage imbalance, this paper begins by investigating the air gap flux density and proceeds to derive expressions for both the air gap flux density and stator parallel branch circulating current of the synchronous condenser in four distinct operating conditions. Additionally, it conducts a comprehensive analysis of the characteristics of the stator parallel branch circulating current under different operating conditions, shedding light on their intricate dynamics: normal operation of the synchronous condenser under balanced voltage, inter-turn short circuit fault in the excitation windings of the synchronous condenser under balanced voltage, normal operation of the synchronous condenser under voltage imbalance, and inter-turn short circuit fault in the excitation windings of the synchronous condenser under voltage imbalance. The degree of voltage imbalance is varied by adjusting the voltage magnitude. Mathematical representations are derived, and a finite element simulation model of the synchronous condenser is constructed to validate the analysis. This study aims to provide a theoretical basis for diagnosing inter-turn short circuit faults in the excitation windings of synchronous condensers under voltage imbalance conditions.

The main structure of this paper is as follows: In Section 2, the expressions for the air gap flux density and stator parallel currents under different fault types of a synchronous condenser are derived, and the impacts of faults are analyzed from a theoretical perspective. Section 3 validates the proposed analytical model's effectiveness and accuracy using finite element analysis, and the changes in air gap flux density and stator parallel currents after faults in the synchronous condenser are analyzed using the finite element model, which is consistent with the theoretical analysis. Finally, Section 4 summarizes the conclusions of this paper.

2. Analysis of Composite Faults in Synchronous Condensers

To compare the variations in air gap magnetic flux density, stator parallel current amplitude, and harmonic content under different operating conditions, this paper focuses on four scenarios: normal operation of the synchronous condenser under balanced voltage, inter-turn short circuit fault in the excitation windings of the synchronous condenser under balanced voltage, normal operation of the synchronous condenser under voltage imbalance, and inter-turn short circuit fault in the excitation windings of the synchronous condenser under voltage imbalance. Compared to symmetrical grid voltage, it is assumed that when the grid voltage becomes asymmetric, the amplitude of one or two phases is reduced. Nevertheless, the derived formulas apply to all cases, demonstrating their universality.

2.1. Analysis of Air Gap Magnetic Field

2.1.1. Air Gap Magnetic Potential during Normal Operation of Synchronous Condenser under Balanced Grid Voltage

Neglecting higher-order harmonics, the air gap magnetic potential of the synchronous condenser during normal operation can be expressed as follows:

$$\begin{aligned} f(\alpha_m, t) &= F_s \cos(\omega t - \alpha_m - \psi - \tfrac{\pi}{2}) + F_r \cos(\omega t - \alpha_m) \\ &= F_1 \cos(\omega t - \alpha_m - \beta) \end{aligned} \quad (1)$$

$$F_1 = \sqrt{F_s^2 \cos^2 \psi + (F_r - F_s \sin \psi)^2} \quad (2)$$

$$\beta = \arctan \frac{F_s \cos \psi}{F_r - F_s \sin \psi} \quad (3)$$

where ω is the stator current angular frequency and rotor rotational angular velocity, α_m is the stator spatial electrical angle, ψ is the internal power factor angle, F_s is the armature magnetic flux amplitude, and F_r is the excitation magnetic flux amplitude.

2.1.2. Air Gap Magnetic Potential during Inter-Turn Short Circuit Fault in the Excitation Windings of Synchronous Condenser under Balanced Grid Voltage

When an inter-turn short circuit occurs in the excitation windings of the synchronous condenser, it generates a counteracting magnetic field [20]. The reverse magnetic MMF (Magneto-Motive Force) generated by the short-circuited winding is given by

$$F_d(\theta_r) = \begin{cases} -\frac{I_{f0}N_{short}(2\pi - \alpha)}{2\pi} & -\frac{\alpha}{2} < \theta_r < \frac{\alpha}{2} \\ \frac{I_{f0}N_{short}\alpha}{2\pi} & \text{other} \end{cases} \tag{4}$$

where I_{f0} is the excitation current, N_{short} is the number of turns in the short-circuited winding in the same slot, α is the mechanical angle between the slot where the short-circuited winding is located and the adjacent slot, θ_r is the mechanical angle of the rotor. The magnetic potential resulting from the short circuit can be expressed in terms of its harmonic components through Fourier decomposition:

$$F_d(\theta_r) = \frac{-2N_{short}I_{f0}}{\pi} \sum_{n=1}^{\infty} \frac{\sin(n\alpha/2)}{n} \cos n\theta_r \tag{5}$$

After performing the Fourier transform on $F_d(\theta_r)$, it can be observed that the main magnetic field in the air gap exhibits various harmonics. When $\alpha \neq 2 k\pi/n$, taking $n = 1,2$, and $\theta_r = \omega t - \alpha_m$, we have

$$\begin{aligned} f(\alpha_m, t) &= F_s \cos(\omega t - \alpha_m - \psi - \frac{\pi}{2}) + F_r \cos(\omega t - \alpha_m) \\ &\quad - F_{d1} \cos(\omega t - \alpha_m) - F_{d2} \cos(2\omega t - 2\alpha_m) \\ &= F_1 \cos(\omega t - \alpha_m - \beta) - F_{d2} \cos(2\omega t - 2\alpha_m) \end{aligned} \tag{6}$$

$$F_{d1} = \frac{2N_{short}I_{f0}}{\pi} \sin \frac{\alpha}{2} \tag{7}$$

$$F_{d2} = \frac{N_{short}I_{f0}}{\pi} \sin \alpha \tag{8}$$

$$F_1 = \sqrt{F_s^2 \cos^2 \psi + (F_r - F_{d1} - F_s \sin \psi)^2} \tag{9}$$

$$\beta = \arctan \frac{F_s \cos \psi}{F_r - F_{d1} - F_s \sin \psi} \tag{10}$$

2.1.3. The Air Gap Magnetic Potential of the Synchronous Condenser under Unbalanced Grid Voltage Conditions without Any Faults Occurring

Under balanced grid voltage, the armature magnetic field of the synchronous condenser exhibits a synchronized circular rotation with the rotor. However, when the grid voltage becomes unbalanced, the armature magnetic field undergoes distortion, assuming an elliptical shape. The symmetrical component method can be applied to characterize the armature magnetic field expression in such scenarios. In this method, the positive-sequence armature magnetic field rotates synchronously with the rotor, while the negative-sequence armature magnetic field rotates in the opposite direction. Meanwhile, the zero-sequence armature magnetic field remains at zero [21]. Consequently, in the presence of an unbalanced grid voltage, the expression for the armature magnetic potential is given by

$$f_s(\alpha_m, t) = F_s^+ \cos(\omega t - \alpha_m - \psi - \frac{\pi}{2}) + F_s^- \cos(\omega t + \alpha_m - \psi - \frac{\pi}{2}) \tag{11}$$

where F_s^+ is the positive-sequence armature magnetic flux amplitude, F_s^- is the negative-sequence armature magnetic flux amplitude. Based on the previous assumption, both F_s^+ and F_s^- are smaller than F_s. The negative-sequence magnetic field caused by the unbalanced grid voltage induces a double-frequency current in the rotor winding. Therefore, the expression for the excitation current is as follows:

$$I_f(t) = I_{f0} + I_{f2}\cos 2\omega t \tag{12}$$

where I_{f0} is the direct current excitation current generated by the generator excitation system, I_{f2} is the amplitude of the twice-frequency excitation current induced. The expression for the excitation magnetic field generated by the excitation current is as follows:

$$\begin{aligned} f_r(\alpha_m, t) &= (I_{f0} + I_{f2}\cos 2\omega t)Nk\cos(\omega t - \alpha_m) \\ &= F_r \cos(\omega t - \alpha_m) + I_{f2}Nk\cos 2\omega t \cos(\omega t - \alpha_m) \end{aligned} \tag{13}$$

where k is the waveform coefficient of the excitation magnetic flux. Therefore, the expression for the synthesized air-gap magnetic potential generated under unbalanced voltage conditions is as follows:

$$\begin{aligned} f(\alpha_m, t) &= f_s(\alpha_m, t) + f_r(\alpha_m, t) \\ &= F_1 \cos(\omega t - \alpha_m - \beta) + F_2 \cos(\omega t + \alpha_m - \gamma) \\ &\quad + \tfrac{1}{2} I_{f2} Nk \cos(3\omega t - \alpha_m) \end{aligned} \tag{14}$$

$$F_1 = \sqrt{F_s^{+2} \cos^2 \psi + (F_r - F_s^+ \sin \psi)^2} \tag{15}$$

$$\beta = \arctan \frac{F_s^+ \cos \psi}{F_r - F_s^+ \sin \psi} \tag{16}$$

$$F_2 = \sqrt{F_s^{-2} \cos^2 \psi + (\tfrac{1}{2} I_{f2} NK - F_s^- \sin \psi)^2} \tag{17}$$

$$\gamma = \arctan \frac{F_s^- \cos \psi}{\tfrac{1}{2} I_{f2} NK - F_s^- \sin \psi} \tag{18}$$

Equations (14)–(18) reveal that unbalanced voltage conditions lead to the induction of a double-frequency current in the excitation winding of the synchronous condenser. As a consequence, third harmonic components are produced in the air gap. It is worth noting that the amplitude of the third harmonic exhibits a direct proportionality to the double-frequency current.

2.1.4. The Air Gap Magnetic Flux in the Synchronous Condenser When There Is an Inter-Turn Short Circuit in the Excitation Winding under Unbalanced Grid Voltage

Under unbalanced grid voltage, a negative-sequence magnetic field is produced, generating double-frequency excitation current in the rotor. Therefore, it becomes essential to account for the influence of this double-frequency current on the excitation magnetic flux when investigating inter-turn short circuits in the excitation winding. The expression for the air gap magnetic flux in the synchronous condenser under these circumstances is as follows:

$$\begin{aligned} f(\alpha_m, t) &= F_s^+ \cos(\omega t - \alpha_m - \psi - \tfrac{\pi}{2}) + F_s^- \cos(\omega t + \alpha_m - \psi - \tfrac{\pi}{2}) \\ &\quad + F_r \cos(\omega t - \alpha_m) + I_{f2} Nk \cos 2\omega t \cos(\omega t - \alpha_m) \\ &\quad - F_{d1} \cos(\omega t - \alpha_m) - F_{d2} \cos(2\omega t - 2\alpha_m) \\ &\quad - F'_{d1} \cos(\omega t - \alpha_m) \cos 2\omega t - F'_{d2} \cos(2\omega t - 2\alpha_m) \cos 2\omega t \\ &= F_1 \cos(\omega t - \alpha_m - \beta) + F_2 \cos(\omega t + \alpha_m - \gamma) \\ &\quad - F_{d2} \cos(2\omega t - 2\alpha_m) + (\tfrac{1}{2} I_{f2} Nk - \tfrac{1}{2} F'_{d1}) \cos(3\omega t - \alpha_m) \\ &\quad - \tfrac{1}{2} F'_{d2} \cos 2\alpha_m - \tfrac{1}{2} F'_{d2} \cos(4\omega t - 2\alpha_m) \end{aligned} \tag{19}$$

$$F'_{d1} = \frac{2N_{short}I_{f2}}{\pi}\sin\frac{\alpha}{2} \tag{20}$$

$$F'_{d2} = \frac{N_{short}I_{f2}}{\pi}\sin\alpha \tag{21}$$

$$F_1 = \sqrt{F_s^{+2}\cos^2\psi + (F_r - F_{d1} - F_s^+\sin\psi)^2} \tag{22}$$

$$\beta = \arctan\frac{F_s^+\cos\psi}{F_r - F_{d1} - F_s^+\sin\psi} \tag{23}$$

$$F_2 = \sqrt{F_s^{-2}\cos^2\psi + \left(\frac{1}{2}I_{f2}NK - \frac{1}{2}F'_{d1} - F_s^-\sin\psi\right)^2} \tag{24}$$

$$\gamma = \arctan\frac{F_s^-\cos\psi}{\frac{1}{2}I_{f2}NK - \frac{1}{2}F'_{d1} - F_s^-\sin\psi} \tag{25}$$

According to Equations (19)–(25), the presence of unbalanced voltage, coupled with an inter-turn short circuit in the excitation winding, gives rise to a multifaceted air gap magnetic flux in the synchronous condenser. This flux encompasses several components, namely even harmonic components induced by the inter-turn short circuit, a DC component introduced by the rotor's double-frequency current, and odd harmonic components.

2.1.5. Air Gap Magnetic Flux Density

During operation, the air gap magnetic flux density of the synchronous condenser, denoted as Λ_0 per unit area, remains constant. The following expression is derived to represent the air gap magnetic flux density:

$$B(\alpha_m, t) = \Lambda_0 f(\alpha_m, t) \tag{26}$$

2.2. Analysis of Parallel Branch Current Circulation in the Stator

The stator winding of a large synchronous condenser is connected in a double Y configuration, employing a three-phase double-layer winding form. This configuration comprises two parallel branches for each phase (A, B, and C), with multiple coil windings connected in series within each branch. Hence, the expression for the instantaneous value of the induced electromotive force in a single branch of the parallel stator winding of the synchronous condenser is as follows:

$$\begin{aligned} e_1(\alpha_m, t) &= N_z k_{w1} B(\alpha_m, t) l v \\ &= N_z k_{w1} B(\alpha_m, t) l (2\tau f) \\ &= 2N_z k_{w1} \tau l f \Lambda_0 F_1 \cos(\omega t - \alpha_m - \beta) \end{aligned} \tag{27}$$

where l is the air gap length, f is the electrical frequency, N_z is the number of conductors connected in series in a single stator branch, and k_{w1} is the fundamental winding factor. Figure 1 illustrates the equivalent circuit of the parallel branch in phase A of the synchronous condenser. The circuit includes R_{a1}, R_{a2}, X_{a1}, and X_{a2}, which represent the resistance and leakage reactance of the two branches. The circulating current is denoted as i_c.

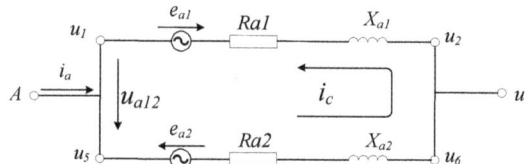

Figure 1. A phase winding equivalent circuit diagram.

When the synchronous condenser is operating normally, applying Kirchhoff's Voltage Law (KVL) to the stator parallel circuit yields the following equation:

$$e_1(\alpha_m, t) + e_2(\alpha_m, t) + i_c(R_{a1} + R_{a2}) + ji_c(X_{a1} + X_{a2}) = 0 \tag{28}$$

The expression for the stator parallel circuit current can be obtained from Equation (28) as follows:

$$i_c = -\frac{e_1(\alpha_m, t) + e_2(\alpha_m, t)}{(R_{a1} + R_{a2}) + j(X_{a1} + X_{a2})} \tag{29}$$

2.2.1. Under the Condition of Balanced Grid Voltage, the Synchronous Condenser Operates without Any Faults

The induced electromotive force in the two parallel branches during the normal operation of the synchronous condenser is given by the following expression:

$$\begin{cases} e_1(\alpha_m, t) = 2N_z k_{w1} \tau l f \Lambda_0 [F_1 \cos(\omega t - \alpha_m - \beta)] \\ e_2(\alpha_m, t) = 2N_z k_{w1} \tau l f \Lambda_0 [F_1 \cos(\omega t - \alpha_m - \pi - \beta)] \end{cases} \tag{30}$$

The stator parallel branch current in this case is

$$i_c = -\frac{e_1(\alpha_m, t) + e_2(\alpha_m, t)}{(R_{a1} + R_{a2}) + j(X_{a1} + X_{a2})} = 0 \tag{31}$$

Based on Equation (31), it can be concluded that during normal operation of the synchronous condenser, the stator parallel branch current is zero, indicating the absence of any current flowing through the stator parallel branches in this case.

2.2.2. Grid Voltage Balance Synchronous Condenser Excitation Winding Inter-Turn Short Circuit Occurs

When a short circuit occurs in the excitation winding of the synchronous condenser, the induced electromotive force in the two parallel branches can be expressed as follows:

$$\begin{cases} e_1(\alpha_m, t) = 2N_z k_{w1} \tau l f \Lambda_0 \\ \qquad [F_1 \cos(\omega t - \alpha_m - \beta) - F_{d2} \cos(2\omega t - 2\alpha_m)] \\ e_2(\alpha_m, t) = 2N_z k_{w1} \tau l f \Lambda_0 \\ \qquad [F_1 \cos(\omega t - \alpha_m - \pi - \beta) - F_{d2} \cos(2\omega t - 2\alpha_m - 2\pi)] \end{cases} \tag{32}$$

The parallel stator current is given by

$$\begin{aligned} i_c &= -\frac{e_1(\alpha_m,t) + e_2(\alpha_m,t)}{(R_{a1}+R_{a2})+j(X_{a1}+X_{a2})} \\ &= \frac{4N_z k_{w1} \tau l f \Lambda_0 F_{d2} \cos 2(\omega t - \alpha_m)}{(R_{a1}+R_{a2})+j(X_{a1}+X_{a2})} \end{aligned} \tag{33}$$

Based on the above analysis, it can be concluded that during an inter-turn short circuit in the excitation winding of the synchronous condenser under balanced grid voltage, the stator parallel current exhibits a second harmonic component that is directly proportional to

F_{d2}. In other words, the magnitude of the parallel current is directly related to the severity of the short circuit.

2.2.3. The Air Gap Magnetic Potential of the Synchronous Condenser under Unbalanced Grid Voltage Conditions without Any Faults Occurring

In the case of unbalanced grid voltage, the induced electromotive force in the two parallel branches of the stator is given by the following expression:

$$\begin{cases} e_1(\alpha_m, t) &= 2N_z k_{w1} \tau l f \Lambda_0 [F_1 \cos(\omega t - \alpha_m - \beta) \\ &\quad + F_2 \cos(\omega t + \alpha_m - \gamma) + \frac{1}{2} I_{f2} Nk \cos(3\omega t - \alpha_m)] \\ e_2(\alpha_m, t) &= 2N_z k_{w1} \tau l f \Lambda_0 [F_1 \cos(\omega t - \alpha_m - \beta - \pi) \\ &\quad + F_2 \cos(\omega t + \alpha_m - \gamma + \pi) + \frac{1}{2} I_{f2} Nk \cos(3\omega t - \alpha_m - \pi)] \end{cases} \quad (34)$$

At this time, the stator parallel current is given by

$$i_c = -\frac{e_1(\alpha_m, t) + e_2(\alpha_m, t)}{(R_{a1} + R_{a2}) + j(X_{a1} + X_{a2})} = 0 \quad (35)$$

Based on the analysis, it can be concluded that during the operation of the synchronous condenser under unbalanced grid voltage conditions, there are no stator parallel currents induced under normal operation.

2.2.4. The Air Gap Magnetic Flux in the Synchronous Condenser When There Is an Inter-Turn Short Circuit in the Excitation Winding under Unbalanced Grid Voltage

In the presence of an inter-turn short circuit in the excitation winding of the synchronous condenser under unbalanced grid voltage, the induced electromotive forces in the two stator branches can be expressed as follows:

$$\begin{cases} e_1(\alpha_m, t) &= 2N_z k_{w1} \tau l f \Lambda_0 [F_1 \cos(\omega t - \alpha_m - \beta) + F_2 \cos(\omega t + \alpha_m - \gamma) \\ &\quad - F_{d2} \cos(2\omega t - 2\alpha_m) + (\frac{1}{2} I_{f2} Nk - \frac{1}{2} F'_{d1}) \cos(3\omega t - \alpha_m) \\ &\quad - \frac{1}{2} F'_{d2} \cos 2\alpha_m - \frac{1}{2} F'_{d2} \cos(4\omega t - 2\alpha_m)] \\ e_2(\alpha_m, t) &= 2N_z k_{w1} \tau l f \Lambda_0 [F_1 \cos(\omega t - \alpha_m - \beta - \pi) + F_2 \cos(\omega t + \alpha_m + \pi - \gamma) \\ &\quad - F_{d2} \cos(2\omega t - 2\alpha_m - 2\pi) + (\frac{1}{2} I_{f2} Nk - \frac{1}{2} F'_{d1}) \cos(3\omega t - \alpha_m - \pi) \\ &\quad - \frac{1}{2} F'_{d2} \cos(2\alpha_m - 2\pi) - \frac{1}{2} F'_{d2} \cos(4\omega t - 2\alpha_m - 2\pi)] \end{cases} \quad (36)$$

At this time, the parallel stator current is given by

$$\begin{aligned} i_c &= -\frac{e_1(\alpha_m, t) + e_2(\alpha_m, t)}{(R_{a1} + R_{a2}) + j(X_{a1} + X_{a2})} \\ &= \frac{4 N_z k_{w1} \tau l f \Lambda_0}{(R_{a1} + R_{a2}) + j(X_{a1} + X_{a2})} \\ &\quad [\frac{1}{2} F'_{d2} \cos 2\alpha_m + F_{d2} \cos(2\omega t - 2\alpha_m) + \frac{1}{2} F'_{d2} \cos(4\omega t - 2\alpha_m)] \end{aligned} \quad (37)$$

Based on the theoretical analysis presented, it can be concluded that the parallel stator current in the synchronous condenser, under the combined conditions of unbalanced grid voltage and inter-turn short circuit in the excitation winding, is predominantly characterized by even harmonics. The magnitude of the parallel stator current is determined by the degree of grid voltage unbalance and the severity of the inter-turn short circuit.

Inter-turn short circuit faults in the excitation winding of the synchronous condenser result in significant modifications to the air gap flux density and parallel stator current. Unbalanced grid voltage conditions further influence these changes in characteristics. Consequently, these distinctive variations can serve as reliable references for fault diagnosis. A comprehensive verification will be conducted through finite element simulation to validate our findings.

3. Simulation Analysis

This paper presents a case study on the TTS-300-2 type novel synchronous condenser at a specific power station. To investigate its behavior, a two-dimensional (2D) finite element simulation model and its corresponding external circuit are developed using Ansys Maxwell 2021 R1 software. This approach enables a comprehensive analysis of the synchronous condensers' performance and facilitates valuable insights into its operation. The utilization of adaptive meshing in condensor modeling offers significant advantages by automatically adjusting the grid density in response to electromagnetic field variations. This adaptive approach enhances the accuracy of simulation results, improves computational efficiency, and optimizes both time and computational resources. In this paper, we employed an adaptive mesh design for the simulation. Details regarding the mesh partition can be found in Table 1. The Finite Element Method is a numerical technique used to solve integral and partial differential equations, offering superior accuracy compared to other analytical analyses. It employs magnetic linearized parameters to accurately model electromagnetic phenomena. In this study, a 2D field-circuit coupled model of the synchronous condenser is developed in ANSYS Maxwell using the Finite Element Method. It is important to note that, for simplicity, the model neglects the effects of skin effects and eddy current losses. The electromagnetic field expression for the electrical machine is represented by Equation (38).

$$\begin{cases} \frac{\partial}{\partial x}(\mu \frac{\partial A}{\partial x}) + \frac{\partial}{\partial y}(\mu \frac{\partial A}{\partial y}) = -J_z \\ A = A_0 \end{cases} \quad (38)$$

where A is the axial components of the magnetic vector potential; A_0 is the magnetic vector potential in the first boundary; J_Z is source current density; μ is material reluctivity.

Table 1. The main information of Mmesh division in 2D finite element model.

Component	Num Elements	Min Edge Length (mm)	Max Edge Length (mm)	Min Element Area (mm^2)	Max Element Area (mm^2)
Stator	13,303	0.0050	0.1044	2.65×10^{-5}	0.00270
Rotor	5745	0.0035	0.0420	1.50×10^{-5}	0.00054
OuterRegion	4914	0.0040	0.0237	1.20×10^{-5}	0.00021
InnerRegion	1836	0.0035	0.0240	1.05×10^{-5}	0.00015
Band	945	0.0073	0.0236	6.24×10^{-5}	0.00020
Shaft	674	0.0079	0.0257	4.09×10^{-5}	0.00023
Stator Coil	23	0.0060	0.0219	4.80×10^{-5}	0.00012
Excitation coil	28	0.0096	0.0224	6.72×10^{-5}	0.00016

Figures 2 and 3 illustrate the model and circuit schematic representations, respectively. The critical parameters of the synchronous condenser are provided in Table 2. In Figure 3, the symbols LPhaseA, LPhaseA1, LPhaseB, LPhaseB1, LPhaseC, and LPhaseC1 represent the windings of the three-phase double parallel branches. The excitation winding is denoted as LField, while the faulty portion responsible for the inter-turn short circuit is referred to as LShortWinding. By manipulating the parameters LField and LShortWinding, precise control over the number of turns in the short-circuited winding can be achieved. Furthermore, the resistances LR and LShortR correspond to LField and LShortWinding, respectively, and must be adjusted accordingly when altering the number of turns.

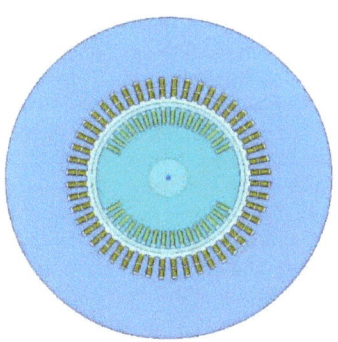

Figure 2. Two-dimensional finite element model of new type synchronous condenser.

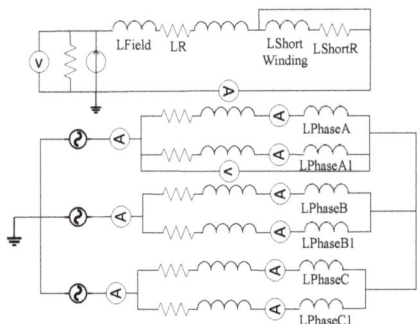

Figure 3. External circuit settings.

Table 2. Parameters of New Synchronous Condenser.

Parameter	Value	Parameter	Value
Rated capacity (Mvar)	300	Number of stator slots	48
Rated exaltation current (A)	1800	Rotor slot number	32
Rated field voltage (V)	407	Stator rated voltage (kV)	20
Number of conductors per slot of stator	2	Stator rated current (A)	8660
Number of turns per slot of rotor	12	Number of parallel branches of stator winding	2
Number of pole-pairs	1	Number of phases	3
Rotor body length (mm)	5950	Inner diameter of stator core (mm)	1240
Air gap length (mm)	70	Frame cushion diameter (mm)	2500
Maximum leading phase operation capability (Mvar)	−200	Rated power factor	0
No-load excitation voltage (V)	137	No-load excitation current (A)	705
Rated speed (rpm)	3000	Rated frequency (Hz)	50

3.1. Model Accuracy Verification

Given the short operating time and the absence of actual on-site fault data for large synchronous condensers, conducting direct short-circuit experiments on-site is impractical. Therefore, to validate the accuracy of the simulation model, simulations were performed under rated operating conditions to analyze the output torque, stator phase voltage, and phase current on the rotor shaft. The rated operating condition of the synchronous condenser refers to its operation at the rated voltage provided in Table 2, carrying the rated load and being connected to the power grid. The specific results are illustrated in Figures 4–6. Additionally, we conducted a comparative analysis between the actual data of stator phase voltage and stator phase current obtained during the synchronous condenser's rated operation and the corresponding simulation data. The detailed results are presented in Table 3.

From the figures, it is evident that the average output torque of the synchronous condenser is zero, which aligns with its expected operational state. Moreover, the phase voltages and currents comply with the rated parameters, and the error falls within an acceptable range. The three-phase currents exhibit symmetry, and there exists a time gap of approximately 5 ms between the stator phase current and the phase voltage. This gap indicates that the stator phase current leads the phase voltage by 90°. These observations confirm the accuracy of the simulation model. By accurately simulating the output torque, stator phase voltage, and phase current under rated conditions, the simulation model has demonstrated its ability to replicate the behavior of the synchronous condenser. Despite the challenges posed by the lack of actual fault data and on-site experiments, the validated simulation results instill confidence in the reliability of the model for further analysis and fault diagnosis.

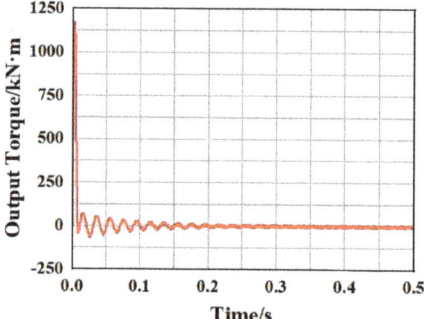

Figure 4. Output torque.

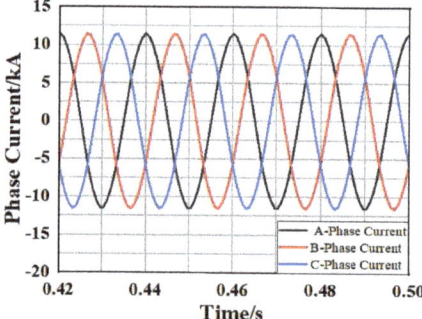

Figure 5. Three-phase current.

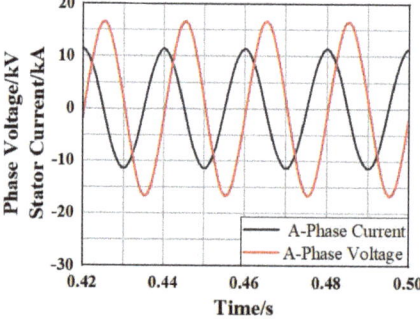

Figure 6. A-phase winding terminal voltage and stator current.

Table 3. Comparison of actual data and simulation data.

Data Type	Actual Data	Simulation Data	Error
Stator phase voltage (kV)	11.547	11.536	0.0954%
Stator phase current (kA)	8.66	8.61	0.5774%

The 2D finite element simulation model of the TTS-300-2 synchronous condenser used in this paper was based on operational parameters primarily extracted from the device's technical manual. Key electrical and structural parameters from Table 2 were accurately applied during the model setup to ensure alignment with the equipment's characteristics. Furthermore, by conducting simulations at the rated voltage and load, we validated the accuracy of torque, phase voltage, and phase current output results under the TTS-300-2 synchronous condenser's rated operating conditions. As a result, the selected parameters effectively mimic the actual device's operational characteristics, guaranteeing the reliability of the simulation analysis.

To simulate the short-circuit of the excitation winding, this study conducted simulation analyses with different numbers of coil turns (1 turn, 3 turns, and 5 turns) short-circuited in the first slot near the large tooth side. In the voltage imbalance simulation, the B-phase and C-phase voltages were maintained at their rated values. In contrast, only the magnitude of the A-phase voltage was adjusted to control the degree of voltage imbalance in the power grid. Specifically, the A-phase voltage values of 96%, 93%, and 90% of the rated voltage were chosen to highlight the simulation results. Due to space constraints, only a partial waveform is presented in this section.

3.2. Analysis of Simulation Results

3.2.1. Analysis of Air Gap Flux Density and Stator Parallel Circulating Current in Faulty Conditions of Synchronous Condenser under Balanced Grid Voltage

- Analysis of Air Gap Magnetic Flux Density

Figure 7a,b depict the analysis results of air gap magnetic flux density under balanced grid voltage with a short circuit fault. Analysis of Figure 7a reveals that a rotor inter-turn short circuit leads to a reduction in air gap magnetic flux density due to a loss of magnetic potential. The severity of the inter-turn short circuit corresponds to a more pronounced decay in the magnetic flux density. Figure 7b demonstrates that in the absence of an inter-turn short circuit, the air gap magnetic flux density is primarily composed of fundamental frequency and odd harmonic components, with only minor influence from slotting effects in the stator and rotor, resulting in a small amount of even harmonic components. The occurrence of a second harmonic in the air gap magnetic density under normal conditions could be attributed to simulation errors resulting in non-uniformity within the air gap magnetic field. However, the presence of an inter-turn short circuit significantly increases the even harmonic magnetic flux density, which intensifies as the severity of the short circuit increases. This observation confirms the accuracy of the derived magnetic field theory.

- Analysis of Parallel Circulation between Stator Branches

Figure 8a,b depict the analysis results of stator parallel branch circulating currents under balanced grid voltage with a short circuit fault. In normal operating conditions, no circulating current or harmonic component is present in the stator parallel branches. However, the occurrence of a short circuit in the excitation winding leads to the generation of circulating currents in the parallel branches, with a predominant presence of even harmonic circulating currents, especially the second harmonic component. Moreover, as the fault severity increases, the amplitude of the harmonic circulating currents also intensifies. This observation confirms the accuracy of the theoretical derivation.

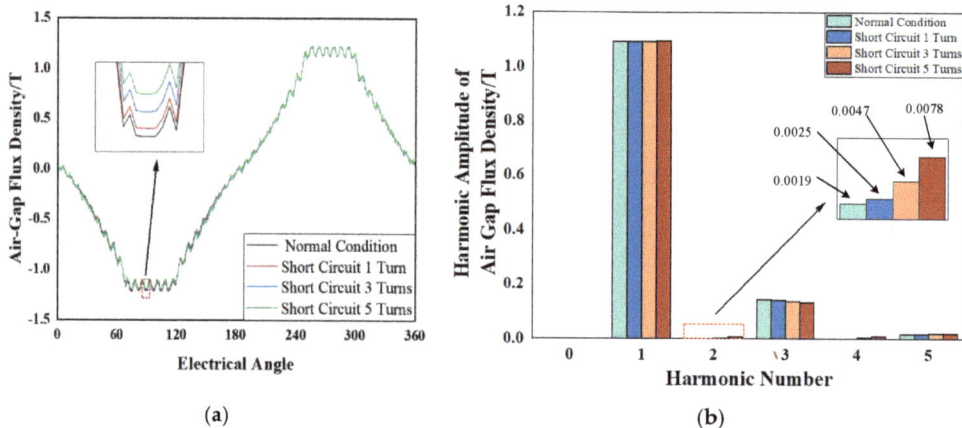

Figure 7. (**a**) Air gap flux density of different short circuit degrees under grid voltage balance; (**b**) Harmonic analysis of air gap flux density with different short circuit degrees under grid voltage balance.

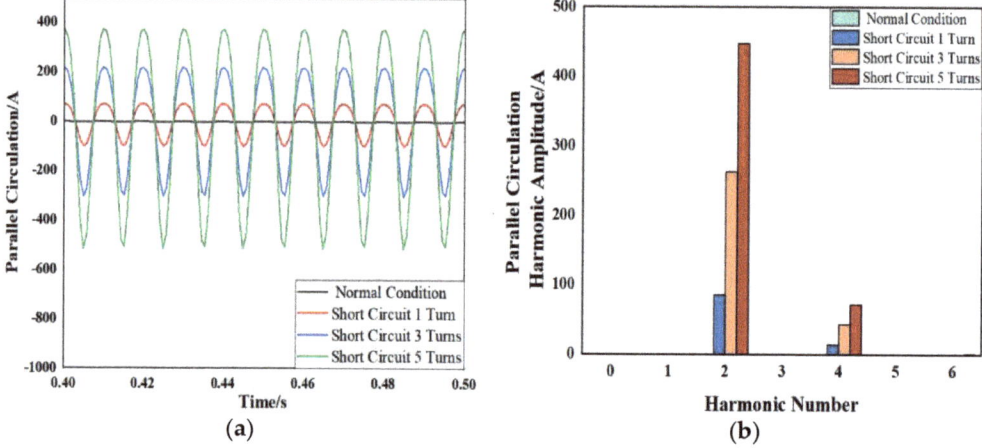

Figure 8. (**a**) Stator parallel circulation with different short circuit degrees when grid voltage is balanced; (**b**) Harmonic analysis of stator parallel circulating current with different short circuit degrees under grid voltage balance.

3.2.2. Analysis of Air Gap Flux Density and Stator Parallel Circulation When the Grid Voltage Is Unbalanced and the Synchronous Condenser Is Fault-Free

- Analysis of Air Gap Magnetic Flux Density

Figure 9a,b showcase the analysis results of air gap flux density under the condition of unbalanced grid voltage. Based on the preceding analysis, it is evident that unbalanced grid voltage causes a reduction in air gap flux density, with a greater degree of voltage imbalance resulting in a more substantial loss of air gap flux density. In the presence of voltage imbalance, the amplitude of the fundamental component exhibits a negative correlation with the degree of voltage imbalance. In contrast, the amplitude of the third harmonic component demonstrates a positive correlation. This behavior can be attributed to the induction of twice the frequency current in the excitation winding by the negative-sequence magnetic fields caused by voltage imbalance. Consequently, third-harmonic magnetic fields are superimposed on the existing third harmonic, leading to an amplified amplitude of the third harmonic and an increased level of magnetic field distortion. According to Equation (14), voltage imbalance in the grid leads to the emergence of third harmonic components in the air gap magnetic field. Consequently, the magnetic field assumes an elliptical shape, deviating from the circular shape predicted by Equation (1). This elliptical magnetic field causes variations in both the amplitude and harmonic content of the air gap magnetic field at different time points, as illustrated in Table 4. Table 4 illustrates the temporal variations of the magnetic field under unbalanced voltage conditions. The primary changes observed are in the amplitude and fundamental amplitude. Although the third harmonic also undergoes changes, its magnitude remains relatively small and inconspicuous due to the low degree of failure.

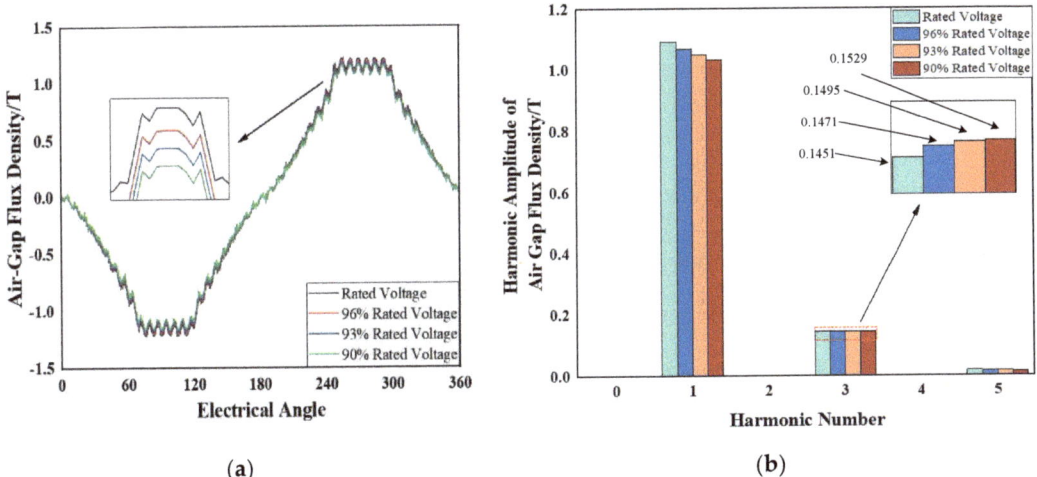

Figure 9. (a) Fault-free air gap flux density of synchronous condenser under unbalanced grid voltage; (b) Harmonic analysis of fault-free air gap flux density of synchronous condenser under unbalanced grid voltage.

Table 4. The values of air gap magnetic flux density at different time instants under 90% rated voltage.

Time/s	Amplitude/T	Fundamental Amplitude/T	Secondary Harmonic Amplitude/T	Third Harmonic Amplitude/T
0.48	1.166	1.034	0.000	0.147
0.485	1.252	1.097	0.000	0.146
0.49	1.191	1.035	0.000	0.147
0.495	1.250	1.097	0.000	0.146
0.5	1.166	1.034	0.000	0.147

- Analysis of Parallel Circulation between Stator Branches

Figure 10 depicts the analysis results of the stator parallel branch current under the condition of unbalanced grid voltage. The theoretical analysis, as indicated by Equations (34) and (35), suggests that the stator parallel branch current remains unaffected by unbalanced grid voltage and maintains a value of 0. However, in the simulation, various factors, such as magnetic saturation and slot effects that occur during the operation of the synchronous condenser, are considered, resulting in a negligible non-zero value for the stator parallel branch current. Although present, the magnitude of this current is minimal and can be disregarded. Thus, during normal operation without faults, the synchronous condenser does not generate any significant stator parallel branch current in unbalanced grid voltage.

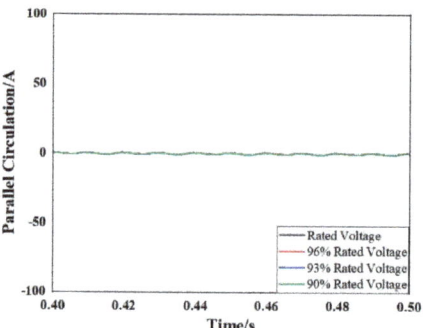

Figure 10. Parallel circulating current of fault-free stator of synchronous condenser under unbalanced grid voltage.

3.2.3. Analysis of Air Gap Magnetic Flux Density and Stator Parallel Circulating Current under Unbalanced Grid Voltage with Excitation Winding Turn-to-Turn Short Circuit

- Analysis of Air Gap Magnetic Flux Density

Figure 11a,b visually represent the air gap magnetic flux density under different degrees of turn-to-turn short circuits in the excitation winding when the grid voltage is unbalanced. Similarly, Figure 12a,b depict the analysis results of air gap magnetic flux density under varying levels of grid voltage imbalance with excitation winding turn-to-turn short circuits. The observations from these figures reveal that unbalanced grid voltage introduces distortions in the air gap magnetic flux density, which are further amplified in the presence of turn-to-turn short circuit faults in the excitation winding. Consequently, there is a significant loss of air gap magnetic flux density. Furthermore, the combined occurrence of faults exacerbates the loss of air gap magnetic flux density compared to a single fault scenario. As the degree of turn-to-turn short circuit in the excitation winding intensifies, even harmonics, particularly the second harmonic, become more prominent in the air gap magnetic flux density. Conversely, grid voltage imbalance primarily affects the fundamental and third harmonic components of the air gap magnetic flux density. The fundamental magnetic flux density decreases while the third harmonic magnetic flux density increases with increasing voltage imbalance. These factors collectively influence the various harmonic components. The simulation results align with the theoretical analysis conducted.

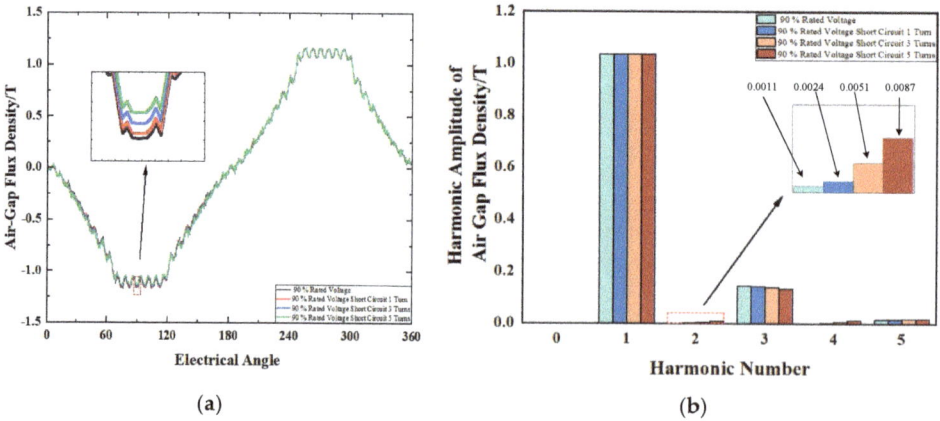

Figure 11. (**a**) A total of 90% rated voltage different short circuit turns air gap flux density; (**b**) Air gap flux density harmonic analysis of 90% rated voltage with different short circuit turns.

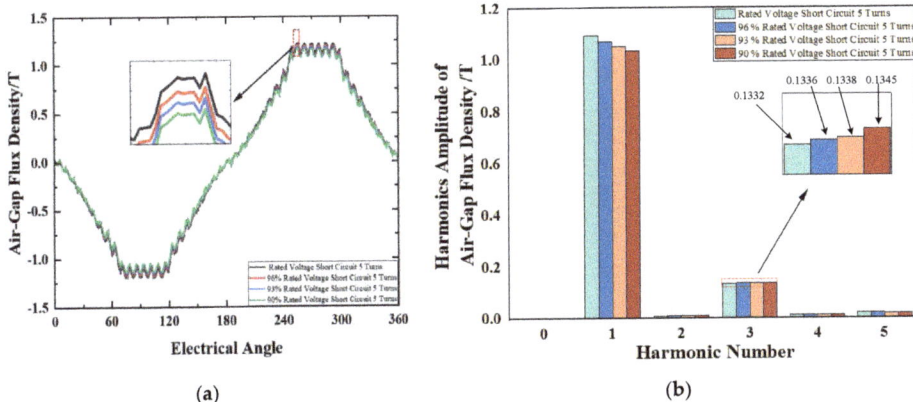

Figure 12. (**a**) Air gap flux density of 5-turn short circuit with different voltage balance; (**b**) Harmonic analysis of air gap flux density of 5-turn short circuit with different voltage balance.

- Analysis of Parallel Circulation between Stator Branches

Figure 13a,b display the analysis results of stator parallel currents under different levels of the rotor winding inter-turn short circuits and voltage imbalance in the power grid. Likewise, Figure 14a,b illustrate the analysis results of stator parallel currents under varying levels of the rotor winding inter-turn short circuit and voltage balance in the power grid. Building upon the simulation results and the theoretical analysis discussed earlier, it is evident that voltage imbalance in the power grid does not impact stator parallel currents in the absence of a short circuit fault. However, following the occurrence of the rotor winding inter-turn short circuit, voltage imbalance in the power grid influences the waveform of stator parallel currents. Consequently, in the case of compound faults, the severity of rotor winding inter-turn short circuits and the degree of voltage imbalance in the power grid affect stator parallel currents. As the severity of rotor winding inter-turn short circuits and voltage imbalance in the power grid intensifies, even harmonics, particularly the second and fourth harmonics, become more prominent in the stator parallel currents. Nonetheless, compared to the severity of rotor winding inter-turn short circuits, the influence of voltage imbalance on stator parallel currents is relatively minor and more susceptible to environmental interference during measurement. Therefore, fault diagnosis of the excitation winding inter-turn short circuit should consider multiple fault characteristics to ensure accurate assessment.

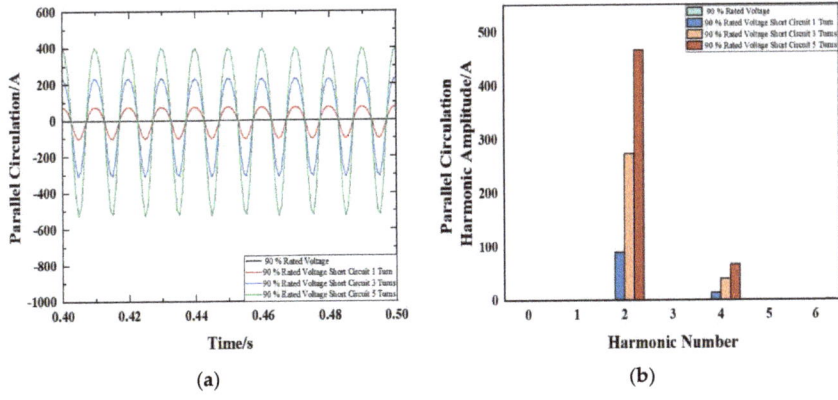

Figure 13. (**a**) A total of 90% rated voltage stator parallel circulation with different short circuit turns; (**b**) 90% rated voltage different short circuit turns stator parallel circulation harmonic analysis.

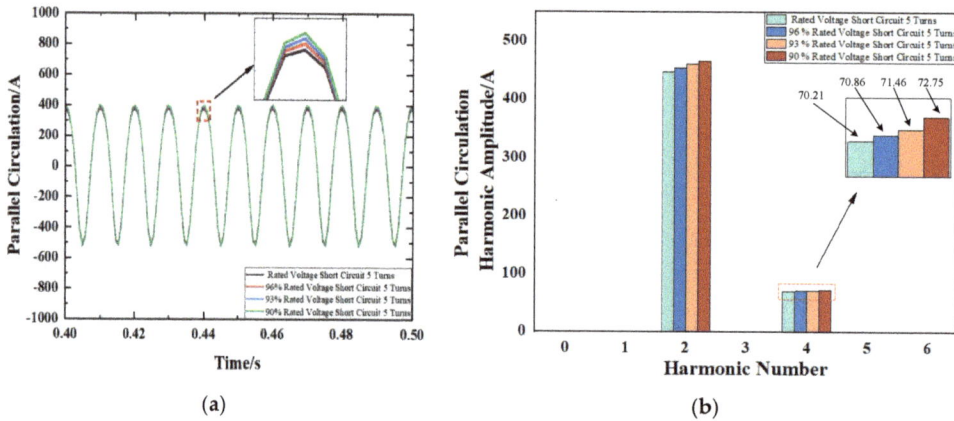

Figure 14. (**a**) Five-turn short circuit stator parallel circulation with different voltage balance; (**b**) Harmonic analysis of five-turn short circuit stator parallel circulation with different voltage balance.

4. Conclusions

This paper examines the impact of inter-turn short circuits in the excitation winding and unbalanced grid voltage on the air gap magnetic flux density and circulating current between stator parallel branches in a synchronous condenser. The study presents derived mathematical expressions for these characteristics and validates them through finite element simulation analysis using a TTS-300-2 synchronous condenser as a case study. The simulation results align with the theoretical derivations, leading to the following conclusions:

1. The occurrence of an inter-turn short circuit in the excitation winding distorts the air gap magnetic flux density. This distortion is further amplified by unbalanced grid voltage, resulting in increased losses in the air gap magnetic flux density. When both unbalanced grid voltage and inter-turn short circuit in the excitation winding are present, the loss of air gap magnetic flux density is even more significant compared to the case of a single fault.
2. In the absence of an inter-turn short circuit in the excitation winding, the impact of unbalanced grid voltage on the circulating current between stator parallel branches can be disregarded. However, when compound faults occur, both the degree of unbalanced grid voltage and the severity of the short circuit can result in fluctuations in the circulating current between stator parallel branches, with the severity of the short circuit having a more significant influence on the circulating current.
3. When diagnosing minor faults in a synchronous condenser, relying solely on features such as the even harmonic component of the circulating current may result in inaccurate fault type determination. To enhance fault diagnosis in future research, it is recommended to incorporate electromechanical information fusion, combining mechanical and electrical characteristics, for more reliable results.

By studying the inter-turn short circuit fault in synchronous condensers under unbalanced voltage conditions, this research provides references and guidance for the safe operation of synchronous condensers. It also establishes a foundation for further research on fault diagnosis in synchronous condensers.

Author Contributions: Conceptualization and methodology, J.L. and C.Z.; software, C.Z. and X.H.; writing—original draft preparation, J.L. and C.Z.; writing—review and editing, J.L., C.Z., J.G. and Y.M.; visualization, J.L. and Y.H.; supervision, resources, project administration, J.L. and Y.H.; funding acquisition, J.L. and Y.H. All authors have read and agreed to the published version of the manuscript.

Funding: This study was financially supported by General Projects of National Natural Science Foundation of China, grant number [52177042].

Data Availability Statement: The data could not be shared due to confidentiality.

Conflicts of Interest: The authors declare no conflict of interest.

References

1. Wu, Q.; Song, P.; Shi, Z.; Zhang, L.; Yan, Y.; Yang, Z.; Shao, L.; Qu, T. Development and Testing of a 300-kvar HTS Synchronous Condenser Prototype. *IEEE Trans. Appl. Supercond.* **2021**, *31*, 1–5. [CrossRef]
2. Wang, P.; Liu, X.; Mou, Q.; Gu, W.; Liu, Y. Dynamic Behaviors and Protection Strategy of Synchronous Condenser Integrated Power System Under Non-Full Phase Fault Conditions. *IEEE Access* **2019**, *7*, 104121–104131. [CrossRef]
3. Wei, C.; Li, H.; Wang, X.; Yang, C.; Wang, W.; Cheng, M. Discrimination Method of Interturn Short-Circuit and Resistive Unbalance Faults for Synchronous Condenser. *IEEE Access* **2021**, *9*, 129706–129717. [CrossRef]
4. Zielichowski, M.; Fulczyk, M. Analysis of operating conditions of ground-fault protection schemes for generator stator winding. *IEEE Trans. Energy Convers.* **2003**, *18*, 57–62. [CrossRef]
5. Gaona, C.A.P.; Blázquez, F.; Frías, P.; Redondo, M. A Novel Rotor Ground-Fault-Detection Technique for Synchronous Machines With Static Excitation. *IEEE Trans. Energy Convers.* **2010**, *25*, 965–973. [CrossRef]
6. Streifel, R.; Marks, R.; El-Sharkawi, M.; Kerszenbaum, I. Detection of shorted-turns in the field winding of turbine-generator rotors using novelty detectors-development and field test. *IEEE Trans. Energy Convers.* **1996**, *11*, 312–317. [CrossRef]
7. Gong, X.; Mao, Q.; Wang, C.; Jiang, B.; Sun, H.; Wang, Z. Analysis of abnormal short circuit between turns of generator rotor windings. In Proceedings of the 2020 IEEE 4th Information Technology, Networking, Electronic and Automation Control Conference (ITNEC), Chongqing, China, 12–14 June 2020; pp. 82–86. [CrossRef]
8. Nadarajan, S.; Panda, S.K.; Bhangu, B.; Gupta, A.K. Hybrid Model for Wound-Rotor Synchronous Generator to Detect and Diagnose Turn-to-Turn Short-Circuit Fault in Stator Windings. *IEEE Trans. Ind. Electron.* **2015**, *62*, 1888–1900. [CrossRef]
9. Polishchuk, V.I.; Gnetova, D.A. Diagnostics system improvement of turn-to-turn short circuits of synchronous generator rotor winding. In Proceedings of the 2016 2nd International Conference on Industrial Engineering, Applications and Manufacturing (ICIEAM), Chelyabinsk, Russia, 19–20 May 2016. [CrossRef]
10. Li, J.; Shi, W.; Li, Q. Research on interturn short circuit fault location of rotor winding in synchronous electric machines. In Proceedings of the 2017 20th International Conference on Electrical Machines and Systems (ICEMS), Sydney, Australia, 11–14 August 2017; pp. 1–4. [CrossRef]
11. Meng, Q.; He, Y. Mechanical Response Before and After Rotor Inter-turn Short-circuit Fault on Stator Windings in Synchronous Generator. In Proceedings of the 2018 IEEE Student Conference on Electric Machines and Systems, Huzhou, China, 14–16 December 2018; pp. 1–7. [CrossRef]
12. Xu, M.-X.; He, Y.-L.; Dai, D.-R.; Liu, X.-A.; Zheng, W.-J.; Zhang, W. Effect of Rotor Intertum Short circuit degree and position on Stator Circulating Current inside Parallel Branches in Generators. In Proceedings of the 2021 IEEE 4th Student Conference on Electric Machines and Systems (SCEMS), Huzhou, China, 1–3 December 2021; pp. 1–7. [CrossRef]
13. Xu, G.; Hu, P.; Li, Z.; Zhao, H.; Zhan, Y. Rotor Loss and Temperature Field of Synchronous Condenser Under Single-Phase Short-Circuit Fault Affected by Different Materials of Rotor Slot Wedge. *IEEE Trans. Ind. Appl.* **2022**, *58*, 7171–7180. [CrossRef]
14. Ma, M.; He, P.; Li, Y.; Jiang, M.; Wu, Y. Analysis of the influence of HVDC commutation failure on rotor winding inter-turn short circuit synchronous condenser. *J. Mot. Control* **2021**, *25*, 1–10. [CrossRef]
15. Zhang, Y.; Wei, C.; Lin, Y.; Ma, H.; Chen, Z.; Jiang, M. Rotor fault diagnosis of synchronous condenser based on wavelet model. *Power Eng. Technol.* **2021**, *40*, 179–184. [CrossRef]
16. Chen, Z.; Li, C.; Ma, H.; Zhao, S.; Tang, X. Rotor fault temperature field analysis of large synchronous condenser. *Power Eng. Technol.* **2022**, *41*, 192–198. [CrossRef]
17. Wei, C.; Sun, L.; Wang, W.; Lin, Y.; Tian, W.; Cheng, M. A Fault Diagnosis Strategy for Rotor Windings Inter-turn Short Circuit of Synchronous Condenser. In Proceedings of the 2019 22nd International Conference on Electrical Machines and Systems (ICEMS), Harbin, China, 11–14 August 2019; pp. 1–5. [CrossRef]
18. Novozhilov, A.N.; Akayev, A.M.; Novozhilov, T.A. Currents in the synchronous condenser windings at turn-to-turn fault in the rotor winding. In Proceedings of the 2016 2nd International Conference on Industrial Engineering, Applications and Manufacturing (ICIEAM), Chelyabinsk, Russia, 19–20 May 2016. [CrossRef]
19. Li, J.; Zhang, L.; Shi, W. Fault characteristics of the DFIG rotor inter-turn short circuit considering inherent imbalance and static eccentricity. In Proceedings of the 2015 IEEE Energy Conversion Congress and Exposition (ECCE), Montreal, QC, Canada; 2015; pp. 971–975. [CrossRef]
20. Ye, L. Research on Fault Diagnosis Algorithm of Condenser Based on RBF Neural Network. Master's Thesis, Huazhong University of Science and Technology, Wuhan, China, 2018.
21. Qiu, J. *Dynamics of Electromechanical Analysis*; Science Press: Beijing, China, 1992; pp. 123–199.

Disclaimer/Publisher's Note: The statements, opinions and data contained in all publications are solely those of the individual author(s) and contributor(s) and not of MDPI and/or the editor(s). MDPI and/or the editor(s) disclaim responsibility for any injury to people or property resulting from any ideas, methods, instructions or products referred to in the content.

Article

Microcontroller-Based Embedded System for the Diagnosis of Stator Winding Faults and Unbalanced Supply Voltage of the Induction Motors

Przemyslaw Pietrzak, Piotr Pietrzak and Marcin Wolkiewicz *

Department of Electrical Machines, Drives and Measurements, Wroclaw University of Science and Technology, Wybrzeze Wyspianskiego 27, 50-370 Wroclaw, Poland; przemyslaw.pietrzak@pwr.edu.pl (P.P.); 259942@student.pwr.edu.pl (P.P.)
* Correspondence: marcin.wolkiewicz@pwr.edu.pl

Abstract: Induction motors (IMs) are one of the most widely used motor types in the industry due to their low cost, high reliability, and efficiency. Nevertheless, like other types of AC motors, they are prone to various faults. In this article, a low-cost embedded system based on a microcontroller with the ARM Cortex-M4 core is proposed for the extraction of stator winding faults (interturn short circuits) and an unbalanced supply voltage of the induction motor drive. The voltage induced in the measurement coil by the axial flux was used as a source of diagnostic information. The process of signal measurement, acquisition, and processing using a cost-optimized embedded system (NUCLEO-L476RG), with the potential for industrial deployment, is described in detail. In addition, the analysis of the possibility of distinguishing between interturn short circuits and unbalanced supply voltage was carried out. The effect of motor operating conditions and fault severity on the symptom extraction process was also studied. The results of the experimental research conducted on a 1.5 kW IM confirmed the effectiveness of the developed embedded system in the extraction of these types of faults.

Keywords: induction motor drive; fault diagnosis; stator winding fault; supply voltage unbalance; ARM Cortex; embedded system

Citation: Pietrzak, P.; Pietrzak, P.; Wolkiewicz, M. Microcontroller-Based Embedded System for the Diagnosis of Stator Winding Faults and Unbalanced Supply Voltage of the Induction Motors. *Energies* **2024**, *17*, 387. https://doi.org/10.3390/en17020387

Academic Editors: Moussa Boukhnifer and Larbi Djilali

Received: 14 December 2023
Revised: 10 January 2024
Accepted: 11 January 2024
Published: 12 January 2024

Copyright: © 2024 by the authors. Licensee MDPI, Basel, Switzerland. This article is an open access article distributed under the terms and conditions of the Creative Commons Attribution (CC BY) license (https://creativecommons.org/licenses/by/4.0/).

1. Introduction

Induction motors (IMs) are widely used in drive systems due to their low production costs, high reliability, and optimal efficiency. In today's industrial landscape, three-phase IMs dominate and account for over 85% of all electric motor utilization [1]. However, despite the high reliability and durability of IMs, they are prone to various types of faults [2]. The most common faults of IMs include bearings, rotor cages, and stator winding faults. Stator winding faults are highly destructive and account for 36% of the total machine failures for low-voltage machines and 66% for high-voltage machines [3].

Stator winding faults are mainly short circuits caused by damage to the winding insulation due to excessive mechanical, thermal, or electrical stresses. There are different types of short circuits: interturn short circuits (ITSCs), short circuits between the coils in one phase, phase-to-phase short circuits, and phase-to-ground short circuits. Most often, a stator winding fault begins with ITSCs and successively spreads from a single turn to subsequent turns and coils in a very short time, leading to a phase-to-phase or phase-to-ground fault. The rapid propagation of ITSC is caused by the very high current flowing in the shorted circuit. For this reason, early and effective detection of this type of fault is crucial and is still a very important research problem [4].

The efficiency and service life of IMs can also be significantly reduced when operating under unbalanced supply voltage conditions, which is quite common in the industrial field [1]. Such conditions can result from a variety of factors, such as unevenly distributed single-phase loads, malfunctioning power factor correction equipment in the same power system, and open circuits in the primary distribution system. An unbalanced supply voltage causes the increased heating of the stator winding, higher losses, vibration, and reduced torque output. Most of these negative effects contribute to the shortened lifespan of IMs [5]. Moreover, supply voltage asymmetry can complicate the process of diagnosing stator winding faults, since even a small supply voltage unbalance results in a large current asymmetry, which also occurs due to short circuits [6]. Therefore, detecting and distinguishing between stator winding faults and an unbalanced supply voltage is extremely important to prevent serious failures and increase the reliability and safety of drive systems. It will also contribute to avoiding motor operation with reduced efficiency, which is crucial to reduce energy consumption and care for the environment.

The requirements for the safety and reliability of modern drive systems are increasing every year. This is due in part to the increasing electrification in many industries and the drive to maximize the lifetime of the equipment. Early detection of a fault can also make it possible to plan motor overhauls accordingly, which will translate into lower repair costs, shorter delays, and reduced potential production losses. For this reason, new diagnostic methods are sought that can be used to diagnose faults at a very early stage of their propagation.

Over the years, several methods for diagnosing IM stator failure have been developed [7–11]. These methods are based on various types of diagnostic signals, such as input voltage [12], stator phase current [13], temperature [14], active and reactive power [15], vibration [16], and axial flux [17]. To extract the fault symptoms from these signals, signal processing methods can be used. Among the most popular and highly effective is the spectral analysis of the signal using the Fast Fourier Transform (FFT). Methods that perform the time-frequency analysis, such as the Short-Time Fourier Transform (STFT) or Continuous Wavelet Transform (CWT) [18], are also attractive in the AC motors stator winding fault diagnosis field. Automation of the AC motors stator winding fault detection and classification process in recent years has most often been implemented using a variety of artificial intelligence techniques [19], such as machine learning algorithms [20,21] and deep learning (DL) [22]. When it comes to computerized diagnostic systems, the fastest growing area in recent years is the application of DL, especially convolutional neural networks (CNN). They are not only applied for fault detection and classification [23] but also for predictive maintenance and remaining useful time estimation [24].

Detection of an unbalanced supply voltage is dominated by methods based on voltage and current signals. The combination of the wavelet transform and principal component analysis of the mains current signal for the detection of unbalanced supply voltage, automated with a support vector machine model, is presented in [25]. The diagnosis and discrimination of the ITSC and unbalanced supply voltage fault method, based on the analysis of the ratio of the third harmonic to the fundamental FFT magnitude component of the three-phase stator line current and voltage, is presented in [26]. An effective approach for the detection of the supply voltage unbalance condition in IM drives, based on a data mining process using the amplitude of the second harmonic of the stator current zero crossing instants as a supply voltage unbalance indicator, is presented in [27]. The online detection method of the unbalanced supply voltage condition, by monitoring a pertinent indicator calculated using the voltage symmetrical components, is shown in [28].

Nevertheless, most of the methods proposed in the literature for stator winding and unbalanced supply voltage diagnosis have been described based on the results obtained using high-end data acquisition (DAQ) equipment and software, such as LabVIEW or MATLAB & SIMULINK [12,27,28], the price of which often exceeds the cost of the machine being monitored. For this reason, the real potential of their industrial deployment is diminishing, as there is no detailed description of the possibilities of low-cost hardware implementation. In this article, special attention is paid to analyzing the possibilities and describing how to apply an embedded system based on a low-budget microcontroller with an ARM Cortex-M core to extract the symptoms of the SI stator winding faults and unbalanced supply voltage.

Embedded systems are used in most modern electronic devices and are a key component of them. An embedded system is a type of system that is designed to perform specific functions, usually in real-time. In recent years, embedded systems that utilize microcontrollers have been the most common. This allows the achievement of a high degree of compactness, responding to increasing demands for the greatest possible miniaturization of devices. The range of applications of embedded systems is very wide, from special-purpose on-board systems, and inverters powering electric motors, to household appliances, HVAC, and many other technical objects [29].

Microcontrollers are small single-chip microcomputers. They are equipped with a variety of peripherals, which include analog-to-digital converters (ADCs), digital-to-analog converters (DACs), comparators, counters, communication interfaces, as well as RAM and Flash memory. In recent years, microcontrollers with the ARM Cortex-M core have been particularly popular. ARM Cortex-M processors are currently one of the best choices for a wide range of applications. In the fourth quarter of 2020, ARM reported a record 4.4 billion chips shipped with Cortex-M processors, confirming their very high popularity [30]. The Cortex M family is a subset of the Cortex family cores, which in turn is a subset of the ARM architecture. Semiconductor manufacturers implement selected versions of the cores, equipping them with peripherals and memory to produce a ready-to-use microcontroller. ARM Cortex-M-based microcontrollers are characterized by high reliability, high performance, and affordability [31].

Currently, there is a noticeable lack of research in the literature that deals with embedded, low-cost implementation of AC motor fault diagnosis methods. In [32], the Arduino board-based system is developed to monitor parameters such as speed, temperature, current, and voltage of the one-phase IM. Authors believe that using these parameters, faults such as over-voltage, over-current, overload, and excessive heating can be detected. In [33], the 8-bit PIC16F877A microcontroller-based system is developed and programmed in C++ language for the detection of under-voltage, over-voltage, over-current, and line-to-ground faults of one-phase IMs. The STM32F4V11VET microcontroller-based fault diagnosis system is proposed in [34] for the detection of the faults of three-phase IMs. This method is based on the stator phase current signals measured using MCR1101-20-5 (ACEINNA, Phoenix, USA) current sensors. It is a promising AI-driven method but requires the current measurement in each of the three phases and it is not strictly defined; the kind of faults the system can detect (broken bearings and misalignment) are mentioned. There are also commercial condition monitoring solutions available on the market that are characterized by a relatively low cost, such as the VB300 G-Force datalogger by EXTECH Instruments, which records a 3-axis shock and vibration and allows the detection of mechanical damage to IMs based on the vibration signal. Nevertheless, it also requires additional PC software. No work has been found that describes a low-cost embedded implementation of an IM stator winding fault and unbalanced supply voltage diagnosis method based on the measurement of the voltage induced by the axial flux.

In this paper, the feasibility of using a low-cost ARM Cortex-M4 core microcontroller (STM32L476RG) to extract the symptoms of ITSCs in IM stator winding and supply voltage unbalance, based on the voltage signal induced in the measuring coil by the axial flux, is discussed. A NUCLEO-L476RG module with a 32-bit STM32L476RG microcontroller (STMicroelectronics, Plan-les-Ouates, Geneva, Switzerland) is used to measure and acquire the diagnostic signal. The method of measuring the diagnostic signal, its acquisitions, the components of the prepared system, and the necessary configuration are presented in detail. Experimental tests were carried out, the results of which confirmed the feasibility of using an embedded system based on a microcontroller with an ARM Cortex-M4 core to extract symptoms of SI stator winding faults and supply voltage unbalance.

The main theoretical and practical contributions of this research are as follows:

(1) An analysis of the possibility of using and the proposal of the concept of an embedded fault diagnosis system based on a low-cost ARM Cortex-M4 core microcontroller to extract the symptoms of IM stator winding faults and unbalanced supply voltage, including the comparison of the results with a high-end solution;
(2) A detailed description of the process of setting up diagnostic signal measurement and acquisition using low-cost microcontrollers, which may serve as a guide for various embedded system applications;
(3) A detailed analysis of the effect of an ITSC in the stator winding and an unbalanced supply voltage of the IM drive on the waveform of the voltage induced in the measuring coil by the axial flux;
(4) A detailed analysis of the effect of an ITSC in the stator winding and an unbalanced supply voltage of the IM drive on the FFT spectrum of the voltage induced in the measuring coil by the axial flux;
(5) An analysis of the possibility of distinguishing between ITSC in the stator winding and an unbalanced supply voltage of the IM drive based on symptoms characteristic of these abnormal conditions;
(6) A proposal for future research and plans to improve and develop the embedded diagnostic system, including reference to current trends related to the Industry 4.0 paradigm.

The rest of the article is organized as follows: Section 2 describes the key parameters of the NUCLEO-L476RG evaluation board, STM32L476RG microcontroller, and motor test bench; Section 3 presents the configuration of the data acquisition system; Then, the results of the ITSC and unbalanced supply voltage symptom extraction based on the voltage inducted by axial flux are presented and discussed. This section also outlines key discoveries and plans for future research; and finally, Section 5 concludes the paper.

2. Experimental Setup

2.1. Characteristics of the Development Board and Microcontroller Used

The development board used in the scope of this research was the NUCLEO-L476RG evaluation board (Figure 1a). It is one of the most popular evaluation boards by STMicroelectronics. The STM32 NUCLEO-L476RG (STMicroelectronics, Plan-les-Ouates, Geneva, Switzerland) is a low-cost and easy-to-use development platform used in flexible prototyping to quickly develop projects based on STM32 microcontrollers. The NUCLEO-L476RG is designed around the 32-bit STM32L476RG microcontroller in a 64-pin LQFP package.

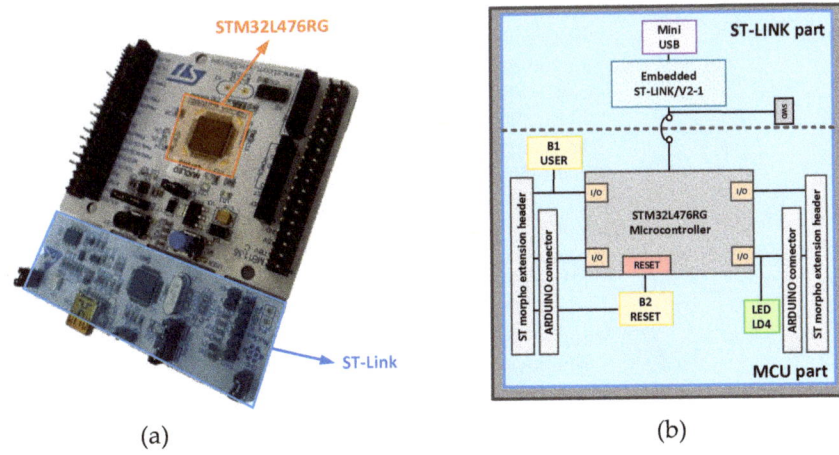

Figure 1. (**a**) Real view and (**b**) simplified hardware block diagram of the NUCLEO-L476RG evaluation board.

The STM32L476RG microcontroller is based on the ARM Cortex-M4 core that operates at a frequency of up to 80 MHz. This microcontroller embeds 1 MB of Flash memory and 128 kB of SRAM memory. The ARM Cortex-M4 core features a single precision Floating Point Unit (FPU) and a Memory Protection Unit (MPU). Its design also features 3 ADC modules and 2 DAC modules with 12-bit resolution. The key parameters of this microcontroller are grouped in Table 1.

Table 1. Key parameters of the STM32L476RG microcontroller.

Parameter	Value
Core	ARM Cortex-M4 (32-bit)
Operating clock frequency	Up to 80 MHz
Flash memory	1 MB
SRAM memory	128 kB
DMA	14-channel
Key communication interfaces	USB OTG, $3\times$ I2C, $5\times$ USART, $3\times$ SPI, CAN
ADC	$3 \times$ 12-bit
DAC	$2 \times$ 12-bit
Comparators	$2\times$ ultra-low power

The NUCLEO-L476RG evaluation board is equipped with ST-Link/V2, which allows flash programming and microcontroller debugging. The power supply of the module is flexible, being possible both via USB and from an external voltage source (3.3 V, 5 V, and 7–12 V). There are also three built-in LEDs on the PCB, which indicate USB communication (LD1), can be programmed by the user (LD2), and indicate the module's power supply (LD3). The ST morpho extension pin headers and ARDUINO connectors are also available on the board for full access to all STM32 inputs and outputs. The simplified hardware block diagram of the board is presented in Figure 1b. A detailed specification of the NUCLEO-L476RG is available on the manufacturer's website [35].

2.2. Motor Test Bench

Experimental tests were carried out on a specially prepared test bench consisting of an IM with a rated power of 1.5 kW (by Indukta) and a DC motor that provided the load torque. The real view of the motor test bench is presented in Figure 2a. The rated parameters of the IM under test are grouped in Table 2. The IM was powered directly from the three-phase grid. The stator winding of the IM under test was wound in such a way that its design allowed for the physical modeling of ITSCs with a certain number of turns. Each phase of the IM under analysis consisted of a coil with 312 turns. During the tests, a maximum of 8 turns were short-circuited, representing 2.6% of all turns in one phase. The schematic diagram of the phase terminals led out to the board of this winding is shown in Figure 2b. The numbers above each piece of winding, visible in Figure 2, are the number of turns that correspond to the taps derived from the winding. The ITSCs were carried out without additional current limiting resistance in the shorted circuit.

The diagnostic signal, the induced voltage in the measurement coil, was measured not only by the NUCLEO-L476RG module described in the previous section but also in parallel for comparison purposes by a high-end DAQ card by National Instruments (DAQ NI PXIe-4492). The DAQ card was placed in an NI PXI 1082 industrial computer. The price of the PXIe-4492 DAQ card was, at the time of article publication, more than 300 times higher ($\approx$USD 8240) compared to the price of the NUCLEO-L476RG board ($\approx$USD 25). Moreover, the PXIe-4492 has an ADC module with a significantly higher resolution (24-bit) compared to the NUCLEO's 12-bit ADC.

 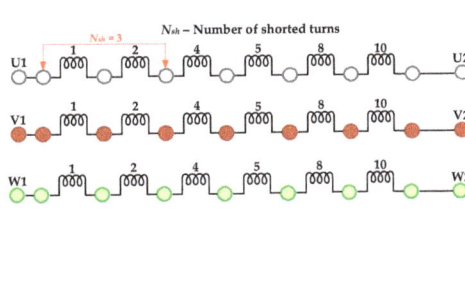

(a) (b)

Figure 2. (a) Real view of the motor test bench: A—SI under test, B—DC motor (load), C—measurement coil, and D—terminal board; and (b) diagram of the derived stator winding terminals.

Table 2. Rated parameters of the tested IM.

Name of the Parameter	Symbol		Units
Power	P_N	1500	[W]
Torque	T_N	10.16	[Nm]
Speed	n_N	1410	[rpm]
Stator phase voltage	U_{sN}	230	V
Stator current	I_{sN}	3.5	[A]
Frequency	f_{sN}	50	[Hz]
Pole pairs number	p_p	2	[-]
Number of stator turns	N_{st}	312	[-]

As presented in Figure 2a, the measurement coil was mounted coaxially with the shaft. Proper mounting of the measurement coil is necessary to apply the proposed method in practice. The design of the used coil and its location coaxially with the shaft does not allow the installation of the cooling fan. Nevertheless, if a fan is necessary, the measurement coil can also be placed on the top or the side of the motor housing. In this experiment, a measurement coil with 300 turns and a DNE copper winding wire cross-section equal to 0.35 mm^2 was used.

2.3. Details of the Developed Microcontroller-Based Fault Diagnosis System

Since preliminary tests have shown that the value of the voltage induced in the coil by the axial flux as a result of the ITSC in the IM stator winding is in the order of mV, to obtain the best possible resolution of the measurement, the signal of this voltage was amplified using an ultra-precise INA241A2 amplifier from Texas Instruments (Dallas, TX, USA). This amplifier has a gain of 20 V/V, a maximum gain error of ±0.01%, a maximum voltage offset error of ±10 µV, and a CMRR (Common Mode Rejection Ratio) of 166 dB (typically). The amplifier was powered directly from the NUCLEO-L476RG module with a voltage of 3.3 V.

The input and output pins of the INA241A2 amplifier, a diagram of its internal connections, and the manufacturer's (Texas Instruments) recommended configuration, as well as a simplified schematic diagram and a photo of the whole prepared system, are shown in Figure 3.

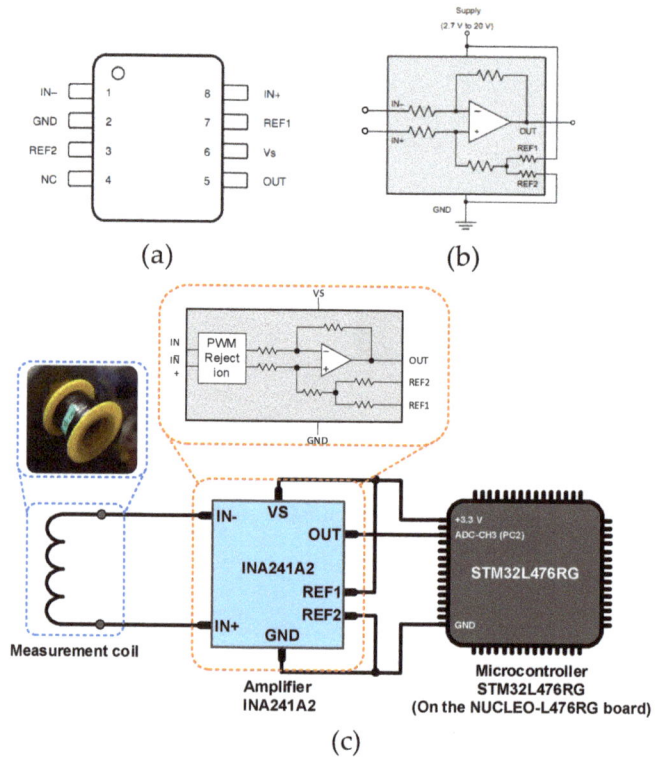

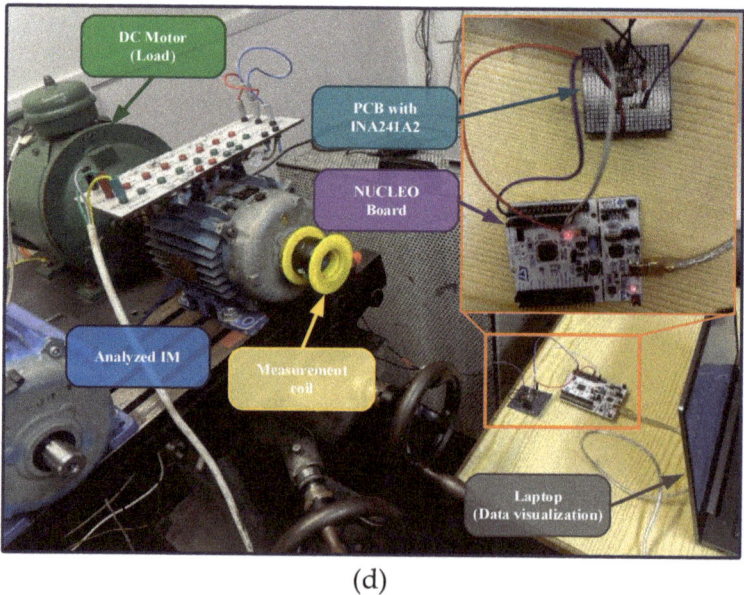

Figure 3. (**a**) Input and output pins of the INA241A2 amplifier, (**b**) a diagram of its internal connections and the manufacturer's recommended configuration, (**c**) a simplified schematic, and (**d**) a photo of the whole system.

The INA241A amplifier, as shown in Figure 3, is located between the STM32L476RG microcontroller and the measurement coil. The IN− and IN+ inputs to which the measurement coil is connected are characterized by a differential mode operation. The REF1 and REF2 pins are reference pins connected to the amplifier supply voltage and GND. The OUT pin is the output voltage, which is connected to the PC3 pin of the microcontroller, which is programmed to operate in ADC mode; it recognizes the conversion of the analog voltage signal induced in the measuring coil into a digital signal. In future research, it is planned to design a customized PCB, which will contain both the microcontroller and the amplifier on a single board. The laptop shown in the lower right corner was needed only for data visualization purposes. In the final system implementation, all functions could be performed by an embedded system that would indicate the information about the stator winding and supply voltage symmetry condition.

The application responsible for the data acquisition and processing was written using the C programming language. Programming and debugging were performed in the STM32CubeIDE environment. Visualization of the values of variables from the microcontroller's memory during real-time operation was carried out using STMStudio (v3.6.0) software.

3. Configuration and Verification of the Data Acquisition Process

3.1. Configuration of the Measurement and Data Acquisition Process

To enable the effective extraction of the IM stator winding fault symptoms from the diagnostic signal, it is necessary to properly configure the measurement and signal acquisition process that is performed by the microcontroller. The microcontroller's pin configuration was done using the Integrated Development Environment (IDE) developed by STMicroelectronics (STM32CubeIDE). STM32CubeIDE is an advanced programming platform for C/C++ languages with the ability to recognize peripheral configuration, code generation, compilation, and debugging for STM32 microcontrollers.

The key task at the stage of preparing the measurement and acquisition of the diagnostic signal is to correctly configure the microcontroller pin that will be associated with the ADC module, which converts the analog voltage signal induced in the measurement coil into a digital signal. Pin two of port C (PC2) was configured as the input of the ADC1 module (channel 3), operated in a single-ended operation mode.

To set the sampling frequency, f_p, when using the microcontroller-based embedded system, it is necessary to configure the timer accordingly so that it generates a cyclic interrupt every specified time, during which the voltage measurement (sampling) will be performed. Cyclic interruptions are used to trigger actions that need to be called at the appropriate frequency. To determine at what frequency the interrupt-generating timer will count, it is necessary to check the microcontroller's specification to verify which bus provides the clocking to the timer and the clocking frequency. In this project, a 16-bit TIM6 timer was used. It was located on the APB1 bus and clocked at 80 MHz. Obtaining this information allows the TIM6 to be appropriately configured for measurement and signal acquisition.

The TIM6 counts from 0 to the value defined in the AutoReload Register (ARR), then generates an interrupt, and resets the counter register to 0 (after the defined time). The frequency of the interrupt triggering (that is the sampling frequency) f_p can be calculated according to the following equation:

$$f_p = \frac{f_{CLK\_CNT}}{TIM\_ARR + 1}, \qquad (1)$$

where f_{CLK_CNT} is the clock frequency of the bus on which the timer is located, and TIM_ARR is the value written in the ARR register.

The f_{CLK_CNT} equals 80 MHz, while 4000 Hz was taken as the desired value of the interrupt trigger frequency (corresponding to the sampling frequency). Hence, the value written in the ARR register is as follows:

$$TIM\_ARR = \frac{f_{CLK\_CNT}}{f_p} - 1 = \frac{80 \cdot 10^6 \text{ Hz}}{4000 \text{ Hz}} - 1 = 19,999. \quad (2)$$

3.2. Verification of the Measurement and Data Acquisition Process

After configuring the peripherals required for the project, it is necessary to verify the correctness of the measurement and signal acquisition before proceeding with the tests. In the first step, it was verified if the TIM6 settings were configured correctly. In the timer interrupt handler function (*HAL_TIM_PeriodElapsedCallback*()), an additional function was added to negate the state of one of the test pins, which was configured as a general-purpose output (GPIO Output). Figure 4 shows the voltage waveform at the output of the test pin recorded using a GW INSTEK MDO-2102A (Montclair, CA, USA) digital oscilloscope. As can be seen, the state change at the output of this pin (state 0 corresponded to ground level, while state 1 corresponded to 3.3 V) occurred every 250 μs. This confirms the correct configuration of TIM6 and corresponds to a frequency of 4 kHz. The interrupt service function will be called to measure (sample) the value measured by the ADC module.

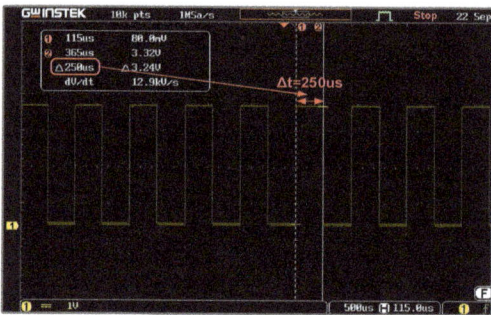

Figure 4. The voltage waveform at the output of the test pin was recorded using a GW INSTEK MDO-2102A digital oscilloscope.

The INA241A2 amplifier (Dallas, TX, USA), which is used in the measurement circuit, not only amplifies the input voltage 20 times but also adds an offset (offset) equal to half the value of its supply voltage. In the case of the supply voltage V_s = 3.3 V, this offset equals 1.65 V. This offset allows the ADC module to measure negative voltages. The output voltage of the INA241A2 amplifier in the absence of input voltage and supply voltage V_s = 3.3 V is shown in Figure 5.

As expected, the value at the output of the amplifier, when there was no input signal connected, was close to 1.65 V. In the next step, the results of signal acquisition recognized by the internal ADC module of the STM32L476RG microcontroller were verified. The waveform of the read-out digital signal after converting the analog voltage signal at the amplifier output by the ADC module (variable *ui16RawADCResult*) is shown in Figure 6a. To convert the raw ADC value to voltage, it is necessary to perform the following calculations [36]:

$$V_{ADC} = \frac{V_{REF+}}{FULL\_SCALE} \cdot ADC\_DATA, \quad (3)$$

where V_{ADC} is the actual voltage measured by the ADC module, V_{REF+} is the reference voltage value of the ADC module equal to 3.3 V, *ADC_DATA* is the digital value converted by the ADC module, and *FULL_SCALE* is the maximum digital value of the ADC output equal to $FULL\_SCALE_{12\text{-}bit} = (2^{12} - 1) = 4095$ for an ADC with 12-bit resolution.

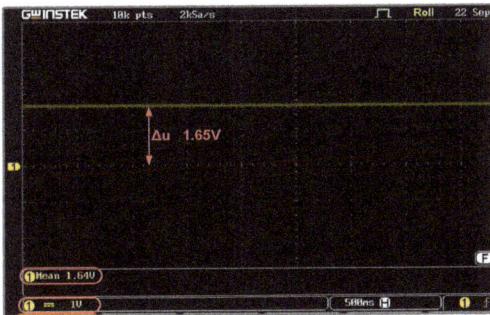

Figure 5. The output voltage of the INA241A2 amplifier in the absence of input voltage (amplifier supply voltage V_s = 3.3 V).

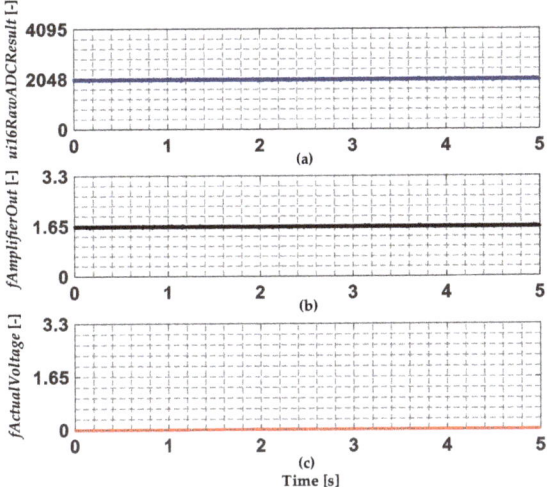

Figure 6. The waveform of (**a**) the measured digital signal after converting the analog voltage signal at the amplifier output by the ADC module (*ui16RawADCResult*), (**b**) the voltage after converting the raw ADC value (*fAmplifierOutput*), and (**c**) the actual amplifier output voltage after offset and gain compensation (*fActualVoltage*).

Figure 6b shows the waveform of the *fAmplifierOut* variable carrying the information about the voltage after converting the raw ADC value according to Equation (3). This value coincides with the waveform recorded using a digital oscilloscope (Figure 5). After taking into account the offset compensation and gain introduced by the amplifier, the waveform of the actual voltage at the output of the amplifier (variable *fActualVoltage*) is shown in Figure 6c. The value was close to 0, confirming that the measurement and acquisition configuration were correct in the absence of input voltage.

The final step in verifying the correct configuration of the ADC module was to measure a sinusoidal voltage signal with an amplitude of 50 mV and a frequency of 50 Hz, generated using the NI myDAQ card and the NI ELVISmx Function Generator environment. The waveform of this signal, recorded using a digital oscilloscope, is shown in Figure 7a, while the measurement at the output of the amplifier by the ADC module of the STM32L476RG microcontroller, after offset and gain compensation (×20), is shown in Figure 7b. The FFT amplitude spectrum is shown in Figure 8. The frequency range of the spectrum includes frequencies from 0 to 2000 Hz, which is due to the adopted sampling frequency of 4000 Hz.

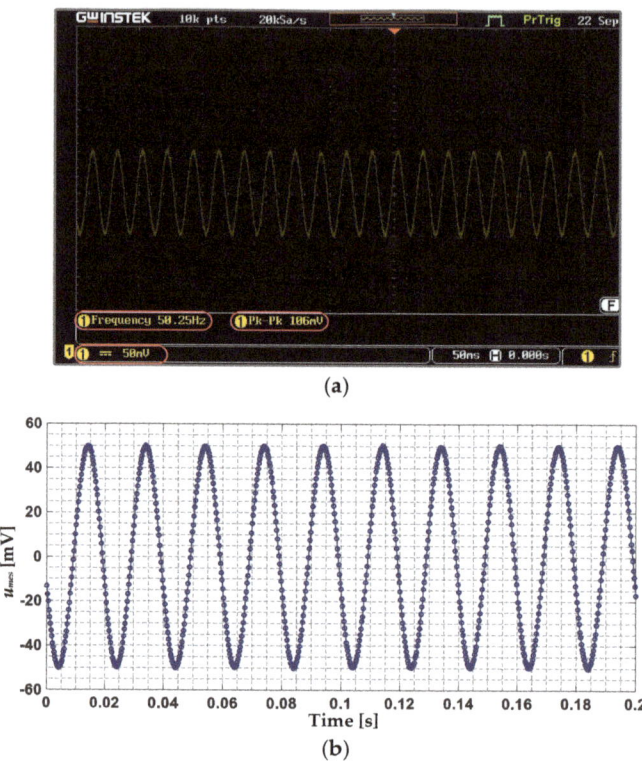

Figure 7. (**a**) The waveform of a sinusoidal voltage signal with an amplitude of 50 mV and a frequency of 50 Hz generated using the NI myDAQ card and the NI ELVISmx Function Generator environment (recorded using a digital oscilloscope), and (**b**) the waveform of the generated signal after conversion and offset compensation measured by the ADC module of the STM32L476RG microcontroller.

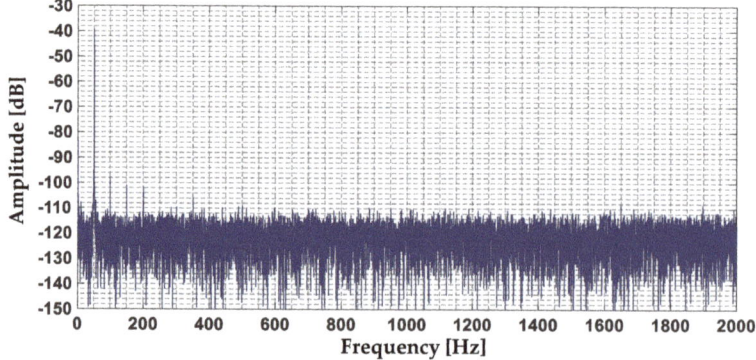

Figure 8. The FFT spectrum of a sinusoidal voltage signal with an amplitude of 50 mV and a frequency of 50 Hz measured by the ADC module of the STM32L476RG microcontroller.

Based on the analysis of the results of the conducted verification of the reliability of the measurement and signal acquisition, it was concluded that the configuration was carried out correctly and the noise level was at a satisfactorily low level of about −125 dB.

4. ITSC and Unbalanced Supply Voltage Symptom Extraction Based on the Voltage Inducted by Axial Flux

Due to the limited accuracy of the technological processing of machine components, IMs are characterized by the presence of certain asymmetries in electrical or magnetic circuits. The effect of these imperfections is the occurrence of leakage fluxes, the value of which will depend on the level of asymmetry of the motor. Since the axial flux finds its source in the currents flowing through the motor windings, faults to the electrical circuits will also be reflected in this signal. In the case of an undamaged motor, the axial flux will have a very low value, close to zero [37]. In the following subsections, the effect of the stator winding faults and unbalanced supply voltage on the waveform of the voltage inducted by the axial flux, as well as on its FFT spectrum, will be analyzed. This will allow the assessment of the possibility of extracting the symptoms of these abnormal motor conditions using a low-cost system.

4.1. Stator Winding Faults (ITSCs)

To verify the validity of the measurement performed with the NUCLEO-L476RG evaluation board, the initial results were compared with those obtained with a high-end NI data acquisition board (DAQ), which has a built-in ADC module with a resolution as high as 24 bits. Figure 9a shows a comparison of the waveform of the induced voltage in the measurement coil by axial flux, u_{mc}, for an unloaded IM with undamaged stator winding ($N_{sh} = 0$, N_{sh}—number of shorted turns), measured using the NUCLEO-L476RG module and the DAQ NI PXI-4492 card. As expected, based on the analysis of these waveforms, the value of the induced voltage in the absence of a stator winding fault was very low; the signal amplitude was about 4 mV. The results obtained for the DAQ NI PXI-4492 measurement card and the STM32L476RG microcontroller were similar. Figure 9b shows a comparison of the u_{mc} waveform for the same drive system operating conditions but with 8 shorted turns ($N_{sh} = 8$) in the IM stator winding. As a result of the damage to the stator winding, the value of the amplitude of the voltage induced by the axial flux increased by about 10 times. Again, both the measurement and data acquisition methods yielded similar results, confirming the correctness of the measurement and signal acquisition performed by the developed low-lost system.

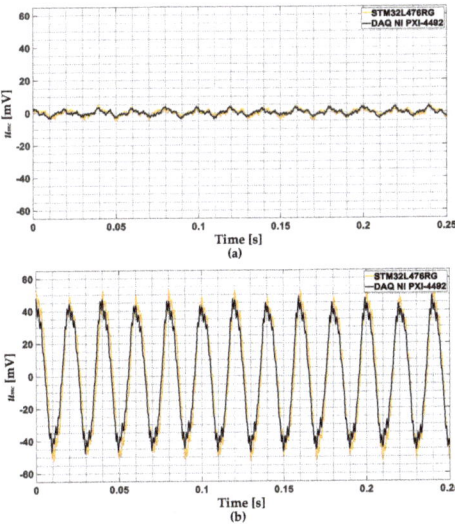

Figure 9. The waveform of the u_{mc} measured using the NUCLEO-L476RG module and DAQ NI PXI-4492 measurement card for an unloaded motor and (**a**) $N_{sh} = 0$, (**b**) $N_{sh} = 8$.

The effect of the stator winding fault can also be seen in the FFT spectrum of the induced voltage signal as an increase of the selected characteristic frequency components. These harmonics are described by the following equation [37]:

$$f_{sp} = kf_s \pm n\frac{1-s}{p_b} = kf_s \pm nf_r, \qquad (4)$$

where:

f_s—fundamental frequency of the supply voltage;
f_r—rotational frequency;
p_p—number of pole pairs;
s—slip;
n—1, 3, 5, ..., $2p_p - 1$;
k—consecutive positive integers (1, 2, 3 ...).

Figure 10 shows the FFT spectra of the u_{mc} measured using the NUCLEO-L476RG module and the DAQ NI PXI-4492 measurement card for an unloaded motor and $N_{sh} = 0$ (Figure 10a) and $N_{sh} = 8$ (Figure 10b). Based on the analysis of these figures, it was concluded that the amplitudes of the harmonics seen in these spectra were similar for both methods of measurement and signal acquisition. The spectrum when the signal was measured with the DAQ NI PXI-4492 card had a lower noise level (by about 30 dB). The higher noise level for the NUCLEO-L476RG did not adversely affect the analysis of the harmonic values. The spectra also show selected stator winding fault-specific frequency components, calculated according to Equation (4). By comparing the spectra shown in Figure 10a,b, it is possible to find the largest increase in the harmonic corresponding to the frequency of the supply voltage (f_s = 50 Hz) as a result of the damage. The results confirmed the correctness of the measurement carried out using the NUCLEO-L476RG. Therefore, further detailed analysis will be performed only for the proposed solution based on the embedded system.

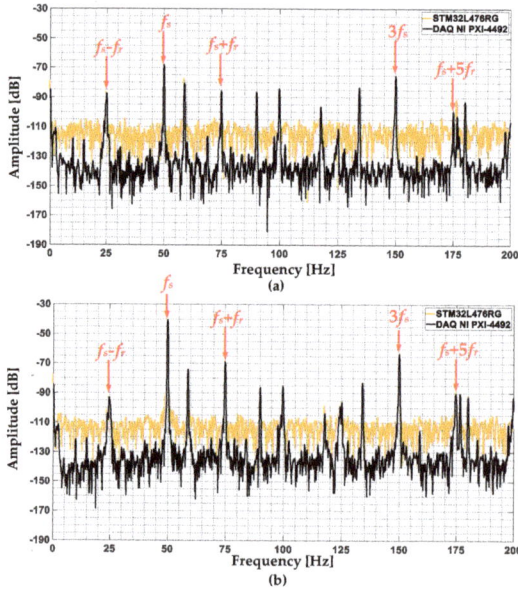

Figure 10. The FFT spectrum of the u_{mc} measured using the NUCLEO-L476RG module and DAQ NI PXI-4492 card for an unloaded motor and (**a**) N_{sh} = 0, (**b**) N_{sh} = 8 ($f_s = f_{sN}$ = 50 Hz, f_r = 24.9 Hz).

Figure 11a shows the waveform of the u_{mc} for the measurement using the NUCLEO-L476RG module, the motor loaded with the rated torque ($T_L = T_N$, T_L—load torque.), and different severities of the stator winding fault ($N_{sh} = 0$, $N_{sh} = 2$, $N_{sh} = 4$, and $N_{sh} = 8$). Based on the analysis of these waveforms, it can be seen that there was a clear trend of increasing u_{mc} amplitude values as the fault deepened. The FFT spectrum for the same operating conditions and degrees of stator winding fault is shown in Figure 11b. Based on the analysis of the amplitudes of characteristic frequencies, it can be concluded that the largest increase in amplitude due to ITSCs was seen for the f_s component. An increase in other harmonics calculated according to Equation (4) can also be observed, especially the $f_s - f_r$, $f_s + f_r$, $f_s + 5f_s$, and $3f_s$ components.

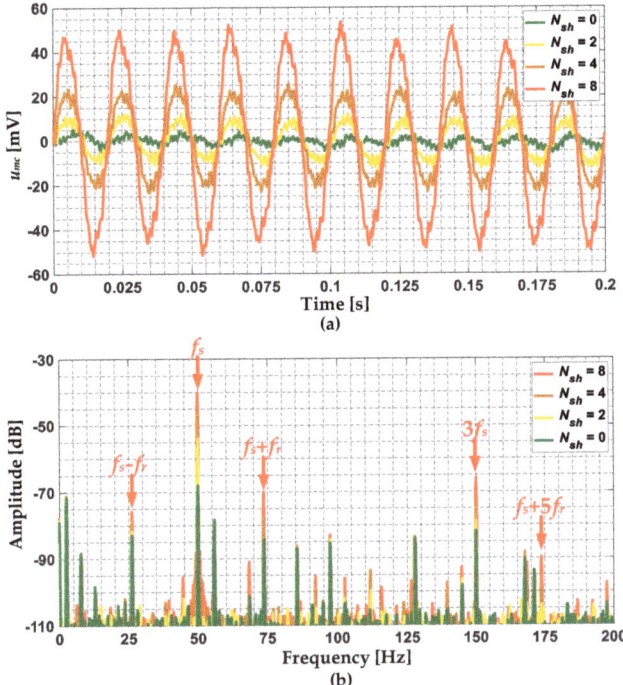

Figure 11. (a) The waveform and (b) the FFT spectrum of the u_{mc} measured using the NUCLEO-L476RG module, for a motor loaded with rated torque and different severities of the stator winding fault.

The effect of N_{sh} in the IM stator winding and the T_L level on the increase in the amplitudes of the characteristic fault frequencies (f_{ITSC}), $f_s - f_r$, $f_s + f_r$, $f_s + 5f_s$, and $3f_s$, in the FFT spectrum of the u_m, is shown in Figure 12. The increase in the value of the A_{diff} (f_{ITSC}) amplitudes is calculated as the difference between the amplitude value for the undamaged winding and a given number of shorted turns. Based on the analysis of the results shown in Figure 12, it can be concluded that the value of the amplitude of the f_s component increased significantly already with 1 shorted turn in the stator winding. In addition, only a very small effect of the load torque was visible. A similar trend was seen for the $f_s + f_r$ (Figure 12c) and $3f_s$ (Figure 12e) components. Nevertheless, the increases in the amplitudes of these components due to ITSCs were lower compared to the amplitudes of f_s. For the $f_s - f_r$, and $f_s + 5f_s$ frequency components, the increase in amplitudes due to the ITSC did not occur for the entire range of analyzed stator winding conditions and T_L levels. Thus, it can be concluded that monitoring the amplitude of the f_s and optionally

$f_s + f_r$ components can allow the detection of the IM stator winding fault at an early stage of its propagation ($N_{sh} = 1$).

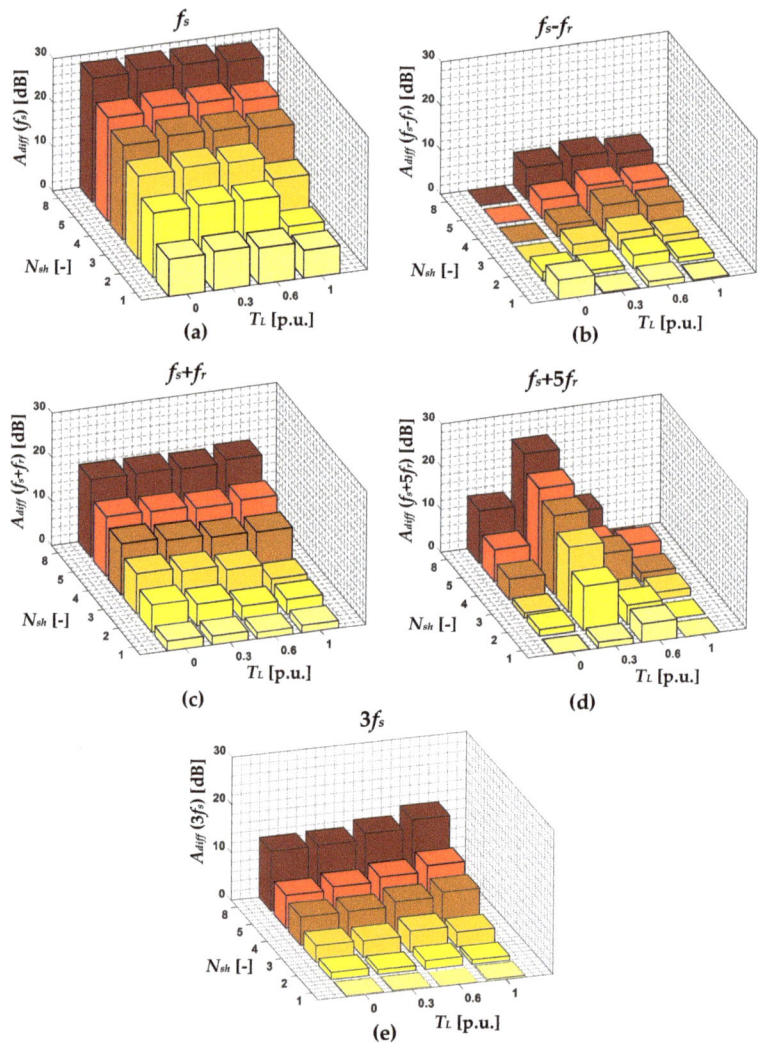

Figure 12. The effect of N_{sh} in the IM stator winding and T_L level on the increase in the amplitude of the selected frequency components: (**a**) f_s, (**b**) $f_s - f_r$, (**c**) $f_s + f_r$, (**d**) $f_s + 5f_r$ and (**e**) $3f_s$ in the FFT spectrum of the u_{mc}.

4.2. Unbalanced Supply Voltage

To introduce a condition of power supply voltage unbalance, each of the phases of the IM under study was supplied by a single-phase autotransformer allowing stepless voltage regulation. The supply voltage value in one of the phases (phase A) was reduced from 230 V to 210 V. The analyzed levels of the supply voltage unbalance are grouped in Table 3. The supply voltage unbalance coefficient, α_{u2}, was calculated as the ratio of the negative sequence supply voltage component to the positive sequence supply voltage component.

Table 3. Analyzed levels of the supply voltage unbalance.

The RMS Value of the Supply Voltage of the Phase C	Supply Voltage Unbalance Coefficient α_{u2}
230 V	0.08%
228 V	0.32%
225 V	0.57%
220 V	1.39%
215 V	2.23%
210 V	2.84%

The waveforms of the voltage induced in the measuring coil by the axial flux for different levels of power supply unbalance are shown in Figure 13. Based on the analysis of these waveforms it can be concluded that the effect of the power supply unbalance on the amplitude of the u_{mc} was significantly less visible compared to the effect of ITSCs. For the analyzed range of supply voltage unbalance levels, the u_{mc} amplitude was close to 5 mV.

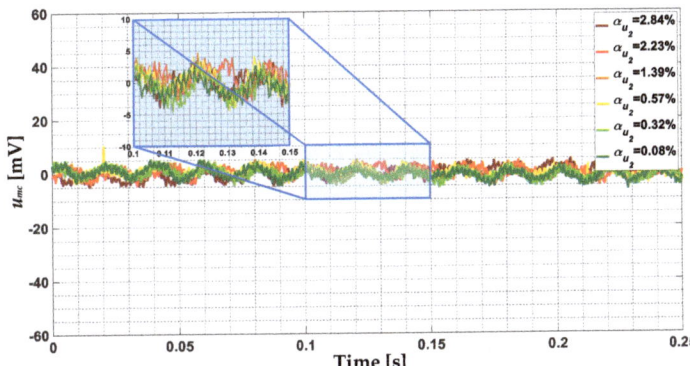

Figure 13. The waveforms of the voltage induced in the measuring coil by the axial flux for different levels of power supply unbalance.

The FFT spectrum of the u_{mc} measured using the NUCLEO-L476RG module for an unloaded motor (Figure 14a), a motor loaded with the rated load torque (Figure 14b), and different power supply unbalance levels is presented in Figure 14. Based on the analysis of these spectra, it can be concluded that there was an increase in the amplitude of the $f_s + 2f_r$ and $3f_s - 2f_r$ frequency components as a result of the supply voltage unbalance.

One of the most important observations is that the frequency components that increased as a result of the ITSC did not change their value due to the unbalanced supply voltage, and the other frequency components ($f_s + 2f_r$ and $3f_s - 2f_r$) appeared. This may allow the distinguishing between these two abnormal conditions (stator winding fault and unbalanced supply voltage). Nevertheless, a more detailed analysis is needed.

The effect of the supply voltage unbalance level and T_L on the increase in the amplitudes of the frequency components that increased the most significantly in the case of the ITSC (f_s and $f_s + f_r$), and the characteristics for supply voltage unbalance (according to Figure 14) ($f_s + 2f_r$, and $3f_s - 2f_r$) are presented in Figure 15. Based on the analysis of these results, it can be concluded that the values of the amplitude of the f_s and $f_s + f_r$ did not increase as a result of the unbalanced supply voltage. The amplitude increase of the $f_s + 2f_r$ component was visible already with an α_{u2} value of 1.39% (the phase voltage RMS value reduced to 220 V). In the case of the $3f_s - 2f_r$ frequency component, the increase as a result of the unbalanced supply voltage was more irregular for different levels of load torque but still visible. Thus, it can be concluded that monitoring the amplitude of the $f_s + 2f_r$, and optionally $3f_s - 2f_r$ components can allow the detection of the unbalanced supply voltage of the IM drive.

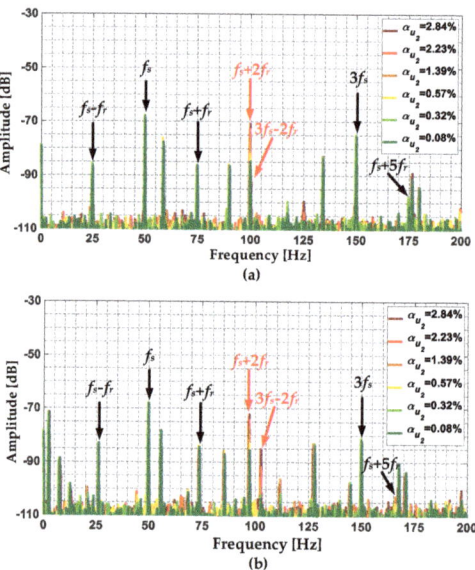

Figure 14. The FFT spectrum of the u_{mc} measured using the NUCLEO-L476RG module, for (**a**) an unloaded motor, and (**b**) a motor loaded with rated load torque, and different power supply unbalance levels.

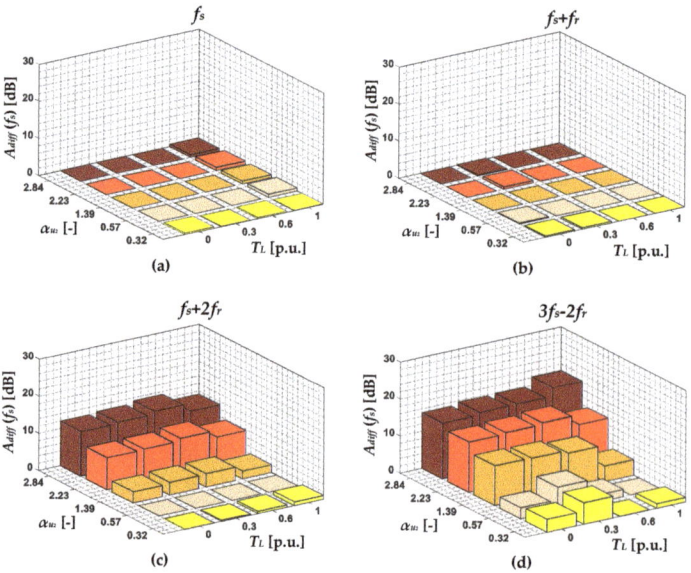

Figure 15. The effect of unbalanced supply voltage level and T_L on the increase in the amplitude of the selected frequency components: (**a**) f_s, (**b**) $f_s + f_r$, (**c**) $f_s + 2f_r$, and (**d**) $3f_s - 2f_r$ in the FFT spectrum of the u_{mc}.

4.3. Influence of the Power Supply Method on the Amplitude Increase of the Selected Harmonics

In the present work, special attention was paid to the IM powered directly from the grid. Nevertheless, the effectiveness of the ITSC symptom extraction was also verified for the IM powered by a Danfoss VLT AutomationDrive FC-302 inverter. The FFT spectra of the u_{mc} measured using the NUCLEO-L476RG module for a motor loaded with rated torque, the different severity of the stator winding fault, and the IM supplied by a Danfoss VLT AutomationDrive FC-302 inverter, for three different values of f_s, are shown in Figure 16. Based on the analysis of the results, it can be concluded that both in the case of the IM powered directly from the grid and a voltage source inverter, there was an increase in the amplitude value of the f_s component as a result of the ITSCs. It confirms the versatility of the proposed solution in terms of the power supply method.

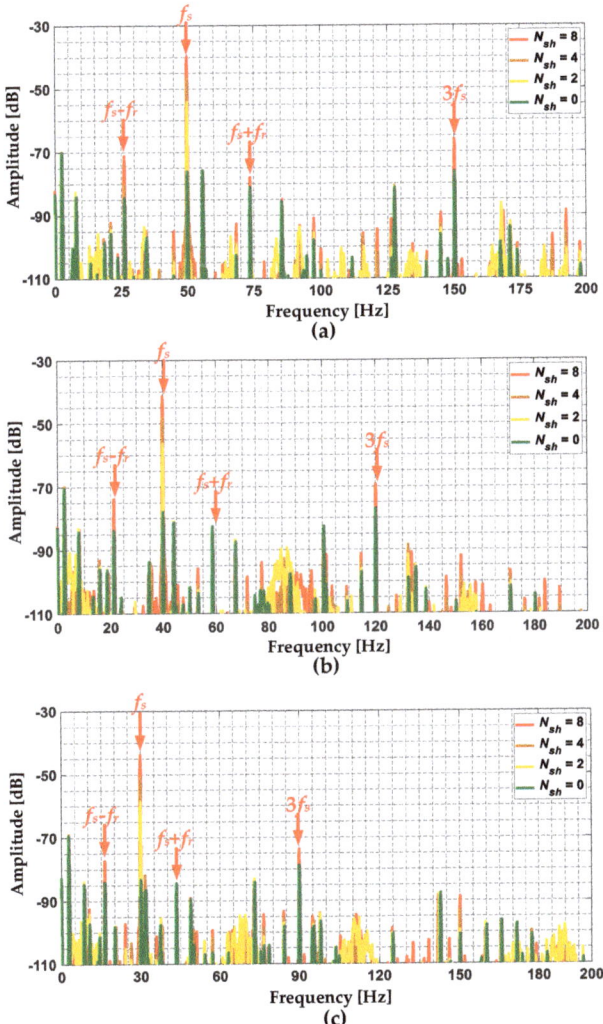

Figure 16. The FFT spectra of the u_{mc} measured using the NUCLEO-L476RG module for a motor loaded with rated torque, different degrees of stator winding fault, and power supply from voltage source inverter, (**a**) f_s = 50 Hz, (**b**) f_s = 40 Hz, and (**c**) f_s = 30 Hz.

4.4. Discussion of the Key Results and Plans for Future Research and Development

Experimental tests made it possible to evaluate the performance of the developed embedded system, analyze the effect of the IM stator winding fault and an unbalanced supply voltage on the voltage induced in the measurement coil by the axial flux, and extract the symptoms that appear in the FFT spectrum of this voltage as a result of these abnormal conditions. The key discoveries are as follows:

- The results of the measurement and signal acquisition process performed using the developed embedded system based on the STM32L476RG microcontroller did not differ from the results obtained using the high-end PXIe-4492 DAQ measurement card by NI;
- The ITSC in the IM stator winding resulted in a significant increase in the amplitude of the u_{mc} regardless of the level of the load torque. The greater the severity of the fault, the greater the increase in amplitude;
- The unbalanced supply voltage of the IM drive did not lead to an increase in the amplitude of the u_{mc};
- The value of the amplitude of the f_s component in the u_{mc} FFT spectrum increased the most as a result of the ITSC already with one shorted turn in the stator winding. The amplitudes of the $f_s + f_r$ and $3f_s$ components also increased, but the increase was smaller compared to f_s;
- The value of the amplitudes of the $f_s + 2f_r$ and $3f_s - 2f_r$ (particularly) components in the u_{mc} FFT spectrum increased the most due to the unbalanced supply voltage;
- The distinguishing between the two abnormal conditions analyzed (stator winding fault and unbalanced supply voltage) were recognized based on the monitoring of the amplitudes of the f_s (characteristic for stator winding fault) and the $3f_s - 2f_r$ (characteristic for unbalanced supply voltage) components;
- The developed method of monitoring the condition of the IM stator winding proved to be effective not only in the case of an IM supplied from the grid but also by the inverter.

Even though the developed system, despite its low cost, already at this stage allows monitoring of the condition of the IM stator winding and the symmetry of the supply voltage, it will be developed in the future and improved with important functions to meet the requirements for modern drive systems that are associated with the Industry 4.0 paradigm. There are many specific areas for future research and development (R&D), including:

- The improvement of the proposed system with the addition of a module that, based on the input vector consisting of statistical information about the voltage signal induced in the measuring coil by the axial flux and the values of harmonic amplitudes, will automatically indicate the state of the stator winding and the symmetry of the supply voltage;
- The integration of the amplifier and microcontroller on a single, specially designed compact PCB that can be mounted at the installer's convenience;
- An extension of the functionality of the developed system to measure other diagnostic signals, such as stator phase currents, and the ability to detect other types of faults, such as broken rotor cage bars, bearing faults, and others;
- An extension of the functionality of the developed system with other mathematical apparatuses that can be used for diagnostic signal processing to extract the symptoms of ITSC and unbalanced supply voltage, such as STFT;
- Adding the function of predicting the possibility of a given failure of the analyzed machine; an extension with the functionality called predictive maintenance;
- The analysis of application possibilities and industry areas where the developed system could be also applied.

From the above points, the first step will be to fully automate the process of monitoring IM stator winding conditions and supply voltage symmetry using AI (machine learning and deep learning) techniques. A simplified block diagram of the flow of diagnostic information processing from the measurement of the signal on the monitored object to the automation of the process of inferring its condition (including the parts that have already been implemented and future R&D plans) is shown in Figure 17.

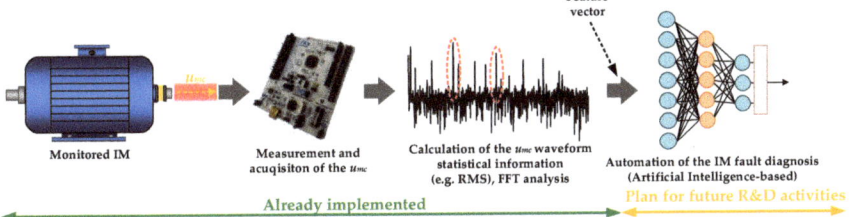

Figure 17. Simplified block diagram of the flow of diagnostic information including the parts that have already been implemented and future R&D plans.

5. Conclusions

The experimental results presented in this paper confirmed the feasibility of using a module based on a low-cost ARM Cortex-M4 core microcontroller to monitor the condition of the IM stator winding based on the voltage induced in the measuring coil by the axial flux. The experimental results also proved that FFT analysis of this signal made it possible to extract symptoms of an incipient ITSC, even with a single shorted turn in the stator winding of IM. The results also showed that it was possible to detect and distinguish from a short circuit an asymmetrical IM supply voltage based on analysis of the amplitudes of selected harmonics in the induced voltage spectrum. This study was carried out over a wide range of operating conditions of the drive system, including the verification of the effect of the power supply method on the increase of individual amplitudes in the analyzed spectrum of the voltage inducted by the axial flux.

The use of the proposed hardware implementation poses several challenges, including those related to the correct configuration of the measurement and acquisition of the diagnostic signal, which is explained in detail and can be helpful in the process of preparing an embedded system for diagnostic purposes for various types of applications. Since the evaluation board used in this work (NUCLEO-L476RG with low-cost STM32L476RG microcontroller) is much cheaper compared to the high-end data acquisition boards used, such as DAQ NI PXI-4492, it has great potential for industrial applications.

Future research will focus on the development and hardware implementation (on a low-cost microcontroller) of an algorithm that will fully automate the process of detecting and classifying an ITSC in the stator winding of an IM, which will use the statistical information of the induced voltage waveform or the amplitude of selected harmonics in the FFT spectrum of this signal. For this purpose, it is planned to use machine learning algorithms whose computational complexity will allow implementation in an embedded system.

Author Contributions: All of the authors contributed equally to the concept of the paper, and proposed the methodology; investigation and formal analyses, P.P. (Przemyslaw Pietrzak), P.P. (Piotr Pietrzak) and M.W.; software and data curation, P.P. (Przemyslaw Pietrzak) and P.P. (Piotr Pietrzak); measurements, P.P. (Przemyslaw Pietrzak), P.P. (Piotr Pietrzak) and M.W.; proposed the paper organization, P.P. (Przemyslaw Pietrzak) and M.W.; validated the obtained results, M.W. All authors have read and agreed to the published version of the manuscript.

Funding: This research received no external funding.

Data Availability Statement: Data are contained within the article.

Conflicts of Interest: The authors declare no conflict of interest.

References

1. Laadjal, K.; Amaral, A.M.R.; Sahraoui, M.; Cardoso, A.J.M. Machine Learning Based Method for Impedance Estimation and Unbalance Supply Voltage Detection in Induction Motors. *Sensors* **2023**, *23*, 7989. [CrossRef] [PubMed]
2. Aguayo-Tapia, S.; Avalos-Almazan, G.; Rangel-Magdaleno, J.D.J.; Ramirez-Cortes, J.M. Physical Variable Measurement Techniques for Fault Detection in Electric Motors. *Energies* **2023**, *16*, 4780. [CrossRef]
3. He, J.; Somogyi, C.; Strandt, A.; Demerdash, N.A.O. Diagnosis of Stator Winding Short-Circuit Faults in an Interior Permanent Magnet Synchronous Machine. In Proceedings of the 2014 IEEE Energy Conversion Congress and Exposition (ECCE), Pittsburg, PA, USA, 15–18 September 2014; pp. 3125–3130.
4. Baruti, K.H.; Li, C.; Erturk, F.; Akin, B. Online Stator Inter-Turn Short Circuit Estimation and Fault Management in Permanent Magnet Motors. *IEEE Trans. Energy Convers.* **2023**, *38*, 1016–1027. [CrossRef]
5. Gonzalez-Cordoba, J.L.; Osornio-Rios, R.A.; Granados-Lieberman, D.; Romero-Troncoso, R.D.J.; Valtierra-Rodriguez, M. Thermal-Impact-Based Protection of Induction Motors Under Voltage Unbalance Conditions. *IEEE Trans. Energy Convers.* **2018**, *33*, 1748–1756. [CrossRef]
6. Lashkari, N.; Poshtan, J.; Azgomi, H.F. Simulative and Experimental Investigation on Stator Winding Turn and Unbalanced Supply Voltage Fault Diagnosis in Induction Motors Using Artificial Neural Networks. *ISA Trans.* **2015**, *59*, 334–342. [CrossRef] [PubMed]
7. Siddique, A.; Yadava, G.S.; Singh, B. A Review of Stator Fault Monitoring Techniques of Induction Motors. *IEEE Trans. Energy Convers.* **2005**, *20*, 106–114. [CrossRef]
8. Niu, G.; Dong, X.; Chen, Y. Motor Fault Diagnostics Based on Current Signatures: A Review. *IEEE Trans. Instrum. Meas.* **2023**, *72*, 1–19. [CrossRef]
9. Riera-Guasp, M.; Antonino-Daviu, J.A.; Capolino, G.-A. Advances in Electrical Machine, Power Electronic, and Drive Condition Monitoring and Fault Detection: State of the Art. *IEEE Trans. Ind. Electron.* **2015**, *62*, 1746–1759. [CrossRef]
10. Garcia-Calva, T.; Morinigo-Sotelo, D.; Fernandez-Cavero, V.; Romero-Troncoso, R. Early Detection of Faults in Induction Motors—A Review. *Energies* **2022**, *15*, 7855. [CrossRef]
11. Gultekin, M.A.; Bazzi, A. Review of Fault Detection and Diagnosis Techniques for AC Motor Drives. *Energies* **2023**, *16*, 5602. [CrossRef]
12. Alloui, A.; Laadjal, K.; Sahraoui, M.; Marques Cardoso, A.J. Online Interturn Short-Circuit Fault Diagnosis in Induction Motors Operating under Unbalanced Supply Voltage and Load Variations, Using the STLSP Technique. *IEEE Trans. Ind. Electron.* **2023**, *70*, 3080–3089. [CrossRef]
13. Cruz, S.M.A.; Cardoso, A.J.M. Stator Winding Fault Diagnosis in Three-Phase Synchronous and Asynchronous Motors, by the Extended Park's Vector Approach. *IEEE Trans. Ind. Appl.* **2001**, *37*, 1227–1233. [CrossRef]
14. Piechocki, M.; Pajchrowski, T.; Kraft, M.; Wolkiewicz, M.; Ewert, P. Unraveling Induction Motor State through Thermal Imaging and Edge Processing: A Step towards Explainable Fault Diagnosis. *Eksploat. Niezawodn. Maint. Reliab.* **2023**, *25*. [CrossRef]
15. Drif, M.; Cardoso, A.J.M. Stator Fault Diagnostics in Squirrel Cage Three-Phase Induction Motor Drives Using the Instantaneous Active and Reactive Power Signature Analyses. *IEEE Trans. Ind. Inf.* **2014**, *10*, 1348–1360. [CrossRef]
16. Kumar, T.C.A.; Singh, G.; Naikan, V.N.A. Sensitivity of Rotor Slot Harmonics Due to Inter-Turn Fault in Induction Motors through Vibration Analysis. In Proceedings of the 2018 International Conference on Power, Instrumentation, Control and Computing (PICC), Thrissur, India, 18–20 January 2018; pp. 1–3.
17. Skowron, M.; Wolkiewicz, M.; Orlowska-Kowalska, T.; Kowalski, C.T. Effectiveness of Selected Neural Network Structures Based on Axial Flux Analysis in Stator and Rotor Winding Incipient Fault Detection of Inverter-Fed Induction Motors. *Energies* **2019**, *12*, 2392. [CrossRef]
18. Pietrzak, P.; Wolkiewicz, M. Fault Diagnosis of PMSM Stator Winding Based on Continuous Wavelet Transform Analysis of Stator Phase Current Signal and Selected Artificial Intelligence Techniques. *Electronics* **2023**, *12*, 1543. [CrossRef]
19. Orlowska-Kowalska, T.; Wolkiewicz, M.; Pietrzak, P.; Skowron, M.; Ewert, P.; Tarchala, G.; Krzysztofiak, M.; Kowalski, C.T. Fault Diagnosis and Fault-Tolerant Control of PMSM Drives–State of the Art and Future Challenges. *IEEE Access* **2022**, *10*, 59979–60024. [CrossRef]
20. Al-Andoli, M.N.; Tan, S.C.; Sim, K.S.; Seera, M.; Lim, C.P. A Parallel Ensemble Learning Model for Fault Detection and Diagnosis of Industrial Machinery. *IEEE Access* **2023**, *11*, 39866–39878. [CrossRef]
21. Ma, J.; Liu, X.; Hu, J.; Fei, J.; Zhao, G.; Zhu, Z. Stator ITSC Fault Diagnosis of EMU Asynchronous Traction Motor Based on ApFFT Time-Shift Phase Difference Spectrum Correction and SVM. *Energies* **2023**, *16*, 5612. [CrossRef]
22. Das, A.K.; Das, S.; Pradhan, A.K.; Chatterjee, B.; Dalai, S. RPCNNet: A Deep Learning Approach to Sense Minor Stator Winding Interturn Fault Severity in Induction Motor under Variable Load Condition. *IEEE Sens. J.* **2023**, *23*, 3965–3972. [CrossRef]
23. Zhou, Y.; Shang, Q.; Guan, C. Three-Phase Asynchronous Motor Fault Diagnosis Using Attention Mechanism and Hybrid CNN-MLP By Multi-Sensor Information. *IEEE Access* **2023**, *11*, 98402–98414. [CrossRef]
24. Guo, J.; Wan, J.-L.; Yang, Y.; Dai, L.; Tang, A.; Huang, B.; Zhang, F.; Li, H. A Deep Feature Learning Method for Remaining Useful Life Prediction of Drilling Pumps. *Energy* **2023**, *282*, 128442. [CrossRef]
25. Sawitri, D.R.; Asfani, D.A.; Purnomo, M.H.; Purnama, I.K.E.; Ashari, M. Early Detection of Unbalance Voltage in Three Phase Induction Motor Based on SVM. In Proceedings of the 2013 9th IEEE International Symposium on Diagnostics for Electric Machines, Power Electronics and Drives (SDEMPED), Valencia, Spain, 27–30 August 2013; pp. 573–578.

26. Refaat, S.S.; Abu-Rub, H.; Saad, M.S.; Aboul-Zahab, E.M.; Iqbal, A. Detection, Diagnoses and Discrimination of Stator Turn to Turn Fault and Unbalanced Supply Voltage Fault for Three Phase Induction Motors. In Proceedings of the 2012 IEEE International Conference on Power and Energy (PECon), Kota Kinabalu, Malaysia, 2–5 December 2012; pp. 910–915.
27. Çakır, A.; Çalış, H.; Küçüksille, E.U. Data Mining Approach for Supply Unbalance Detection in Induction Motor. *Expert Syst. Appl.* **2009**, *36*, 11808–11813. [CrossRef]
28. Laadjal, K.; Sahraoui, M.; Alloui, A.; Cardoso, A.J.M. Three-Phase Induction Motors Online Protection against Unbalanced Supply Voltages. *Machines* **2021**, *9*, 203. [CrossRef]
29. Vassiliev, A.E.; Ivanova, T.Y.; Cabezas Tapia, D.F.; Luong, Q.T. Microcontroller-Based Embedded System Equipment Development for Research and Educational Support. In Proceedings of the 2016 International Conference on Information Management and Technology (ICIMTech), Bandung, Malaysia, 16–18 November 2016; pp. 219–223.
30. Zachary Lasiuk, P.V.J.A. *The Insider's Guide to Arm Cortex-M Development. Leverage Embedded Software Development Tools and Examples to Become an Efficient Cortex-M Developer*; Packt Publishing: Burmingham, UK, 2022.
31. Saha, S.; Tyagi, T.; Gadre, D.V. ARM(R) Microcontroller Based Automatic Power Factor Monitoring and Control System. In Proceedings of the 2013 Texas Instruments India Educators' Conference, Bangalore, India, 4–6 April 2013; pp. 165–170.
32. Waswani, R.; Pawar, A.; Deore, M.; Patel, R. Induction Motor Fault Detection, Protection and Speed Control Using Arduino. In Proceedings of the 2017 International Conference on Innovations in Information, Embedded and Communication Systems (ICIIECS), Coimbatore, India, 17–18 March 2017; pp. 1–5.
33. Sutar, P.P.; Panchade, V.M. Induction Motor Faults Mitigation Using Microcontroller. In Proceedings of the 2017 International Conference on Energy, Communication, Data Analytics and Soft Computing (ICECDS), Chennai, India, 1–2 August 2017; pp. 489–493.
34. Gargiulo, F.; Liccardo, A.; Schiano Lo Moriello, R. A Non-Invasive Method Based on AI and Current Measurements for the Detection of Faults in Three-Phase Motors. *Energies* **2022**, *15*, 4407. [CrossRef]
35. STM32 Nucleo-64 Development Board with STM32L476RG MCU Product Overview. Available online: https://www.st.com/en/evaluation-tools/nucleo-l476rg.html (accessed on 10 December 2023).
36. Reference Manual RM0351: STM32L47xxx, STM32L48xxx, STM32L49xxx and STM32L4Axxx Advanced Arm®-Based 32-Bit MCUs. Available online: https://www.st.com/resource/en/reference_manual/rm0351-stm32l47xxx-stm32l48xxx-stm32l49xxx-and-stm32l4axxx-advanced-armbased-32bit-mcus-stmicroelectronics.pdf (accessed on 10 December 2023).
37. Skowron, M. Application of Deep Learning Neural Networks for the Diagnosis of Electrical Damage to the Induction Motor Using the Axial Flux. *Bull. Pol. Acad. Sci. Tech. Sci.* **2020**, *68*, 1031–1038. [CrossRef]

Disclaimer/Publisher's Note: The statements, opinions and data contained in all publications are solely those of the individual author(s) and contributor(s) and not of MDPI and/or the editor(s). MDPI and/or the editor(s) disclaim responsibility for any injury to people or property resulting from any ideas, methods, instructions or products referred to in the content.

MDPI
St. Alban-Anlage 66
4052 Basel
Switzerland
www.mdpi.com

Energies Editorial Office
E-mail: energies@mdpi.com
www.mdpi.com/journal/energies

Disclaimer/Publisher's Note: The statements, opinions and data contained in all publications are solely those of the individual author(s) and contributor(s) and not of MDPI and/or the editor(s). MDPI and/or the editor(s) disclaim responsibility for any injury to people or property resulting from any ideas, methods, instructions or products referred to in the content.

www.ingramcontent.com/pod-product-compliance
Lightning Source LLC
LaVergne TN
LVHW070738100526
838202LV00013B/1261